OPERATING SAFELY
in Hazardous Environments

Second Edition

Joseph A. Cocciardi, PhD, MS, CSP, CIH, REHS/RS
Cocciardi and Associates

JONES & BARTLETT
LEARNING

World Headquarters
Jones & Bartlett Learning
5 Wall Street
Burlington, MA 01803
978-443-5000
info@jblearning.com
www.jblearning.com

Jones & Bartlett Learning books and products are available through most bookstores and online booksellers. To contact Jones & Bartlett Learning directly, call 800-832-0034, fax 978-443-8000, or visit our website, www.jblearning.com.

> Substantial discounts on bulk quantities of Jones & Bartlett Learning publications are available to corporations, professional associations, and other qualified organizations. For details and specific discount information, contact the special sales department at Jones & Bartlett Learning via the above contact information or send an email to specialsales@jblearning.com.

Copyright © 2013 by Jones & Bartlett Learning, LLC, an Ascend Learning Company

All rights reserved. No part of the material protected by this copyright may be reproduced or utilized in any form, electronic or mechanical, including photocopying, recording, or by any information storage and retrieval system, without written permission from the copyright owner.

Operating Safely in Hazardous Environments, Second Edition is an independent publication and has not been authorized, sponsored, or otherwise approved by the owners of the trademarks or service marks referenced in this product.

Some images in this book feature models. These models do not necessarily endorse, represent, or participate in the activities represented in the images.

Production Credits
Chief Executive Officer: Ty Field
President: James Homer
SVP, Editor-in-Chief: Michael Johnson
SVP, Chief Marketing Officer: Alison M. Pendergast
Executive Publisher: Kimberly Brophy
Executive Acquisitions Editor: William Larkin
Associate Editor: Olivia MacDonald
Associate Production Editor: Nora Menzi
Vice President of Sales, Public Safety Group:
   Matthew Maniscalco
Senior Marketing Manager: Brian Rooney
V.P., Manufacturing and Inventory Control: Therese Connell
Composition: Laserwords Private Limited, Chennai, India
Director of Photo Research and Permissions: Amy Wrynn
Rights & Photo Research Assistant: Gina Licata
Cover Design: Kristin E. Parker
Cover Images: Top Right Image: Courtesy of Mass Communications Specialist 2nd Class Paul Honnick/U.S. Navy
Printing and Binding: Courier Companies
Cover Printing: Courier Companies

Library of Congress Cataloging-in-Publication Data
Cocciardi, Joseph A.
 Operating safely in hazardous environments / Joseph A. Cocciardi. -- 2nd ed.
    p. ; cm.
 Includes bibliographical references and index.
 ISBN 978-1-4496-0966-5
 I. Title.
 [DNLM: 1. Safety Management--methods. 2. Environmental Exposure--prevention & control. 3. Hazardous Substances. 4. Occupational Health. WA 485]

604.7--dc23

2012019086

6048

Printed in the United States of America
16 15 14 13 12   10 9 8 7 6 5 4 3 2 1

Contents

CHAPTER 1 Public and Occupational Safety and Health: A Historical Perspective 1

CHAPTER 2 Current Regulations, Regulatory Agencies, and Professional Associations: Roles and Responsibilities 11

CHAPTER 3 Recognizing and Identifying Hazardous Environments 26

CHAPTER 4 Recognizing and Identifying Hazardous Materials 40

CHAPTER 5 Quantifying Hazardous Environments 61

CHAPTER 6 Protective Planning 101

CHAPTER 7 Personal Protective Equipment 125

CHAPTER 8 Decontamination 146

CHAPTER 9 Responding to Emergencies 156

CHAPTER 10 Medical Monitoring and Occupational Health 180

CHAPTER 11 Personal Safety 196

APPENDIX A Bloodborne Pathogen Information 214

APPENDIX B Asbestos and the OSHA Asbestos Standards 216

APPENDIX C LEAD 220

APPENDIX D New York City Department of Health and Mental Hygiene: Fungal Remediation Protocols ... 229

APPENDIX E DOT Chart 14: Hazardous Materials Markings, Labeling and Placarding Guide 231

Index 235

Photo Credits 247

Preface

After more than 30 years of learning, teaching, and evaluating health and safety education offerings, I have identified a need for an organized, interdisciplinary teaching and learning guide on how to work and play safely. In partnership with Jones & Bartlett Learning, we have published two titles which includes: *Operating Safely in Hazardous Environments* and *Operating Safely in Hazardous Environments: A Review and Refresher*.

These books are designed as "Safety and Health" primers for individuals in the educational, workplace, and public sectors. They are designed to facilitate optimal information flow (and, subsequently, learning) for students. Areas covered include public and workplace hazard detection, identification, and mitigation, as well as operations in these dangerous areas. The texts also serve the purpose of instructional books and workbooks in classroom settings, and act as technical reference material for students in workplace, home, and recreational environments.

Operating Safely in Hazardous Environments, the first book to be published in this series, initially focuses on mechanisms for recognizing and identifying hazardous atmospheres, environments, and operations. It then focuses on quantifying the identified hazards or potential hazards, and finally identifies mechanisms for protecting people in these situations. Emergency response procedures and emergency action plans are also covered.

This book can serve as the required text for training and educating workers who enter a variety of hazardous atmospheres (e.g., hazardous waste operations; permit required confined spaces; emergency response situations; toxic materials work, such as asbestos containing materials and lead or lead based paint removal; work at heights; and work within other immediately dangerous or hazardous areas). After years of developing and providing educational sessions in these areas, I have identified commonalties that exist across these occupational boundaries. *Operating Safely in Hazardous Environments* provides information on these common characteristics and operations (e.g., proper use of a respirator, or use of toxic materials monitoring equipment). It pinpoints key learning for presentation to students and can minimize repetitive training. Use of this text by instructors may qualify students for multiple hazardous occupations and meet multiple training needs with minimal time expenditures, a true benefit in today's industrial society.

The text begins with a historical perspective of public and industrial health and safety organizations and efforts and contains a chapter describing present governmental and professional health and safety organizations and their roles and responsibilities. Hazardous atmospheres, and mechanisms for recognizing and identifying them are explained, as well as means of quantifying hazards. The book concludes with sections describing the proper use of engineering and administrative safety and health controls and personal protective equipment. Emergency response plans and procedures for the development of these plans are given.

Other Titles

Operating Safely in Hazardous Environments: A Review and Refresher (ISBN 978-1-4496-0967-2) is designed to lead students through an annual review and practice of skills associated with operating safely in each type of hazardous environment. Use of a text such as this is required by regulations governing training requirements for hazardous work.

Acknowledgments

Countless individuals have assisted with the development of this text. These include my instructors at both the Masters and Doctoral levels, my peers in the public and occupational safety and health fields who continuously challenge me to excel (in particular Steve Strayer and Rocco DiPietro, who daily teach the fundamentals covered in these books and Tim Davis who has broadened my experience to include medical consequences of hazardous environments, students who have attended 30 years of training in the fields described in the book, and those who have operated safely in hazardous environments over the years. I am always amazed at how these individuals continue to perform these dangerous tasks in a safe and routine manner. Their knowledge of their particular fields is immense and it is this knowledge that I have attempted to distill into the material.

I also need to acknowledge and thank Shelly Mixell, who put endless hours into the typing and electronic mailings associated with this Edition of the manuscript.

Finally, I could not have produced these works without the support of my wife and family (Jolene, Dan, Steve, Joey, and Jenny). They have helped me to live a "safe and healthy life" which makes writing about these characteristics easy, and instructing about the concepts even easier.

1 Public and Occupational Safety and Health: A Historical Perspective

Key Concepts

- Safety and Accident Prevention
- Health Promotion and Illness Prevention
- Property and Resource Protection
- Hazardous Environments
- Barriers to Hazardous Environments
- The Historical Perspective

Introduction

Before examining the **hazardous environments** that exist today and before examining how individuals operate safely in and around these environments, it is beneficial to review the historic development of public and occupational safety and health programs.

Individuals and organizations dealing with public and occupational safety and health traditionally addressed the areas of both property protection and injury and illness prevention. Although individuals at first provided public and occupational safety and health services, organizations subsequently developed services that affected larger areas or groups.

Today, public and occupational safety and health organizations provide a wide variety of property and resource protection, **accident** and **illness** prevention programs, and hazard mitigation services. These are provided to protect and minimize threats to both individuals—such as workers—and communities from unwanted unhealthy events such as sickness, physical accidents, fires, and other disasters.

Historical Activities: Personal Health and Property Protection

Historical records document the development of healthy practices, such as the presence of physicians, in ancient cultures. Although in some cultures physicians assumed the role of the administrators of medicines and potions (e.g., Greek and Roman "Physicians"), in others there was a moral or religious overtone to their activities (e.g., religious activities during the Middle Ages).

Other cultures, such as those found in Central and South American native populations (e.g., the Southern Mexican Mayan ancestors), practiced an integration of religious, physiological, and physical healing practices and preventative activities.

During ancient times communities practiced property conservation as well. Specific examples include the following.

Light Towers

The first use of lighthouses (c. 300 B.C.E.) is traced to the placement of lighting and alarm towers at the entrances to harbors during ancient times (e.g., by the Ptolemaic Pharaohs of Alexandria, Egypt). Interestingly, many health- and safety-related events occurred around these port cities, such as the construction of light towers, the formation of mutual insurance groups (Rhodes, Greece), and the organization of life-safety groups to rescue mariners and cargo in distress. A variety of light towers were built during the Roman 1st century C.E. and the Italian middle ages. The British "Dover Tower Light," the oldest "lighthouse" in Britain, was built around the third century C.E.

Roman "Vigils" Corps

After a disastrous Roman fire in 6 C.E., the Emperor Augustus instituted the use and staging of fire hoses (made of animal intestines) in certain Greek and Egyptian cities (anecdotally reported), and the organization of servants and slaves into "Vigils," an organization formed for the purpose of Roman fire watch and night patrol. They would serve in this capacity for 500 years. In addition, the Roman Vigilante Corps were used at this time to patrol Rome from dusk to dawn, warn individuals to extinguish fires, and sound alarms in the event of major fires. This corps was in use until approximately 500 C.E., and numbered 16,000 at the height of power of the Roman Empire.

The Great London Fire (1666)

Codes and organizations to enforce these codes were developed after times of disaster, such as the great London fire of 1666. Subsequent to this fire, organizations were formed and legislation codified to prevent conflagrations and to attack fires once they occurred. Specific activities after the great London fire (which burned for a week and destroyed approximately four-fifths of the wooden city) included prohibition of thatched roofs within the city of London, provision for wider streets, requirements to use brick and stone for new construction, and the organization and training of the Metropolitan (London) Fire Brigade.[1]

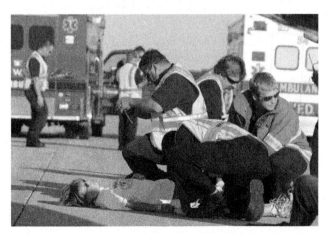

PHOTO 1-1 Public and Occupational Safety and Health Programs focus on the prevention of unwanted events, while preparing to minimize their consequences in case they do occur.

---

[1] James Braidwood, originally a surveyor specializing in buildings and structures, founded the municipal fire service in Edinburgh, Scotland, and subsequently was appointed the first Director of the London Fire Brigade (London Fire Engine Establishment). He was succeeded by Sir Eyre Massey Shaw.

PHOTO 1-2 Subsequent to the 1666 London fire, organizations were formed to attack conflagrations and legislation codified to prevent fires.

By the mid-18th century, systems such as the fire brigade were truly developing to protect people and property in hazardous environments.

The Beginnings of Epidemiological Study

Activities to prevent the spread of illnesses can also be traced through these times, again most notably after large-scale public health events, such as outbreaks of the Bubonic Plague. The movement of river access points for water companies to areas up river after outbreaks of cholera in London in the first half of the 19th century is a specific example of a health-related response to illness outbreaks.

The roots of modern epidemiological investigations can be traced to the cholera investigations at this time in London (e.g., John Snow's investigation of contaminated drinking wells in the Golden Square area of London is considered the first example of field epidemiology).

Industrialization

The industrial revolutions introduced technologies and events to the world that increased personal safety and health concerns. As countries became more populated and industrialized, the amount of hazardous environments to which individuals and property were exposed occurred with greater frequency. To address the increasing number of hazardous exposures, individuals and organizations again concerned themselves with accident and illness prevention and property protection. However, a more aggressive role became necessary as hazardous areas or events became more commonplace.

Fire Service Organizations

A pioneer in this area was Benjamin Franklin who, in the mid-18th century, recommended the use of noncombustible materials in cities and lightning protection to prevent fires in buildings (Sir Christopher Wren had similar requirements for the rebuilding of London after the great fire of 1666). Franklin was reportedly the first individual to install a lightning rod—an iron rod that rose 5 feet above his Philadelphia home and was connected to another rod that was buried 5 feet into the ground.

Although not the first fire protection organization in America,[2] Franklin is generally credited with organizing and developing fire protection societies (fire suppression services) in the United States for the purposes of conserving property. His Union Volunteer Fire Company served this purpose in Philadelphia. Before Franklin's "Philadelphia Contributorship" (Fire Insurance Company) was formed in 1752, the Boston Mutual Fire Society had been established in 1718 to assist society members in removing and conserving property during fires. These fire suppression societies matured through the 19th century into the fire service organizations we see today.

Lifesaving Maritime Organizations

In 1795, under the administration of Alexander Hamilton, the U.S. Treasury Department organized the Revenue Cutter Service. This predecessor of both the U.S. Coast Guard and the U.S. Public Health Service was tasked with the responsibility of conserving property and saving lives at maritime disasters. In addition to rescues, their health-related duties included fumigation and quarantine of foreign ships prior to harbor entrance.

PHOTO 1-3 American fire marks were used to identify the insuring company and policy number of members. In many towns, insurance companies paid the first arriving fire society. This system is blamed for the fierce competition that occurred among American volunteer fire societies.

---

[2]In 1647 the Dutch Governor of New Amsterdam had founded a "Rattle Watch" to warn citizens in the event of fire, and in 1678 Boston received a "fire engine" and paid 13 men to maintain it and use it at fires; the Boston Fire Department was officially organized in 1717.

PHOTO 1-4 By 1860, the U.S. Lifesaving Service staffed more than 1000 Lifesaving Stations. This service, as well as the Revenue Cutter Service, the Maritime Inspection Bureau, and the U.S. Lighthouse Service, were organized into the U.S. Coast Guard over time.

Other predecessors of the U.S. Coast Guard included the United States Lighthouse Service and the United States Lifesaving Service, which were formalized in the 19th century. As maritime trade and related disasters increased, these groups began a basic regimentation of public safety services. For example, by the year 1860, the United States Lifesaving Service had been governmentally organized into more than 1000 lifesaving stations. Full- and part-time personnel stationed roughly five miles apart on both coastal and inland rivers drilled weekly in lifesaving rescue practices and first aid.

Still, industrial development and population growth continued into the later 19th century, and with it came the increased need for public health and safety services.

First Aid and Field Medical Services

In 1864, with the passage of the Ambulance and Stretcher Bearer Act, the U.S. government formalized requirements for military ambulances and stretchers. The purpose of this equipment and service was to transport traumatic injury victims quickly and safely from hazardous environments to locations where health care could be provided. Other attempts to bring safe environments to hazardous areas through the provision of field medical services also occurred during this time. These include the development of first aid practices for miners by Dr. Matthew J. Shields in the minefields of Northeastern Pennsylvania in 1899. This practice was the forerunner of first aid services, as they exist in hazardous industrial settings today.

Professional Organizations

In the late 19th and early 20th century, organizations such as the Association of Iron and Steel Electrical Engineers (forerunner of the National Safety Council) and the National Board of Fire Underwriters were formed. These groups began to codify procedures for safe work in hazardous environments. Disasters such as the Triangle Shirt Waist Company factory fire in New York City—which killed 92 workers in 1911 as a result of blocked exits—and the great Chicago fire of 1871—which killed 250 citizens—had placed the hazards of both workplaces and communities in the minds of the public.

TABLE 1-1 identifies the time line of key American health and safety events.

Modern Safety and Health Organization

The wide use of chemicals, in particular explosives and hydrocarbon fuels, continued the legacy of disasters into the 20th century.

By the mid-20th century, local response organizations were well trained in the area of fire protection. Consensus Codes, such as those promulgated by the National Fire Protection Association, were published for use by insurance groups, as well as municipal organizations and the public.

Codes and code enforcement to reduce hazards were now common. Rudimentary illness prevention activities were affected by the public health organizations (in particular, the public health services) and organizations providing life safety in hazardous areas (such as the U.S. Coast Guard protecting life at sea) were established and regimented.

With mechanization and industrialization, the development of safety equipment occurred to assist work in hazardous environments. **TABLE 1-2** lists select safety equipment development.

TABLE 1-1 American Health and Safety Time Table	
1647	New Amsterdam "Rattle Watch"
1678	Boston receives "fire engine" and pays 13 men to maintain and use it. The Boston Fire Department was officially organized in 1717.
1716	Boston Light built: first American lighthouse
1734	Fire services organized: Contribution Fire Society, Philadelphia
1752	Fire services organized: Green Tree Fire Society, Philadelphia
1790	U.S. Revenue Cutter Service authorized (maritime rescue)
1794	Journeymen Corkwainers Society–Boston (Mutual Worker's Insurance Group)
1799	Boston: First Board of Health levies fines under the direction of Paul Revere
1832	Revenue Cutter Service initiates "Winter Patrol"
1847	U.S. Lifesaving Service initiated; monies given to fund 18 lifeboat and rope throwing stations along the Massachusetts coast; 2,900 documented lives saved during the first five years of operation
1869	Pennsylvania mine safety legislation, after various mine field explosions
1871	Chicago Fire (250 Killed) Peshtigo (Wisconsin) Fire (1,552 Killed, however the total deaths may have approached 2,400).
1877	Massachusetts "Guarding" Laws passed
1890	U.S. Lifesaving Service trains 3,000 surfmen in physician-developed rescue protocol
1894	Underwriter Laboratories, Chicago
1896	National Fire Protection Association, Boston
1899	Safe Harbors and Waters Act: initial regulation of dangerous goods in transport
1903	Iroquois Theater Fire, Chicago (575 Killed)
1904	General Slocum Fire, New York (1,021 Killed)
1906	San Francisco Earthquake (452 Killed), lifesaving response by U.S. Army
1909	Lakeview Grammar School Fire, Chicago (175 Killed)
1910	U.S. Bureau of Mines organized, with mine safety responsibilities
1911	Triangle Shirt Waist Company Factory Fire, New York City (92 Killed)
1912	Safety Congress, Association of Iron and Steel Electrical Engineers (National Safety Council)
1936	Walsh-Healy Act, initiated government labor safety rules
1945	Safety Harbors and Waters Act, revisions enacted to 1899 Legislation
1955	Our Lady of Angels School Fire, Chicago (95 Killed)
1966	Highway Safety Act–U.S. Department of Transportation formed
1970	EPA formed by presidential directive OSHA formed–Williams Steiger Occupational Safety and Health Act of 1970
1972	Consumer Product Safety Commission formed
1974	55 mph speed limit–required across the United States
1976	Resource Conservation and Recovery Act (RCRA) Toxic Substances Control Act (TSCA)

(continues)

TABLE 1-1 (continued)

1979	Three Mile Island, Nuclear incident, Pennsylvania
	Love Canal incident, New York
1980	Comprehensive Environmental Response Compensation and Liability Act (CERCLA)
1984	Bhopal, India, hydrocyanic acid gas release (3,000 immediate deaths, up to 18,000 eventual deaths.)
1986	Chernobyl Nuclear disaster, Russia
1986	Superfund Amendments and Reauthorization Act
1993	First terrorist attack on World Trade Center, New York City
2001	Terrorist attacks on World Trade Center, Pentagon, and crash of fourth plane in Pennsylvania
2003	U.S. Department of Homeland Security Formed

TABLE 1-2 Select Safety Equipment Development

Date	Device	Comment
1778	Rudimentary fire extinguisher	This device was a modified rifle, used to "shoot" water at burning roofs; it was patterned after European "Fire Squirts" of the same principle
1815	Miners' safety lamp	Used by Sir Walter Davy in England to detect gas in mines
1822	Fresnell lens	Used seven orders of lamp lenses that were developed to intensify lighthouse beacons
1837	Soda acid fire extinguisher	Patent denied by U.S. Patent Office as a useless item
1847	Breeches Buoy	Extensively used with rigging by the U.S. Life Saving Service for ship-to-ship and ship-to-shore rescues
1852	Coston flare	Used for night illumination and warning
1855	First fire alarm telegraph	Dr. William Gomoy, inventor; by 1859 J.N. Gamewell had patent rights for the entire country (Gamewell Fire Alarm Systems); telegraph communications had also been introduced to communicate between U.S. lifesaving stations, replacing flag communications
1865	U.S. Ambulance and Stretcher specifications issued	Extensively used by the Union forces during the Civil War
1910	Carbon tetrachloride fire extinguisher	Used to extinguish fires; was eventually outlawed as a safety hazard for generating poison gas in the process of extinguishing fire
1918	Gibbs rebreathing respirator	Used in mines; McCaa respirator introduced in 1926
1920	Type N respirator	"Hopcalite" cartridge in this respirator converted CO-CO_2, making it a useful respirator for certain situations such as firefighting
1939	Chem-Ox respirator	Oxygen Breathing Apparatus followed in 1946 by the self-contained compressed air breathing apparatus; developed by the military
1942	Splint	Developed by Sir Hugh Owen Thomas, reportedly reduced mortality from 80% to 20%
1947	Placard identifiers used for dangerous goods	Use initiated after Texas City, Texas, explosion and fires

Toxic Materials

Industrialization also spread the use of toxic (illness causing) materials in both occupational and public settings. Lead, a naturally occurring metal whose use had been widespread for thousands of years, and asbestos, a series of fibrous minerals used in the insulating process, were two common toxic materials. The use and maladies related to these materials were widespread in the early twentieth century, while controls and restrictions were prevalent in the latter half of the century. Timelines identifying the use of and restrictions for these materials are found in **TABLES 1-3** and **1-4**. Information about the health effects of these materials is found in Appendices B and C.

TABLE 1-3 Chronological Timeline: Asbestos-Containing Materials	
1920	Unregulated use of asbestos-containing materials
1970	Asbestos Occupational Exposure Standard set at 5 f/cc of air
1971	Asbestos listed as a hazardous air pollutant and prohibition of sprayed-on asbestos. "No Visible Emissions" standard promulgated
1976	Occupational exposure standard lowered to 2 f/cc
1979	EPA "Schools Technical Assistance Rules"–requiring inspection of Asbestos ceilings in Schools
1986	Asbestos Hazards Emergency Response Act (AHERA)
	OSHA PEL lowered to 0.2 f/cc
1990	PEL lowered to 0.1 f/cc
1991	EPA extends AHERA to public and commercial buildings (ASHARA)

TABLE 1-4 Chronological Timeline: Lead-Containing Materials	
6500 B.C.E.	Mining of lead ore begins (for smelting purposes)
4500 B.C.E.	Egyptian use of lead as eye salve, astringents, and cooling agents
3200 B.C.E.	Indian use of lead as eye salve and makeup; also recorded in the Old Testament, Mesopotamian cultures, and found in the ancient graves of North American Indians
	Chinese elixir formula using lead
	Aristotle reports on uses of lead as a spermicide.
1500 B.C.E.	Ancient Romans: 80,000 lbs. of lead produced at peak production
	Use as an internal medicine, a contraceptive, in wine production, paint pigments, food flavoring, and cooking utensils (dramatic rate of birth defects and abnormalities in children, sterility, and psychological deficits in Roman aristocracy)
	Hindu (Sanskrit) medical compendia lists use of lead
	Egyptian medical compendia lists 30 uses of lead
200 B.C.E.	Physician Nikander of Colophon wrote of the poisonous effects of lead, but medicinal uses of lead continue up to 1000 C.E. (Hippocratic collection: 30 medicines containing lead)
	Algerian and Arabic medical uses continue
1450 C.E.	Paracelsus identifies pharmacological change, "treat disease with drugs producing similar, not opposite, effects"
1497	Jean Fernell warns against internal uses of lead and documents lead poisoning
1600	Other medical practitioners reporting on overuse problems with lead
	Prescription as an antiaphrodesiac leads to lead poisoning among monks (lead had previously been used as an aphrodisiac)

(continues)

TABLE 1-4 (continued)	
1786	Ben Franklin noted that the hazards of lead had been known for at least 60 years
1887	Mass poisoning in Philadelphia due to lead dye in cake
1890	Brisbane (Queensland) identifies four clinical lead poisoning groups and identifies painted railings and walls as the source; dusts and soils identified as other exposures
1920	Australian legislation introduced to preclude lead painted surfaces within reach of children Lead poisoning begins to be recognized among children
1932	Burning of discarded battery casings linked to lead poisoning in Baltimore (Baltimore health department halted use of battery casings for fuel)
1966	Chicago: first mass screening for children (60 micrograms/deciliter [mg/dl] is considered normal blood lead level)
1967	"Lead Poisoning in Children" published (first federal acknowledgment of preventable lead poisoning)
1969	Lead poisoning linked to improperly glazed earthenware
1970	40 mg/dl used to identify childhood lead poisoning
1971	200 micrograms per cubic meter of air (mg/m^3) OSHA air (lead) standard
1972	Federally assisted mass screening for children (1971–Lead-Based Paint Poison Prevention Act) (20-45% of children tested had blood/lead levels of > 40 mg/dl)
1973	L.B.P.P.P.A. Amendments: eliminates lead-based paint in federal housing
1977	Consumer Product Safety Commission–0.06% lead maximum in paint (for housing constructed after 1978)
1978	OSHA general industry standard reduced to 50 mg/m^3 of lead in air
1980	National Academy of Health Sciences publishes "Lead in the Human Environment" (60,000 tons of lead released into the environment annually, 90% from auto emissions) It is noted that U.S. mean blood lead levels reductions parallels reductions in consumption of leaded gasoline
1987	EPA proposal: 0.005 mg/liter of lead in water standard
1988	ATSDR: 47 million homes with lead-based paint; 12 million children exposed to lead paint at toxic levels
1990	NIOSH goals published: elimination of exposures resulting in >25 mg/dl of blood lead Interim HUD guidelines issued (McKinnley Homeless Act)
1992	Housing and Community Development Act: Title X, within 180 days: OSHA to issue interim final rule, as protective as the interim HUD guidelines for workers
1993	OSHA standard 29 CFR 1926.62: Lead Exposure in Construction issued
1995	HUD: final document issued: "The Guidelines for the Evaluation of Lead-Based Paint Hazards in Housing"
1996	40CFR745: requirements for lead-based paint activities in target housing and child occupied facilities issued
1997	Chapter 7 of the HUD Guidelines revised and published
2006	EPA issued the Proposed Lead, Renovation, Repair, and Painting (RRP) Rule that requires all contractors and home professionals to become certified, receive training, and use lead-safe work practices, controls, and cleaning procedures when working in pre-1978 housing. Rule would be later issued in 2008
2008	EPA reduced the National Ambient Air Quality Standards (NAAQS) criteria for lead by a factor of ten. As of November 2008, the NAAQS criteria for lead is now 0.15 ug/m^3 averaged over a 24-hour period
2009	Consumer Product Safety Commission reduces allowable amounts of lead in consumer products to 0.009% (previously 0.06%)
2010	The RRP Rule took effect on April 22, 2010. All work performed in pre-1978 housing for compensation (including payment, rent, tuition, etc.) was required to be performed in compliance with the RRP Rule

Recent Safety and Health Organizational Development

Within recent times (i.e., since World War II), industrial and population expansion have necessitated both an increased organization of health and safety groups as well as the further development of equipment and procedures to protect populations in the modern hazardous environments.

An example is the U.S. Centers for Disease Control and Prevention (CDC), which is a component of the U.S. Department of Health and Human Services. Organized after World War II to address the threat of malaria, the Communicable Disease Center was the forerunner to the Centers for Disease Control and Prevention of the U.S. Public Health Service. After success in the American south, the CDC addressed and expanded into multiple areas of occupational safety and health, infectious disease management, injury and illness statistics, epidemiology, and food and drug safety.

Since 1966, additional federal agencies have addressed dangerous situations. In 1966, the U.S. Department of Transportation (DOT) was organized. The Research and Special Programs Administration, now known as The Research and Innovative Technology Administration, part of the DOT, promulgates standards for the safety of hazardous materials in transport.

In 1970, the Williams Steiger Occupational Safety and Health Act facilitated the formation of both the Occupational Safety and Health Administration (OSHA) and the National Institute for Occupational Safety and Health (NIOSH). OSHA has specific responsibilities for codifying and enforcing engineering, administrative, and other protective controls on businesses to prevent accidents, illnesses, and injuries. NIOSH investigates, records, and recommends optimal safety practices, and is currently one of the Centers for Disease Control and Prevention.

In 1970, the U.S. Environmental Protection Agency was formed with the mission of protecting the country's air, water, and soil.

The U.S. Department of Homeland Security was formed in 2003, subsequent to the 9/11 (2011) attacks on the World Trade Center, the Pentagon, and the crash of a fourth hijacked plane in a field in Pennsylvania. The U.S. Department of Homeland Security has a broad mission of securing the American homeland.

These current regulatory agencies are further discussed in Chapter 2.

Chapter Summary

Hazardous environments have been seen historically in many cultures. As society developed, however, increased populations and expanding industrialization brought more hazardous environments to more people. To prevent illness and accidents, codes identifying the best safety and health practices emerged, as well as organizations (private and governmental) to propagate and enforce them. Finally, response groups and procedures emerged to mitigate the effects of hazardous atmospheres once they occurred.

By the mid- to late 20th century, a variety of organizations had developed to protect individuals in both public and occupationally hazardous areas. These organizations now attempt to work through the following:

1. *Enforceable codes* aimed at preventing illness and injuries both in communities and work forces, and organizations that promulgate these codes and attempt to integrate healthy and safe practices into daily events
2. *Safety related equipment* to detect and prevent the hazardous environments as they emerge, as well as to protect individuals in the hazardous environments
3. *Private and public organizations* that respond to health and safety situations as they occur for the purpose of mitigating damages or severity

Although health and safety initiatives have historically followed disastrous events, today that chain of events may be changing. Currently, positive safety and health characteristics are identified with the expectation that they will be mimicked. In today's environment, health, and safety traits and characteristics cross both occupational boundaries and occupational/nonoccupational lines.

Terms

Accident: An unforeseen or unplanned event or circumstance affecting persons or property.

Hazardous Environment: An operation or area with an increased prevalence of accidents or illnesses affecting nonprotected individuals.

Illness: A sickness or unhealthy condition of body or mind, usually with impairment of function.

References

Carway, W.F. *Firefighting Lore—Strange but True Stories from Firefighting History*. New Albany: FBH, 1996.

CDC's 50th Anniversary, July 1, 1996. *MMWR* 45 (1996):525–545.

Centers for Disease Control and Prevention. *Principles of Epidemiology*. Atlanta: CDC, 1990.

Cocciardi, J.A. A Comparison of Characteristics of Optimal Health and Safety Programs for Occupations Entering Hazardous Environments (Ph.D. Diss., The Union Institute, Cincinnati, 1999).

Holland, F.R. *America's Lighthouses: An Illustrated History*. New York: Dover, 1981.

———. *Lighthouses*. New York: Metrobooks, 1995.

International Fire Service Trainers Association. *Fire Service Orientation and Terminology* (3rd ed). Stillwater, OK: Fire Protective Publications, 1995.

Johnson, R.E. *History of the United States Coast Guard 1915–Present*. Annapolis, MD: Naval Institute Press, 1987.

Kuk, M.L. Markt Saint Florian: Burial Site of a Martyred Fire Chief. *Firehouse Magazine*, 25(2000): 72–74.

Mobley, Joseph A. *Ship Ashore: The U.S. Lifesavers of Coastal North Carolina*. North Carolina University of Archives and History, Raleigh, 1994.

Murphy, Jim. *The Great Fire*. New York: Scholastic, 1995.

Toxic Substances Control Act. U.S.C. (September 28, 1976) 2601 et seq.

World Book Encyclopedia (1961). s.v. "fire."

2 Current Regulations, Regulatory Agencies, and Professional Associations: Roles and Responsibilities

Key Concepts

- U.S. Department of Homeland Security
- U.S. Department of Transportation (DOT)
- U.S. Occupational Safety and Health Administration (OSHA)
- U.S. Environmental Protection Agency (EPA)
- The National Institute for Occupational Safety and Health (NIOSH)
- The National Safety Council
- The Agency for Toxic Substances and Disease Registry (ATSDR)
- The National Fire Protection Association (NFPA)
- The American National Standards Institute (ANSI)
- The American Conference of Governmental Industrial Hygienists (ACGIH)
- The Resource Conservation and Recovery Act (RCRA) (1976)
- The Superfund (Comprehensive Environmental Response Compensation and Liability Act of 1980 (CERCLA)
- The Toxic Substances Control Act (TSCA) (1976)
- The Superfund Amendments and Reauthorization Act of 1986 SARA Standard
- Hazardous Waste Operations and Emergency Response (HAZWOPER)
- Process Safety Management (OSHA) and Risk Management Programs (EPA)

Introduction

Public and occupational safety and health guidelines are promulgated by both professional associations and governmental agencies. Before 1970, few regulatory safety and health standards existed as we know them today.

The formation of the **U.S. Department of Transportation (DOT)** (1966), the **U.S. Environmental Protection Agency (EPA)** (1970), the **Occupational Safety and Health Administration (OSHA)** within the **U.S. Department of Labor** (1970), and the **U.S. Department of Homeland Security** (2003) changed this regulatory stance. These federal enforcement agencies, as well as other professional and governmental groups discussed in this chapter, are the current producers and enforcers of health and safety standards.

Professional Associations

Professional associations, with missions related to public and occupational safety, were generally precursors to today's regulatory agencies. These professional associations include the **National Safety Council**, **American National Standards Institute (ANSI)**, the **National Fire Protection Association (NFPA)**, and the **American Conference of Governmental Industrial Hygienists (ACGIH)**.

The National Safety Council was founded in 1913 as a privately supported public service organization with a mission of accident and occupational disease reduction. Since its founding, the National Safety Council has expanded from workplace safety to highway, community, and recreational safety issues.

ANSI, founded in 1918 by civil engineering societies and three governmental agencies, facilitates the Consensus Standard making process. The institute itself does not develop standards, but rather establishes consensus among qualified groups. Today, there are nearly 15,000 American National Standards, including such safety-related standards as those concerning personal protective equipment (hard hats, safety glasses, and foot protection).

The NFPA, organized in 1896, has as its mission the reduction of the worldwide burden of fire and other hazards on the quality of life by providing scientifically based consensus codes, standards, research, and training and education. The NFPA provides consensus codes on such everyday safety items as the fire extinguisher.

ACGIH, founded in 1938, was initially a body representing federal, state, and municipal industrial hygienists. In 1946, the organization offered full membership to all industrial hygiene personnel. Today the group has 12 standing committees and is best known for its publication of *Threshold Limit Values for Chemical Substances and Physical Agents and Biological Exposure Indices*.

Governmental Agencies

Current safety, health, and environmental regulatory agencies and regulations have developed from initiatives to protect the environment (EPA [1970], Agency for Toxic Substances and Disease Registry [1986]) as well as to ensure both public and occupational safety and health (U.S. DOT, [1966], OSHA, National Institute for Occupational Safety and Health (NIOSH) [1970]). OSHA and the EPA were well organized as enforcement agencies by the mid 1970s.

In addition, consensus standards organizations (both governmental and private) had matured by the 1970s and had begun to codify health and safety, illness, and injury prevention recommendations.

The U.S. DOT, established by Congress on October 15, 1966, is tasked with the development and maintenance of a safe and efficient transportation system. It consists of an Office of the Secretary, and 11 operating administrations, boards, or corporations. The U.S. DOT Research and Innovative Technology Administration publishes the *Emergency Response Guidebook* for first responders to hazardous material releases. In addition, DOT publishes and enforces procedures for the safe transport of hazardous materials.

Since its inception, the U.S. DOT has promulgated hazardous materials safety regulations. These requirements identify appropriate packing containers, labeling, and vehicle placarding, manifesting, and training (commercial drivers and affected employees) for individuals who transport hazardous materials. Further information concerning these transportation safety requirements promulgated by the U.S. DOT is found in **TABLE 2-1**.

OSHA was formed in 1970 by the Williams–Steiger Occupational Safety and Health Act and is a component of the U.S. Department of Labor. The specific mission of OSHA is to enforce workplace safety and health regulations. OSHA requires employers to post notification of worker safety and health rights, record and investigate accidents and illnesses (and immediately report multiple occupational injuries or a single occupational fatality to the agency), and follow published safety rules.

NIOSH, also formed under the 1970 OSH Act, operates under the U.S. Public Health Service, and is one of the Centers for Disease Control and Prevention (CDC), within the Department of Health and Human Services. The mission of NIOSH includes health hazard investigations culminating in recommendations and alerts to both agencies of occupational safety and the working public. NIOSH recommendations are generally considered to be the best available technological works

TABLE 2-1 Select U.S. DOT Hazardous Materials Transporters Requirements

	U.S. DOT Hazmat Employee Training Requirements:

• Determine if a hazardous material is to be transported
• Package and identify the product (Hazard Communication) properly
• Train Hazmat employees:
Hazmat Employee: Person employed by a Hazmat employer, who in the course of employment directly affects hazardous materials transportation safety. This includes self-employees. A Hazmat employee is one who:
  • Loads, unloads, or handles hazardous materials
  • Repairs, modifies, marks, or otherwise represents containers, drums, or packaging as qualified for use in the transportation of hazardous materials
  • Prepares hazardous materials for transportation
  • Is responsible for safety of transporting hazardous materials
  • Operates a vehicle used to transport hazardous materials
  • Completes an appropriate manifest
• Ship according to commercial regulations
• Report all incidents/accidents/releases to the U.S. DOT
	• Awareness Training
	  • DOT regulation information
	  • Recognizing hazardous materials
	  • Identifying hazardous materials
	  • Security Awareness or Security Specific Training
	• Function-Specific Training
	  • Use of Hazmat table
	  • Shipping papers/requirements
	  • Packaging requirements
	  • Manifesting and labeling requirements
	  • Placarding requirements
	• Safety Training
	  • Emergency response information
	  • Employee protection
	  • Emergency Response Guidebook use
	  • Handling packages safely
	• Commercial Drivers Training requirements
	• Training must recur every three years
	• Required manifest components for hazardous materials include the names of shippers, transporters, and receivers; designation as a hazardous material; the proper shipping documentation; an emergency contact number and the shipper's certification. The manifest must be signed.
	• Form U.S. DOT F 5800.1 is used for accident reporting purposes.

Source: U.S. Department of Transportation, 49 CFR 172 (Subpart H) to 177.

in the occupational safety and health field. In addition, NIOSH tests and certifies respiratory protection equipment. Nonoccupational accident and illness prevention and investigation are the responsibility of the National Center for Injury Prevention and Control, which is also a component of the CDC.

The EPA was formed by Presidential Order (Presidential Reorganization Plan #3 of 1970) and combined environmental duties (air, water, and soil conservation functions) of 16 agencies and groups from the U.S. Department of the Interior; Agriculture; Health, Education and Welfare; the Atomic Energy Commission; the Federal Radiation Council; and the U.S. Council on Environmental Quality. Today, the EPA enforces environmental protection rules and regulations.

The Agency for Toxic Substances and Disease Registry (ATSDR) was established in 1986 by the Superfund Amendments and Reauthorization Act (SARA). ATSDR, an independent operating division within the U.S. Department of Health and Human Services, originally studied the public health effects of toxic substances around the nation's superfund sites. Today the agency has branched into diverse public health issues and recommends environmental health guidelines in these areas.

The U.S. Department of Homeland Security (DHS) was established in 2003, subsequent to the 9/11 terrorist attacks on the United States and the passage of the Homeland Security Act of 2002, which combined 22 federal agencies with protective missions. The DHS coordinates national homeland security efforts.

Within the "environmental" decade of the 1970s, several key pieces of legislation were enacted that shaped public and occupational behavior in hazardous environments, and these are described in the next section.

Legislation

Resource Conservation and Recovery Act of 1976 (40 CFR 240-271)

The Resource Conservation and Recovery Act (RCRA) of 1976 fully introduced the concept of liability into the public and environmental health fields. This specific legislation, designed to protect the environment from inappropriate waste disposal practices, further implemented the concept promulgated earlier of the protection of clean air, water, and soil.

Specifically, the RCRA empowered the EPA to develop regulations for the safe disposal of all waste material, including hazardous waste. It also gave the EPA enforcement powers in this area. Underground storage tanks and medical wastes have been regulated under RCRA.

In addition, "cradle-to-grave responsibility" (i.e., liability) was assigned to individuals for inappropriate disposal practices. This cradle-to-grave responsibility has subsequently been interpreted to apply to all individuals handling wastes in manners that may cause damage to public health and the environment, including manufacturers, users, and transporters/disposers of materials.

Specific waste definitions and practices (including definitions and practices for hazardous wastes) to prevent inappropriate discharges to the environment are issued under the RCRA. **TABLE 2-2** provides RCRA definitions of hazardous wastes, a subcategory of solid wastes.

Toxic Substances Control Act of 1976 (40 CFR 700-799), Federal Insecticide, Fungicide, and Rodenticide Act

The Toxic Substances Control Act (TSCA), also enacted in 1976, assigned to the EPA the task of regulating toxic substances in commerce. In addition, import and export of certain toxic materials were regulated. Although the use of toxic materials, in some cases, was controlled by previous legislation, TSCA prohibited manufacture, and in certain cases, disposal and use of toxic substances.

A variety of substances, such as asbestos-containing materials, lead and lead-based paint, polychlorinated biphenyls, ozone-depleting chemicals, and certain herbicides and pesticides are regulated under the TSCA. A polychlorinated biphenyl label is required by TSCA on certain configurations of electrical equipment. Registration of the equipment is also a TSCA requirement.

PHOTO 2-2 Application of pesticide is regulated by FIFRA.

Additional regulation of pesticides is found in the Federal Insecticide, Fungicide and Rodenticide Act (FIFRA). FIFRA regulations identify registration and application practices for these materials. FIFRA is enforced by the EPA.

The EPA has taken emergency actions in some cases to require abandonment of use, collection, and appropriate disposal of hazardous substances, under the authority of the TSCA (e.g., Agent Orange—one of the notorious herbicides, named for the orange striped barrels in which it was transported—was addressed by the U.S. EPA in this manner in 1979).

Comprehensive Environmental Response Compensation and Liability Act of 1980

A variety of environmentally related health disasters occurred throughout the 1970s and 1980s. In August 1978, toxic substances found seeping and volatilizing into homes in upstate New York produced the evacuation of approximately 264 homes in the Love Canal area of Niagara Falls. Toxic substances, previously buried (legally at the time), had been released from containers and had found their way into home basements and an area school.

The outcome of the immediate actions taken by the New York State Health Department identified the need for funding qualified personnel to remediate environmentally hazardous and toxic sites where responsible parties could not be identified.

Subsequently, sites similar to Love Canal were identified throughout the country. The U.S. government itself (notably the Department of Energy and the Department

PHOTO 2-1 An RCRA-licensed hazardous waste facility.

TABLE 2-2 RCRA Hazardous Wastes

• Ignitable Wastes	• A liquid containing less than 24% alcohol, with a flash point <140°F • A non liquid capable of spontaneous and sustained combustion • An ignitable compressed gas, per DOT • An oxidizer, per DOT • Class "A" or "B" explosive
• Corrosive Wastes	• A material with pH less than or equal to 2, or greater than or equal to 12.5 • A liquid that corrodes steel, at a rate greater than 0.25 inch per year
• Reactive Wastes	• A normally unstable material that reacts violently without detonating • Water reactive wastes • A material that forms an explosive mixture with water • A material that generates toxic gases when mixed with water • A material that contains cyanide or sulfide and generates toxic vapors at a pH between 2 and 12.5 • A material listed by DOT as a Division 1.1, 1.2, or 1.3 explosive

• Toxic Wastes (Toxic by Toxicity Concentration Leaching Procedures [TCLP])	*Toxic Wastes*	*mg/Liter*
	Arsenic	5
	Barium	100
	Cadmium	1
	Chromium	5
	Lead	5
	Mercury	0.2
	Selenium	1
	Silver	5
	Benzene	0.5
	Carbon tetrachloride	0.5
	Chlorobenzene	100
	Chloroform	6
	1,4-Dichlorobenzene	7.5
	1,2-Dichloroethane	0.5
	1,1-Dichloroethylene	0.7
	Methyl ethyl ketone	200
	Tetrachloroethylene	0.7
	Trichloroethylene	0.5
	Vinyl chloride	0.2
	2,4-Dinitrotoluene	0.13
	Hexachlorobenzene	0.13
	Hexachlorobutadiene	0.5
	Hexachloroethane	3
	Nitrobenzene	2
	Pyridine	5.0

(continues)

TABLE 2-2 (continued)		
	o-Cresol	200.0
	m-Cresol	200.0
	p-Cresol	200.0
	Cresol	200.0
	Pentachlorophenol	100
	2,4,5-Trichlorophenyl	400
	2,4,6-Trichlorophenyl	2
	Chlordane	0.03
	Endrin	0.02
	Heplachlor	0.008
	Lindane	0.4
	Methoxychlor	10
	Toxaphene	0.5
	2,4-D	10
	2,4,5-TP (Silvex)	1
• Listed Hazardous Wastes	Nonspecific source listed wastes Specific source listed wastes Commercial and chemical products listed wastes	

of Defense) produced many sites falling into this category. In 1980, Congress enacted the Comprehensive Environmental Response Compensation and Liability Act (CERCLA), better known as the Superfund. This congressional act allocated monies for qualified individuals to investigate, characterize, control, and remediate sites meeting certain hazardous criteria. Since this time, many states have implemented legislation to fund a state Superfund or hazardous-sites actions.

In 1979, a release of radioactive waters from primary to secondary containment at the Three Mile Island Nuclear Power Generating Station outside of Harrisburg, Pennsylvania, and a subsequent release of radioactive materials into the environment occurred (**FIGURE 2-1**). This potential disaster identified to emergency planners that adequate preparations, warning and evacuation, and response plans were not in place around some sites defined as potentially hazardous. In 1985, another nuclear incident, the Chernobyl Nuclear Power Station explosion and fire during a maintenance test activity, further identified that large areas could be impacted by potential disasters. Again, both first-responders' activities and adequate emergency preparedness had not occurred, placing responders and the public at increased risk in these hazardous areas.

Finally, the Bophal, India, release of hydrocyanic acid gas into the air in 1985 resulted in 2000 immediate fatalities and a reported total of 10,000 fatalities over the course of this chemical disaster. This situation again verified that appropriate emergency plans were not in place to protect workers, emergency responders, and the public living in areas surrounding chemical use and manufacturing facilities.

Superfund Amendments and Reauthorization Act (1986)

Consequently, Congress enacted SARA in 1986. Although SARA initially re-funded governmental responses to hazardous waste sites, additional requirements were initiated to preclude concerns raised by situations such as Three Mile Island, Chernobyl, and Bhopal.

These requirements included a directive to OSHA to establish stringent safety regulations that would apply to individuals operating at hazardous waste sites; hazardous waste treatment, storage, and disposal facilities; or emergency responses to releases of hazardous substances. OSHA responded in 1987 with the publication of the Hazardous Waste Operations and Emergency

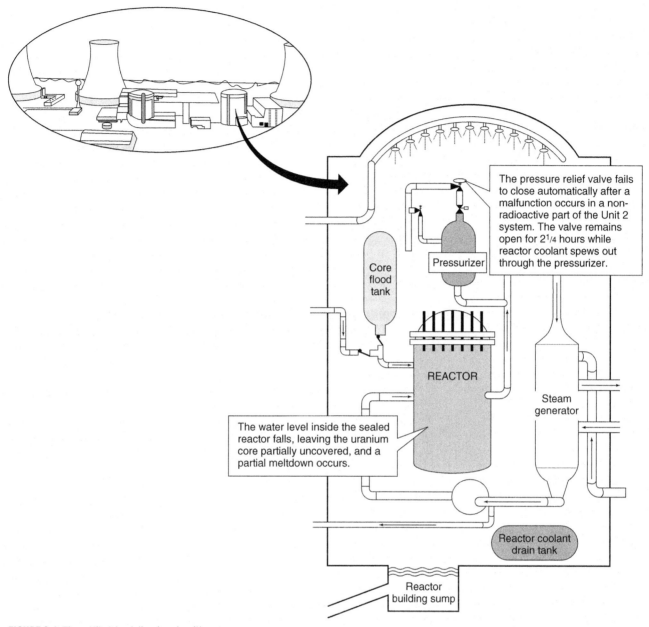

FIGURE 2-1 Three Mile Island disaster algorithm.

Response (HAZWOPER) Standard, affording a high level of protection to these three groups.

An outline of HAZWOPER-covered activities, as well as other hazardous-environments regulation coverage is found in **TABLE 2-3**. **TABLE 2-4** identifies codified requirements of the HAZWOPER Rule. In Title III of the SARA legislations, known as the Emergency Planning and Community Right-to-Know Act (EPCRA), Congress directed states to establish local emergency planning commissions (LEPCs).

Such commissions are required to receive information concerning threshold (reportable) quantities of extremely hazardous substances and their locations, as well as any hazardous substances stored in quantities of more than 10,000 pounds at fixed sites. Hazardous substances include potentially harmful chemical, biological, and radiological materials.

Subsequently, emergency notification, evacuation, and response procedures were required to be developed and coordinated by the LEPCs for releases or potential releases of hazardous substances. In many areas, Hazmat Response Teams have developed specialized equipment, training, and procedures based on the SARA-reported hazards in their areas. The personal safety of responders has also been regulated by the SARA requirements.

PHOTO 2-3 Response actions by trained personnel are required under CERCLA.

PHOTO 2-4 Chemical facilities with reportable quantities of hazardous substances are required to notify LEPC.

TABLE 2-3 Select Hazardous Environments Regulation Coverage

Regulation	Responsible Agency	Coverage
• Hazardous Waste Operations and Emergency Response (HAZWOPER) Standard (29 CFR 1910.120) (29 CFR 1926.65)(40 CFR 311)	OSHA (private employers) EPA (public employers)	• Clean-up operations required by a governmental body, whether federal, state, local, or other, that involve hazardous substances and are conducted at uncontrolled hazardous waste sites (including, but not limited to, the EPA's National Priority Site List [NPL], state priority site lists, sites recommended for the EPA NPL, and initial investigations of government identified sites that are conducted before the presence or absence of hazardous substances has been ascertained) • Corrective actions involving clean-up operations at sites covered by the RCRA as amended (42 U.S.C. 6901 et seq.) • Voluntary clean-up operations at sites recognized by federal, state, local, or other governmental bodies as uncontrolled hazardous waste sites • Operations that involve hazardous wastes and are conducted at treatment, storage, and disposal (TSD) facilities regulated by 40 CFR 264 and 40 CFR 265 pursuant to RCRA; or by agencies under agreement with the EPA to implement RCRA regulations • Emergency response operations for releases of, or substantial threats of releases of hazardous substances without regard to the location of the hazard
• Process Safety Management Standard (29 CFR 1910.119)	OSHA	• Facilities where toxic, reactive, flammable, or explosive chemicals are found above threshold quantities, including listed highly hazardous chemicals and flammable liquids or gases in quantities of 10,000 lbs. or more (Note: Certain flammable fuel gases are exempt from this standard)
• Risk Management Programs (RMP) Requirements (40 CFR 68)	EPA	• Facilities that use or store identified toxic substances or flammable liquid gases above threshold quantities; three levels of programs (Program 1, 2, and 3) are available to facilities dependent on characteristics, such as location and frequency of releases
• Hazardous Materials Transport (49 CFR 170 *et seq*.	DOT	• Shippers of hazardous materials, drivers and transporters of hazardous materials, and individuals who affect the transport of hazardous materials; training required biennially or triennially for Hazmat employees

TABLE 2-4 Hazardous Waste Operations and Emergency Response Program Components

Regulations Codified at:	
29 CFR 1910.120 HAZWOPER	
29 CFR 1926.65 HAZWOPER	
40 CFR 311 WORKER PROTECTION RULE	

• Written Health and Safety Program	• Organizational structure for site activities and chain of command required
	• Site Supervisor, Health and Safety Officer required
	• Comprehensive Work Plan
	  • Training and information
	  • Medical surveillance
	• Site Specific Health and Safety Procedures
	  • Air monitoring component
	  • Personal protective equipment (PPE) program
	  • Site control program
	  • Decontamination procedures
	  • Spill containment program
	  • Emergency response plan
	  • Confined space entry plan
	  • Pre-entry briefing and safety effectiveness component

• Site Characterization	• Preliminary evaluation, risk identification, monitoring, and employee notification of findings
• Site Control	• Map/zones of sites, "buddy system" use, communications, alarming/alerting systems, identification of medical facility for treatment, if required, development and implementation of Standard Operating Procedures (SOPs)

• Training	*Initial Training*
	
	• Site Workers
	  • 40 hours (24 hours for workers occasionally on-site who do not use respirators)
	  • Three day on-the-job training, supervised
	• Site Supervisors
	  • 8 hours additional safety training
	• Site Refresher Training
	  • 8 hours annually
	• Treatment Stage or Disposal Facility Workers
	  • 24 hours initial
	• Treatment Stage or Disposal Facility Supervisors
	  • 8 hours initial
	• Treatment Stage or Disposal Facility Refresher Training
	  • 8 hours annually

	Emergency Responders
	
	• Hazmat Awareness, or
	• Hazmat Operations
	  • 8 hours initial, or
	• Hazmat Technician
	  • 24 hours initial, or
	• Hazmat Specialist
	  • 24 hours initial, or
	• Hazmat Incident Command
	  • 24 hours initial, and
	• Hazmat Refresher
	  • Annually
	• Training Certification Required by the employer

(continues)

TABLE 2-4 (continued)

• Medical Surveillance	• Coverage initiates at 30 days employee exposure above PELs, or
	• Employees who use respirators, or
	• Symptomatic employees, or
	• Hazmat team members
	• Frequency
	  • Initial
	  • Annual
	  • Reassignment/termination
	  • Symptomatic exposures

• Engineering Controls/Work Practices/PPE	• Based on initial evaluation • Level B equipment required if Immediately Dangerous to Life or Health (IDLH) conditions exist or are suspect • Level A equipment required for IDLH skin contact situations • Written PPE program required
• Monitoring	• Initial and periodic monitoring required
• Handling Procedures	• Incipient Fire Requirements in effect (OSHA) • Bulging drums must be considered explosive • Crystallized lab chemicals must be considered explosive • Over-pressurized cylinders must be considered explosive • Radioactive hazards must be assessed • Confined Space Entry Procedures must be in effect
• Decontamination	• Written protocol, prior to entry work • Emergency decontamination capabilities
• Sanitation Requirements	• Potable water on-site • Identification of nonpotable water on-site • Washing and shower requirements for sites • Lighting requirements (minimum 5 ft. candles for work areas)
• TSD Requirements	• Hazard communication • Medical surveillance • Decontamination program • Materials handling program • Emergency response program • PPE program

• Emergency Response	• Specific plan for sites
	• Specific plan for TSDs
	• Subsection "Q" requirements
	  • Plans written, available, and able to handle anticipated emergencies
	  • Elements: Pre-emergency planning/mutual aide
	    • Personnel roles, lines of authority, and communications
	    • Emergency recognition and prevention
	    • Safe distances and places of refuge
	    • Site security and control
	    • Evacuation routes
	    • Decontamination procedures
	    • Emergency medical and first aid procedures
	    • Emergency alerting and response procedures
	    • Critiquing of events required
	    • PPE and equipment program
	• Procedures for Handling Emergency Response
	  • Incident command system (ICS) (site specific)

TABLE 2-4 (continued)	

• Emergency Response (cont'd)	• Incident command (IC) shall
	  • ID all Hazmats involved
	  • Analyze the site
	  • Use engineering controls to reduce exposures
	  • Address PELs and PPE requirements
	  • Develop handling procedures
	  • Ensure PPE Level B is used unless air monitoring occurs/identifies safe atmospheres
	• Use of the "buddy system" in dangerous areas
	• Ensure back-up personnel are available for rescue with first aid and transport capabilities when IDLH environments are present.
	• Designate Safety Officer to
	  • ID and evaluate hazards
	  • Direct safety operations
	  • Alter, suspend, or terminate operations, if IDLH conditions present hazards
	• Develop decontamination procedures
	• Skilled support personnel (special skills) are exempt from training if given an initial briefing concerning PPE and chemical hazards on site.

PHOTO 2-5 Community response resources have been planned subsequent to SARA.

The SARA legislation further increased occupational and public safety in and around areas thought to be the most hazardous to populations.

Clean Air Act Amendments (1990)

In 1990, with the passage of the Clean Air Act Amendments, Congress additionally directed both OSHA and the EPA to afford further protection to employees (OSHA) and communities (EPA) in and around highly hazardous chemical sites. This occurred through the requirements of the Process Safety Management Standard (OSHA), protecting workers onsite with specific quantities of hazardous substances present, and Risk Management Programs (EPA), protecting the communities around these sites. Summaries of required activities under each program are found in **TABLES 2-5** and **2-6**.

TABLE 2-5 Risk Management Program (RMP) Components

- Determine applicability and which RMP applies to the facility.
  Program 1: Worst case release would not affect the public, no significant release in last 5 years, emergency response coordinated with local first responders
  Program 2: Facilities that do not involve chemical processing, and/or do not meet Program 1 or 3 eligibility
  Program 3: Facilities with higher risk, complex chemical processes, such as those covered by OSHA's Process Safety Management Standard (Programs 2 and 3), or other program facilities that have had releases
- Perform off-site consequence analysis, including worst-case scenario
- Develop Integrated Prevention and Emergency Response Program to manage risks, submit to EPA, and hold public information hearings, if required.
- Update at five-year intervals.

Homeland Security Act (2002)

The Homeland Security Act of 2002 was signed into law in November of 2002, in response to the September 11, 2001, terrorist attacks on the United States. The act combined 22 disparate federal agencies and identifies the DHS primary missions of:

1. The prevention of terrorist attacks within the United States
2. The reduction of vulnerability of the United States to terrorism
3. The minimization of damage from and the assistance in the recovery process from attacks which do occur

TABLE 2-6 Process Safety Management Program Components

• Applicability	• 130 specific toxic and reactive substances, or • Flammables in quantities greater than 10,000 lbs.
• Employee Participation	• Employee participation in planning required • Action plan required for employee participation.
• Process Safety Information	• Collect information relative to highly hazardous chemicals in the process, technology of the process, and equipment in the process.
• Perform Process Hazard Analysis (PHA)	• Process hazard team must be assembled and perform PHA. • Mechanism to address findings of PHA • Update at five-year intervals

• Operating Procedures	• Develop written SOP for initial startup.
	  • Normal operations
	  • Temporary operations
	  • Emergency shutdowns
	  • Emergency operations

• Training	• Initial training on process operations • Refresher training at three-year intervals.
• Contractors	• Inform contract employees of hazards of the process and Emergency Action Plan. • Develop and evaluate safe work practices to control contract employees.
• Pre Startup Review	• Perform pre-startup safety review.
• Mechanical Integrity	• Develop written Process Maintenance Procedures and train employees in these procedures.
• Hot Work Permitting	• Develop and use a hot-work permit for operations conducted on or near a PSM process.
• Management of Change	• Develop written procedures for all but "replacements-in-kind" changes. • Update employees and contractors, and PSM information as changes occur.
• Accident Investigation	• Promptly investigate all incidents/accidents and report to PSM team.
• Incident Investigation	• Investigate within 48 hours, an incident that resulted in or could reasonably have resulted in a catastrophic release of a highly hazardous chemical in the workplace. Identify an incident investigation team.
• Emergency Planning and Response	• Establish and implement an Emergency Action Plan and/or an Emergency Response Plan.
• Compliance Audits	• Conduct Compliance Audits at three-year intervals. Audit/update PHA at five-year intervals.

Within 10 years, the Act had been amended more than 30 times, including a major reorganization subsequent to Hurricanes Katrina and Rita, which enhanced the all-hazard response structure of the nation through the Federal Emergency Management Agency (FEMA). In addition to FEMA, diverse agencies such as the U.S. Coast Guard, the U.S. Customs and Border Protection Service, the Domestic Nuclear Detection Office, the Counternarcotics Enforcement Office, and the U.S. Secret Service are housed in DHS, which is currently the third largest agency of the U.S. government.

As a component of the 2007 Homeland Security Appropriations Act, Congress mandated the establishment by DHS of risk-based performance standards for the security of high-risk chemical facilities. The DHS published the Chemical Facility Anti-Terrorism Standards (CFATS) in December of 2007. These standards identified chemicals and quantities of interest, and placed a requirement on facilities to report these to DHS. Originally enacted through 2010, the ability for DHS to regulate high-risk chemical facility security has now been extended.

The CFATS requires the following four actions to occur:

1. A "Top Screen," or review/ranking, of reporting facilities to identify those posing most risk to populations from a terrorist attack
2. A security assessment completed online, to further define/confirm risk
3. A Site Security Plan, addressing DHS risk-based performance standards with 18 required components, and
4. Implementation and confirmatory inspections

Chapter Summary

Although professional organizations have promulgated public and occupational safety and health guidelines, federal agencies such as the EPA, OSHA, U.S. DOT, and DHS have developed their roles as enforcers of safety and environmental health standards. The EPA is the lead agency with air, water, and soil protection responsibilities. OSHA maintains responsibilities for employee safety and health. The U.S. DOT safeguards people and the environment through enforcing rules and regulations that affect the various modes of transporting hazardous materials. The DHS maintains the country's capabilities to prevent, respond to, and recover from terrorist attacks.

Terms

American Conference of Governmental Industrial Hygienists (ACGIH): Founded in 1938, it is a body representing and offering full membership to all industrial hygienist personnel.

National Fire Protection Association (NFPA): Organized in 1896, its mission is the reduction of the worldwide burden of fire and other hazards on the quality of life by providing scientifically based consensus codes, standards, research, training, and education.

National Safety Council, American National Standards Institute (ANSI): Founded in 1918, it facilitates the Consensus Standard making process.

Occupational Safety and Health Administration (OSHA): Formed in 1970, the mission of OSHA is to enforce workplace safety and health regulations. OSHA is a component of the U.S. Department of Labor.

U.S. Department of Labor: Prepares the American workforce for new and better jobs, and ensures the adequacy of workplaces in America.

U.S. Environmental Protection Agency (EPA): Enforces environmental protection rules and regulations.

U.S. Department of Homeland Security: Founded in 2003, it coordinates national homeland security efforts.

U.S. Department of Transportation (DOT): Established by Congress on October 15, 1966, is tasked with the development and maintenance of a safe and efficient transportation system.

References

American Conference of Governmental Industrial Hygienists. *Threshold Limit Values for Chemical Substances and Physical Agents and Biological Exposure Indices*. Cincinnati: ACGIH, 2011.

American National Standards Institute: http://www.ansi.org, 2011.

Clean Air Act, Public Law 91-604, 42 U.S.C. 7401, et seq., December 31, 1970.

Clean Water Act, Public Law 92-500, 33 U.S.C. 1251, et seq., October 18, 1972.

Comprehensive Environmental Response, Compensation and Liability Act, Public Law 96-510, amended at Public Law 99-499, December 11, 1980 (amended 1982, 1986).

Department of Homeland Security: *Chemical Facility Anti-Terrorism Standards Regulations*; 6 CFR 27.

Federal Insecticide, Fungicide and Rodenticide Act, Public Law 92-516, 7 U.S.C. 136, et seq., October 21, 1972.

Homeland Security Act of 2002, Public Law 107-296, November 25, 2002.

Homeland Security Appropriations Act, Public Law 109-295, November 2006.

Interim Final Rule, Chemical Facility Anti-Terrorism Standards (CFATS), Washington, D.C., April 9, 2007.

National Fire Protection Association *Codes and Standards for a Safer World: National Fire Protection Association*. Batterymarch Park, Quincy, MA: NFPA, 2011.

National Safety Council. *Maximize Your Membership: York Desktop Guide to the National Safety Council*. Hasa, IL: NSC, 2006.

Occupational Safety and Health Administration, U.S. Department of Health and Human Services, U.S. Public Health Service, Centers for Disease Control, National Institute for Occupational Safety and Health. *Healthy People 2000: National Health Promotion and Disease Prevention Objectives*. Washington, D.C.: Government Printing Office, 1999.

Resource Conservation and Recovery Act, Public Law 94-580, 42 U.S.C. 6901 *et seq.*, October 21, 1976.

Toxic Substances Control Act, 15 U.S.C., 2601 et seq., September 28, 1976.

U.S. Department of Health and Human Services: http://www.hhs.gov, 2011.

U.S. Department of Health and Human Services, Public Health Service, Centers for Disease Control, National Institute for Occupational Safety and Health. *NIOSH Health Hazard Evaluation Program*. Washington, D.C.: Government Printing Office, 1993.

U.S. Department of Homeland Security: http//www.dhs.gov, 2011.

U.S. Department of Labor: Occupational Safety and Health Administration:http://www.osha.gov, 2011.

U.S. Department of Transportation: http://www.fhwa.gov, 2011.

U.S. Environmental Protection Agency: http://www.epa.gov, 2011.

The Williams-Steiger Occupational Safety and Health Act, Public Law 91-596, 29 U.S.C. et seq., December 29, 1970.

Exercise 2

1. Identify the federal agency (EPA; DOT; OSHA; DHS) with primary jurisdiction for the following events:

    a) Recovery operations subsequent to an earthquake: _____

    b) Transportation of hazmat in cargo aircraft: _____

    c) Identification of locations for hazardous waste disposal: _____

    d) Publication of appropriate regulations for the protection of workers: _____

2. Identify the federal agency with specific jurisdiction over hazardous substance protection from the product gasoline in the following specific situations:

    a) Requirements for the production/maintenance of Material Safety Data Sheets (MSDS) at a work location: _____

    b) Requirements for the "clean-up" of a gasoline spill, at a work location: _____

    c) Requirements for the protection of large quantities of gasoline from potential terrorist acts, at a work location: _____

    d) Publication of appropriate regulations for the transport of large bulk quantities of gasoline through pipelines or by truck transport: _____.

3. The federal regulation that placed "cradle to grave" responsibility on all parties manufacturing, using or disposing of a hazardous substance is _____.

4. _____ is the acronym identifying the federal safety legislation protecting workers at hazardous sites, TSD facilities, and involved in emergency response to releases or substantial threats of release of hazardous substances.

5. Occupational protection of facilities with 130 specific toxic/reactive substances, large quantities of flammables, is provided under which OSHA Standard: _____. Protection for adjacent communities is provided under which EPA Program: _____.

3 Recognizing and Identifying Hazardous Environments

Key Concepts

- Thermal Hazards
- Radiological Hazards
- Asphyxiant Hazards
- Chemical Hazards
- Etiological Hazards
- Mechanical Hazards

Introduction

Hazardous environments may occur within buildings, outside structures, or may exist in the air, water, and soil. In general, six environmentally hazardous situations may be encountered:
1. Thermal hazards
2. Radiological hazards
3. Asphyxiant hazards
4. Chemical hazards
5. Etiological hazards
6. Mechanical hazards

The mnemonic TRACEM can be used to recall these six primary hazards.

Thermal Hazards

The inner body is designed to work efficiently under ordinary temperatures (i.e., temperatures at which the body can heat or cool itself to maintain its core body temperature of 98.6°F [37°C]). Any situations that may prevent maintenance of body temperature should be professionally evaluated and appropriate actions taken to assist with body cooling or warming. In addition to ambient temperatures, the addition or lack of protective clothing and equipment, or the increase or decrease in workload may substantially affect the body's ability to heat or cool itself. Severe **thermal hazards** result in immediate physical problems, such as burns (the destruction of skin tissue). Older individuals or those with circulatory system problems are particularly susceptible to thermal-related problems.

Heat-Related Disorders

Many individuals spend time in hot environments. The human body maintains a fairly constant internal temperature even though it is exposed to varying degrees of environmental heat.

To maintain body safety, the body must rid itself of excess heat, primarily through varying the rate and amount of blood circulation through the skin (as heart rate increases and blood flows closer to skin surfaces, excessive heat is lost) and releasing fluid onto the skin through sweating and subsequent evaporation (cooling). These are automatic responses. As environmental temperatures approach normal skin temperatures, cooling the body becomes more difficult. Blood brought to the body surface cannot lose its heat, and sweating takes over as the primary body cooling mechanism. During conditions of high humidity (or when sweat cannot evaporate adequately through body coverings), the body's ability to maintain acceptable temperatures may become impaired. An individual's alertness and ability to work may also be affected.

Excessive exposure to hot work environments may induce a variety of heat-related disorders, such as heat stroke. Heat stroke is the most serious health problem associated with working in hot environments and is a medical emergency. Heat exhaustion, heat cramps, fainting, or transient heat fatigue may also occur. There are four environmental stresses in a hot environment:
1. Temperature
2. Humidity
3. Radiant heat
4. Air velocity (movement of air over the body surface)

When these factors increase (or decrease, in the case of air velocity) at a rate that cannot be handled by the body's natural mechanisms for shedding heat, physical heat disorders may occur.

Various organizations such as the National Institute for Occupational Safety and Health (NIOSH), the Occupational Safety and Health Administration (OSHA), and the American Conference of Governmental and Industrial Hygienists (ACGIH) have published recommendations for minimizing stress in hot environments. Some states, notably California, have passed legislation to protect workers in hot environments. General recommendations for individuals who must work in hot environments supplied by OSHA can be found in **FIGURE 3-1**.

Typically, safety protocol for hot environments should be considered when temperatures reach 77°F. Symptoms of heat strain should never be ignored. Factors such as work regimen, area ventilation, ambient sunlight, and worker health and fitness affect the initiation of hot environment monitoring. As referenced environmental conditions change, protocol may change as well.

PHOTO 3-1 Hazardous environments include thermal, radiological, asphyxiant, chemical, etiological or mechanical hazards.

Protecting Workers from Heat Stress

Heat Illness

Exposure to heat can cause illness and death. The most serious heat illness is heat stroke. Other heat illnesses, such as heat exhaustion, heat cramps and heat rash, should also be avoided.

There are precautions your employer should take any time temperatures are high and the job involves physical work.

Risk Factors for Heat Illness

- High temperature and humidity, direct sun exposure, no breeze or wind
- Low liquid intake
- Heavy physical labor
- Waterproof clothing
- No recent exposure to hot workplaces

Symptoms of Heat Exhaustion

- Headache, dizziness, or fainting
- Weakness and wet skin
- Irritability or confusion
- Thirst, nausea, or vomiting

Symptoms of Heat Stroke

- May be confused, unable to think clearly, pass out, collapse, or have seizures (fits)
- May stop sweating

To Prevent Heat Illness, Your Employer Should

- Provide training about the hazards leading to heat stress and how to prevent them.
- Provide a lot of cool water to workers close to the work area. At least one pint of water per hour is needed.

For more information:

 Occupational Safety and Health Administration
U.S. Department of Labor
www.osha.gov   (800) 321-OSHA (6742)

OSHA 3154-09-11R

FIGURE 3-1 OSHA Quick Card: Protecting Workers from Heat Stress.

TABLE 3-1 Recognition and Identification Actions for Heat Related Physiological or Other Events

Event	Action
"Heat Wave" (Multiple days where the temperature is at or above 95°F (35°C) or 90°F (32°C), if 9°F (–13°C) or more above the previous daily high). Heat Indices combine the effects of temperature and humidity.	Develop acclimatization schedule and provide Safety Officer to monitor work. [Note: The U.S. EPA recommends monitoring workers in vapor protective clothing at temperatures above 70°F (21°C); however, good industrial hygiene practices dictate monitoring of all workers in impervious clothing for heat-related symptomology. Monitoring or safety planning for all workers at temperatures above 80°F (26.5°C) is recommended.]
"Heat Stress" (Profuse sweating, weakness, nausea and headache, heat syncope [fainting])	Facilitate cooling, hydrate, and reduce work regimen.
"Heat Cramps" (Painful muscle spasms due to failure to replace salts)	Facilitate cooling, hydrate, and seek medical attention.
"Heat Stroke" (Mental confusion, loss of consciousness, coma, body temperature above 106°F (41°C), hot red/blue dry skin)	HEAT STROKE IS A LIFE THREATENING MEDICAL EMERGENCY. SEEK EMERGENCY MEDICAL ATTENTION.

TABLE 3-2 National Oceanic and Atmospheric Administration (NOAA) Heat Index Table

NOAA's National Weather Service
Heat Index

Relative Humidity (%) \ Temperature (°F)	80	82	84	86	88	90	92	94	96	98	100	102	104	106	108	110
40	80	81	83	85	88	91	94	97	101	105	109	114	119	124	130	136
45	80	82	84	87	89	93	96	100	104	109	114	119	124	130	137	
50	81	83	85	88	91	95	99	103	108	113	118	124	131	137		
55	81	84	86	89	93	97	101	106	112	117	124	130	137			
60	82	84	88	91	95	100	105	110	116	123	129	137				
65	82	85	89	93	98	103	108	114	121	126	130					
70	83	86	90	95	100	105	112	119	126	134						
75	84	88	92	97	103	109	116	124	132							
80	84	89	94	100	106	113	121	129								
85	85	90	96	102	110	117	126	135								
90	86	91	98	105	113	122	131									
95	86	93	100	108	117	127										
100	87	95	103	112	121	132										

Likelihood of Heat Disorders with Prolonged Exposure or Strenuous Activity
☐ Caution   ☐ Extreme Caution   ■ Danger   ■ External Danger

Source: U.S. Department of Commerce: National Oceanic and Atmospheric Administration: National Weather Service.

In addition, time of year, time of day, or other environmental factors specific to sites may affect the need for a heat stress monitoring program, as well as the acclimatization level of workers (i.e., how their bodies have been trained to compensate for heat). The recognition, identification, and actions for heat-related physiological events are found in **TABLE 3-1**. Heat indices—a quick measure of how hot it really feels—are published by the National Oceanic and Atmospheric Administration's National Weather Services (**TABLE 3-2**). Many Site Safety Plans identify protective actions based on heat index calculations—in particular, where more elaborate sensing equipment is not present. The heat index chart was designed for use in light wind and shady conditions; full sunshine and strong winds can increase heat index values. A heat index above 105°F (41°C) may cause increasingly severe heat disorders with continued exposure or physical activity. The NWS issues a Heat Advisory, when a heat index of 105–110°F is expected for three hours in a day, and an Excessive Heat Warning when a heat index

TABLE 3-3 Recognition and Identified Actions for Cold-Induced Events	
60.8°F (16°C)	• Initiate and record temperature measurements.
39.2°F (4°C)	• Provide gloves for stationary workers. • Provide gloves for light work and appropriate total body protection. • Provide additional protection to individuals exposed to evaporating liquids.
35.6°F (2°C)	• Treat wet workers and remove any wet clothing. • Provide eye protection, including UV protection for work in snow- or ice-covered terrain.
30.2°F (–1°C)	• Initiate dry bulb temperature measurements every four hours; record wind speed if in excess of 5 mph. These readings are necessary to calculate equivalent chill temperature (ECT) (see Table 2-3). At these temperatures, a safety officer should be designated to determine ECT and applicable actions. • Review/exclude workers with body temperature regulation concerns. • Prevent contact with cold surfaces.
19.4°F (–7°C) ECT	• Provide gloves for moderate work. Provide warming shelters and warming fluids; record measurements every four hours.
10.4°F (–12°C) ECT	• Use buddy system and acclimatization; provide safety/awareness training to workers.
0°F (–17°C)	• Provide mittens for hand warming.
–11.2°F (–24°C) ECT	• Ensure medical approval for workers at these temperatures (this may be reduced to –11.2°F if wind speeds are below 5 mph).
–25.6°F (–32°C) ECT	• At this ECT, prohibit continuous skin exposure.

of >105°F is expected for three hours in two consecutive days or any heat index of 115°F or greater.

In extreme situations, personal protective equipment, such as layered thermal protective clothing (e.g., firefighters' gear) or reflective layered protective clothing (e.g., aluminized glass suits used for work within the proximity of hot areas), is recommended. Further information concerning thermal protective equipment is found in Chapter 7.

Cold-Related Disorders

Events that may cause the body's core temperature to drop below 96.8°F (36°C) should be avoided (i.e., long periods of time in cold, nonprotected environments, such as air or water). Hypothermia can be caused by routine exposure to water, wind, and moisture in air, or adjacent solid materials whose temperature is below human body temperature. The combination or synergistic effects of these materials may cause cooling with increased rapidity.

In general, hypothermia is recognized by a dull tone to the skin, loss of mental alertness, a feeling of numbness in the extremities, and eventually the loss of consciousness. This is caused by the body's internal reaction to cold. In its attempt to warm the heart, the body shuts down the flow of blood to the extremities. When severe shivering becomes evident, all work should be terminated and body warming initiated.

Actions to warm individuals suffering from hypothermia include removing them from the cold environment and raising the body's core temperature. The recognition and identification of cold-related events are found in **TABLE 3-3**, which describes recommended actions at both ambient and *equivalent chill temperatures* (ECT). ECT includes the cooling effect of wind on exposed flesh. It was originally developed by the U.S. Army Research Institute of Medicine, Natick, Massachusetts, and is now identified by the National Weather Services as the wind chill. **TABLE 3-4** presents the National Weather Service wind chill chart. Layered clothing, in addition to acclimatization to cold events, is recommended for cold-related recreational or work environments. As with hot environments, remaining outside the environment where cold temperature extremes are present is always advisable.

Radiological Hazards

Radiological hazards can be classified as ionizing and nonionizing radiation. Although radioactive decay continuously occurs around us, levels above normal background levels may cause personal harm.

Ionizing Radiation

Ionizing radiation has sufficient energy to break the chemical bonds of atoms. This potentially strips

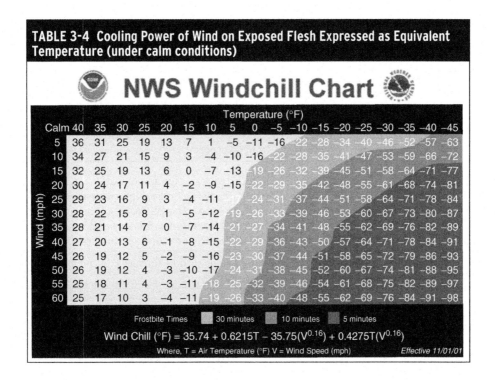

TABLE 3-4 Cooling Power of Wind on Exposed Flesh Expressed as Equivalent Temperature (under calm conditions)

electrons from atomic structures and subsequently may damage cellular materials. Items emitting ionizing radiation (e.g., a pair of positively charged protons and neutrons spontaneously emitted from the nucleus of an atom) may deposit this energy in their path of travel, thus forcing changes (ionizing the chemical make-up) within the body's system, which they affect or pass through.

The characteristics of various types of ionizing radiation—alpha, beta, and gamma radiation (radioactive materials)—are found in **TABLE 3-5**. The effects of ionizing radiation are found in **TABLE 3-6**. Additional types of ionizing radiation (x-ray, photon radiation) may produce hazardous environments as well. Worker exposure to ionizing radiation is generally regulated by the U.S. Nuclear Regulatory Commission (NRC), although the Occupational Safety and Health Administration (OSHA) publishes protective requirements for workers who fall outside the NRC purview.

The amount or strength of the **radioactivity** is related to both the amount of material and the half-life (the time it takes for the radioactive material to reduce [decay] to half its original strength).

Nonionizing Radiation

Nonionizing radiation does not have sufficient energy to break chemical bonds and is therefore considered nonionizing. The primary effect of nonionizing radiation (e.g., radio frequency radiation, microwave emissions, radar) is heating caused by the increased movement and collisions of molecules.

Various recommended exposure limits are published by the National Council on Radiation Protection and Measurements and have been adopted by the U.S. Federal Communication Commission. The American National Standards Institute, the American Conference of Governmental Industrial Hygienists, the U.S. Department of Defense, and the Institute of Electrical and Electronics Engineers have also published recommended exposure limits.

Asphyxiant Hazards

Asphyxiant hazards are those hazards associated with the lack of oxygen in the atmosphere, the introduction

TABLE 3-5 Characteristics of Ionizing Radiation	
Alpha	Two protons/two neutrons charged (+2); heavy, short range in air (< 4"); 0.01 mm in tissue (stopped by dead skin) (4 MeV – 8 MeV energy)
Beta	Electron emitted upon creation from the nucleus; usually a (–1) charge (positions are +1 charge); mid range, (0-20' in air); .2-0.5 cm in tissue, less energy (1 KeV – 1 MeV)
Gamma	No charge/mass, (1 KeV – 10 MeV energy) large range, small energy deposition
X-Ray	10 eV-120 KeV energy, originates in the electron field of the atom

TABLE 3-6 Effects of Ionizing Radiation	
50 R (5 Sv)	Decrease in white blood cells
200 R (2 Sv)	Nausea, vomiting, dizziness, hair loss, diarrhea, and infection
450 R (4.5 Sv)	$LD_{50/30}$: 50% of those exposed are expected to die within 30 days without medical intervention
600 R (6 Sv)	LD_{100}: 100% of those exposed are expected to die without medical intervention
10,000 R (100 Sv)	Death results from central nervous system disorder
Note 1:	Biological half life decreases total body burden
Note 2:	Individuals exposed to 5 rem are considered irradiated (individuals evidencing 0.5 mR/Hr [surface] or 0.1 mR/Hr [thyroid] are considered contaminated). The OSHA/EPA Worker Protection Guideline is 5 rem/year. The NRC Public Exposure Limit is 0.5 rem/year. Background levels of ionizing radiation are generally 0.01–0.02 mR/Hr, and as high as 0.05 mR/Hr in some jurisdictions.
Note 3:	The International System of Units (SI) is used to quantify exposures. In the system, 1 rem = 0.01 sievert (Sv); 1 rad = 0.01 gray (Gy).

of particulate matter into air that may obstruct respiration, or the introduction of certain chemicals into breathable air that may prohibit the up-take of oxygen by the blood.

The first situation described (low oxygen) is considered immediately dangerous and may be found in many occupational environments. Standard breathing air contains approximately 20.9% oxygen and 79% nitrogen by volume. Nitrogen, the inert component, is not used by the body. The oxygen is circulated and used by the body's cardiopulmonary system and used in cellular respiration, where it is converted to carbon dioxide, transported back to the lungs, and expelled. **TABLE 3-7** lists typical air composition.

Human beings must breathe oxygen in order to survive, and they will begin to suffer adverse health effects when the oxygen level of their breathing air drops below the normal atmospheric level. **TABLE 3-8** lists some physical effects of oxygen volume/pressure changes. This is a difficult situation to detect because observable changes are generally not noted in oxygen deficient atmospheres (i.e., there is no color, odor, or visible indicator). Below 19.5% oxygen by volume, air is considered oxygen-deficient. At concentrations of 16% to 19.5%, workers

TABLE 3-7 Typical Air Composition			
Nitrogen	78.1%	Helium	0.0005%
Oxygen	20.9%	Krypton	0.0001%
Argon	0.9%	Xenon	0.000009%
Carbon dioxide	0.03%	Radon	0.000000000000007%
Neon	0.002%		

TABLE 3-8 Physical Effects of Oxygen Volume Changes	
Oxygen Level	Effects
23%	Fire hazard
21%	Normal volume
19.5%	Immediately dangerous to life and health (IDLH) level
16%	Nausea, headaches, and symptoms occur; self-rescue may not be possible
12%	Intermittent respiration and exhaustion
10%	Intermittent respiration, lethargic movement, and unconsciousness
6%	Convulsions and cardiac arrest

engaged in any form of exertion can rapidly become symptomatic as their tissues fail to obtain the oxygen necessary to function properly. Increased breathing rates, accelerated heartbeat, and impaired thinking or coordination occur more quickly in an oxygen-deficient environment. Even a momentary loss of coordination may be devastating to a worker if it occurs while the worker is performing a potentially dangerous activity, such as climbing a ladder. Concentrations of 12% to 16% oxygen cause increased breathing rates; accelerated heartbeat; and impaired attention, thinking, and coordination even in people who are resting.

At oxygen levels of 10% to 14%, faulty judgment, intermittent respiration, and exhaustion can be expected even with minimal exertion. Breathing air containing 6% to 10% oxygen results in nausea, vomiting, lethargic movements, and perhaps unconsciousness. Breathing air containing less than 6% oxygen produces convulsions, then cessation of breathing, followed by cardiac arrest. These symptoms occur immediately. Even if a worker survives the low oxygen event, organs may show evidence of hypoxic damage (injury to the human anatomy due to lack of oxygen), which may be irreversible.

A number of workplace conditions can lead to oxygen deficiency. Simple asphyxiants, or gases that are physiologically inert, can cause asphyxiation when present in high enough concentrations to lower the oxygen content in the air.

At nonstandard atmospheric pressures, the quantity of oxygen that the body may absorb is reduced (i.e., at altitudes above 8,000 feet, the partial pressure of oxygen may be below 100 mm/hg). In addition, at increased pressures such as conditions underwater, nitrogen may be forced into the bloodstream, producing situations such as nitrogen narcosis (divers' euphoria) in which adequate oxygen is not found in the bloodstream. Once introduced, bubbles of nitrogen cannot easily be removed from the bloodstream, producing painful pressure (the "bends"). **TABLE 3-9** identifies situations that may produce low oxygen levels.

Conversely, high levels of oxygen may be considered immediately hazardous. High oxygen concentrations may increase the burning rate and temperature of some materials or reduce the temperature necessary to ignite substances. Consequently, high oxygen levels (higher than 23%) are generally considered to be a fire or explosion hazard. In addition to the hazards of displaced oxygen, particles such as dust may block the respiratory system, or certain chemicals can poison hemoglobin or enzyme systems that permit the transfer and transport of oxygen from the lungs to the blood.

Chemical Hazards

Chemicals may produce hazardous environments. Several classification systems are used to recognize and identify potentially hazardous atmospheres caused by chemicals. These include the United States Department of Transportation, (Categories of Hazardous Materials), the National Fire Protection Association (Code #704: The Identification of Hazards of Materials for Emergency Response), and the OSHA Hazard Communication Standard. Chapter 4 provides further information concerning the recognition and identification of **chemical hazards.**

Etiological Hazards

Etiological hazards are posed by bacteria, viruses, or other human or zoonotic pathogenic agents.

Etiological agents may be found in a variety of packages, many times identified by location. Although many etiological agents coexist naturally, inappropriate introduction of these biological agents into the human body may cause illness or disease. Etiological agents include those of plant, animal, and microbial origin, as well as their aerosols (i.e., fungi, bacteria, molds, viruses, allergens).

Molds and fungi are etiological agents of concern, as they produce adverse effects in some human exposures. Guidelines for safe operations around molds and fungi as promulgated by the New York City Department of Health and Mental Hygiene are found at Appendix D. The body's natural defenses to certain etiological agents can be increased through vaccinations. Over time, the Public Health System has eliminated the etiological hazard of these agents through the vaccination system and through combined sanitation and medical management programs. Individuals with compromised immune systems are at particular risk from etiological agents.

Mechanical Hazards

Hazardous mechanical environments are generally recognized by site inspections. Sharp edges, fragmented equipment, or the chance to be struck by or against

TABLE 3-9 Situations Producing Low Oxygen Pressures or Volumes		
Situation	Effect	Typical Cause
Inert Gases or Gases Heavier than Air	(Oxygen displacement)	Releases of nitrogen, argon, carbon dioxide
Oxygen below 19.5%	(Oxygen deficient)	Fires, sewer work, confined space activities
Chemical Asphyxiation	(Oxygen present but not absorbed into the bloodstream)	CO poisoning, chemical release of cyanides
Dusts, Mists, Aerosols in High Concentrations	(Mechanical blockage of oxygen uptake)	High concentrations of dusts or aerosols

TABLE 3-10 Health Recommendations for Noisy Environments	
Noisy Environment	Assess exposures and individual sound pressure doses.
Environments Measuring Greater than 85 dBA TWA_8 (decibels on the "A" weighted scale, taken as an 8 hour Time Weighted Average)	Hearing Conservation Program required by OSHA for general industry settings and recommended by NIOSH. Hearing Conservation Program required by MSHA for mining sites. This Hearing Conservation Program includes area monitoring, baseline audiograms, optional use of hearing protectors, and annual training.
	The NIOSH-recommended standard requires noise monitoring, engineering, or administrative controls of noise, mandatory hearing protectors, medical surveillance, hazard communication, program evaluation, and record keeping.
Environmental Noise Greater than 90 dBA TWA	Mandatory OSHA and MSHA Noise Control Program, which, in addition, includes mandatory engineering and administrative controls and use of hearing protection when the engineering or administrative controls will not sufficiently reduce employee doses. Impact noise above 140 dB is prohibited.

Adapted from: U.S. Department of Labor, OSHA: Occupational Noise Exposure, 29 CFR 1910.95, 1996; and U.S. Department of Health and Human Services, Centers for Disease Control and Prevention, National Institute for Occupational Safety and Health, "Occupational Noise Exposure, Revised Standard," Cincinnati, OH, 1998.

items should be reviewed when **mechanical hazards** may present themselves. Guarding is generally used as a protective mechanism against mechanical hazards. Excess pressures and overpressures may cause mechanical problems as well. Excessive pressure on the eardrums from loud noise may cause hearing damage.

The U.S. Department of Labor and OSHA require guards for the protection of employees from hazardous mechanical energy, such as moving parts or cylinders that may release pressure.

In addition, the OSHA Control of Hazardous Energy Standard requires lockout of newer energy storage or distribution equipment (e.g., electrical sources) and older equipment incapable of accepting lockouts prior to servicing or maintenance operations. Employee training in these procedures is required. Specific written lockout procedures are required for equipment with multiple energy sources or stored or residual energy. A variety of regulations apply to individuals who work at heights, on scaffolds, and in trenches that may produce mechanical hazards. Further information concerning protection from these hazards is found in Chapter 7.

Noisy Environments

Noisy work environments are regulated by OSHA or other regulatory agencies with appropriate jurisdiction (e.g., the Mine Safety and Health Administration of the U.S. Department of Labor). "Noisy" environments must be evaluated using a calibratable sound level meter. Many sound levels may cause hearing losses. However, those above 85 decibels on the "A" weighted scale are generally considered capable of producing temporary or permanent hearing impairment. The U.S. Department of Labor's hearing conservation regulations apply to most environments with sound pressure levels above 85 decibels. This is also the recommendation of the NIOSH, although their measurement calculations differ from those of OSHA. At 90 decibels, OSHA requires mandatory hearing protection. General requirements for noisy environments are found in **TABLE 3-10**. Typical noisy environments are listed in **TABLE 3-11**.

Community noise protection, a function of public health, is designed to protect individuals from both nuisance and hazardous noise pressures. Most community noise standards analyze noise by octave band, prohibiting excessive pressures in each octave of the sound environment. Community noise standards are typically

TABLE 3-11 Typical Noisy Environments	
Sound Pressure Level (in decibels)	
30	Very soft whisper
55	Conversation, at 3 feet
60	Air conditioner (window unit)
65	Passenger car; 55 mph at 35 feet
70	Vacuum cleaner
80	Garbage disposal at 3 feet
85	Diesel truck, 40 mph at 35 feet
90	Fire alarm system
95	Power lawn mower; Some manufacturing and printing equipment
115	Pneumatic chipper; Wood-working equipment
140	Jet engine

applied at airports, transportation corridors, and mining operations.

Chapter Summary

Workers and the public must recognize and identify hazardous environments to which they may be exposed. Thermal, radiological, asphyxiant, chemical, etiological, and mechanical hazards present hazardous environments. Quantification of these environments allows specific selection of engineering, administrative, or protective equipment controls. The initial recognition and identification of the environment allows individuals the opportunity to avoid, replace, or remove the hazard; remain distant; or review health and safety requirements prior to entry into the hazardous environment.

Terms

Asphyxiant Hazards: Problems occurring due to the lack of oxygen, the constituent in air required for human respiration. In addition to displacing or removing oxygen in/from air, asphyxiant hazards may be caused by particles or substances in air that either block the human respiratory system (e.g., dust particles small enough to inhale) or prohibit the blood from acquiring and distributing oxygen at standard temperatures and pressures (e.g., cyanide).

Chemical Hazards: Exposure to hazardous chemical substances intentionally, unintentionally, or in emergency situations. Chemical hazard exposure may arrive from direct injection into the body, inhalation, ingestion, or absorption through the skin.

Etiological Hazards: Hazards posed by bacteria, viruses, or other human or zoonotic pathogenic agents (see Chemical Hazards for routes of exposure).

Mechanical Hazards: These hazards include the absorption of pressure from sudden events such as falls, explosions, or incidents in which individuals are struck by or against objects. In addition, mechanical hazards include sharp or hard surfaces, uneven or improperly balanced work environments, and the excess pressures of agents such as noise.

Radioactivity: The process by which unstable atoms "try" to become stable and, as a result, emit radiation (energy). This process is called disintegration or decay.

Radiological Hazards: Exposure above background levels of ionizing or nonionizing radiation.

Thermal Hazards: Exposure to temperature extremes that may cause hypothermic or heat-related disorders.

References

American Conference of Governmental Industrial Hygienists. *Threshold Limit Values for Chemical Substances and Physical Agents and Biological Exposure Indices.* Cincinnati: ACGIH Worldwide, 2010.

American Industrial Hygiene Association. *Field Guide for the Determination of Biological Containments in Environmental Samples.* Fairfax, VA: ACGIH Worldwide, 1996.

Cocciardi, J.A. The Development and Validation of Procedures for Working in Hot Environments. *American Society of Safety Engineers, Industrial Hygiene Division Newsletter.* June 2000.

———. The Development and Validation of Procedures for Working in Hot Environments in an Industrial Cleanup Situation During the Summer of 1997. *Paper presented at the American Industrial Hygiene Conference and Exhibition.* Atlanta, May 1998.

———. *Emergency Planning: Medical Services-1: A Manual for Medical Service Personnel Addressing Contaminated or Irradiated and Otherwise Physically Injured Persons and Equipment Following a Nuclear Power Plant Accident.* Commonwealth of Pennsylvania, Emergency Management Agency: Mechanicsburg, PA, 2004.

Department of Defense, Instruction Number 6055.11. Washington, D.C.: Government Printing Office, 2004.

Federal Communications Commission. Evaluating Compliance with FCC Guidelines for Human Exposure to Radiofrequency Electromagnetic Fields. *Office of Engineering & Technology Bulletin 65.* Washington, D.C.: Government Printing Office, 1997.

———. *Guidelines for Evaluating the Environmental Effects of Radiofrequency Radiation.* Washington D.C.: Government Printing Office, 1997, 96–326.

Federal Communications Commission, Office of Engineering & Technology. *Information on Human Exposure to Radiofrequency Fields.* Washington, D.C.: Government Printing Office, 1998.

Federal Register. Hazardous Waste Operations and Emergency Response. *U.S. Department of Labor, Occupational Safety and Health Administration.* March 8, 1989.

Federal Register. Respiratory Protection, Final Rule. *Department of Labor, Occupational Safety and Health Administration.* January 8, 1998.

Hazardous Materials Information Guides. *Laboratory Safety: Industrial Supplies.* Janesville, WI: Lab Safety Supply Company, 1998.

"Heat Related Mortality, U.S., 1997." *MMWR* 47(1998): 23, 473.

National Fire Academy, National Emergency Training Center. *Hazardous Materials Incident Analysis*. Emmittsburg, MD: NFA-SM-HMIA/TTT, February 1, 1985.

National Fire Protection Association. *Code #704: Identification of the Hazards of Materials*. Quincy, MA: NFPA, 2007.

National Institute of Occupational Safety and Health, U.S. Public Health Service, Centers for Disease Control and Prevention. *Chemical Control Corporation, Health Hazard Evaluation Report #TA80 77-853*. Elizabeth, NJ: NIOSH, U.S. Public Health Service, CDC, April 1981.

Occupational Safety and Health Administration. "Heat Stress." In *OSHA Technical Manual*. Washington, D.C.: Government Printing Office, 1986.

Rom, W. *Environmental Occupational Medicine, 2nd ed.* Boston: Little, Brown, 1992.

U.S. Department of Health and Human Services, Centers for Disease Control and Prevention, National Institute for Occupational Safety and Health. *Criteria for a Recommended Standard . . . Occupational Exposure to Hot Environments*. [DHHS-No. 86-113], Washington, D.C.: Government Printing Office, 1986.

———. *Occupational Noise Exposure, Revised Standard*. Cincinnati, OH: NIOSH, 1998.

U.S. Department of Labor, Occupational Safety and Health Administration. *Hazardous Energy, 29 CFR 1910.147*. Washington, D.C.: Government Printing Office, 1994.

———. *Occupational Noise Exposure, 29 CFR 1910.95*. Washington, D.C.: Government Printing Office, 1996.

U.S. Department of Transportation. *Hazardous Materials Table, 49 CFR 171.101*. Washington, D.C.: Government Printing Office, Current.

Exercise 3

Indicate the potential thermal, radiological (ionizing or nonionizing) asphyxiant, chemical, etiological, or mechanical hazards depicted.

1.

PHOTO 3-2

2.

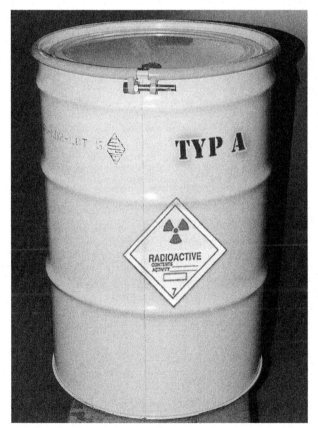

PHOTO 3-3

3.

PHOTO 3-4

4.

PHOTO 3-5

5.

PHOTO 3-6

6.

PHOTO 3-7

7.

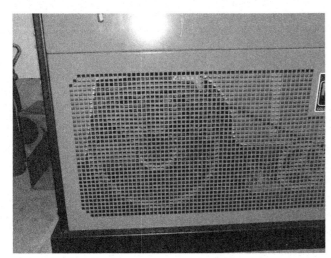
PHOTO 3-8

8.

PHOTO 3-9

CHAPTER 3: Recognizing and Identifying Hazardous Environments

4 Recognizing and Identifying Hazardous Materials

Key Concepts

- U.S. Department of Transportation: Categories of Hazardous Materials
- NFPA: Hazardous Materials Identification System
- The United Nations Globally Harmonized System of Classification and Labeling of Chemicals: HAZCOM 2012
- The Material Safety Data Sheet/Safety Data Sheet
- The Emergency Response Guidebook
- The Hazardous Material Shipping Paper
- The Hazardous Waste Manifest

Introduction

The release of **hazardous materials, substances, or wastes** into the environment, either intentionally or accidentally, may create a hazardous work environment. Recognizing and identifying hazardous materials, substances, or wastes as early as possible at a work site, community event, or emergency incident allows the individual optimal time to plan protection for workers, the public, and the environment.

Individuals who work around these potential hazards need to be aware of the characteristics of the different categories of hazardous materials; the container shapes and sizes used to transport the materials; the signs, symbols, and colors used to identify the materials; and the means for interpreting associated documentation maintained with hazardous chemicals (i.e., materials, substances, and wastes).

The Primary Dangers of Hazardous Materials

Hazardous materials may create two types of hazardous atmospheres:

1. **Fire and Pressure Releases:** Releases of pressure or energy, such as through a fire, which is fueled/oxidized by materials in the U.S. DOT hazard classes:
   - explosives
   - compressed gases
   - flammable and combustible liquids
   - flammable solids
   - oxidizers

   These categories cover the 16 physical hazard categories currently identified in the OSHA Hazard Communications Standard (The United Nations Globally Harmonized System of Classification and Labeling of Chemicals (GHS), which additionally includes materials corrosive to metals. The "Fire Triangle" (**FIGURE 4-1**) identifies the elements necessary for the oxidation/reduction reaction, which we call "fire," to occur. Today, the Fire Triangle has been redefined as the Fire Tetrahedron (**FIGURE 4-2**), with a fourth element included in the fire reaction being the ability for the initial three components to react or combine, chemically. Some oxidation reactions occur rapidly (e.g., an explosion), whereas some occur at a much slower rate (e.g., rusting).

   For a fire to occur, gaseous oxygen generally combines with gaseous fuel in the proper proportions. When mixed appropriately (not too lean/not too rich), the material—which has reached its **flash point** and is in its **flammable range**—may find a source of ignition and oxidize or reduce with the release of pressure, energy, heat, light, gases, and vapors or partial products of combustion (**FIGURE 4-3**).

   Identification of potential materials present in their flammable ranges is essential to protect

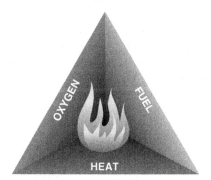

FIGURE 4-1 The Fire Triangle.

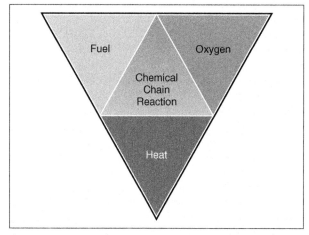

FIGURE 4-2 The Fire Tetrahedron.

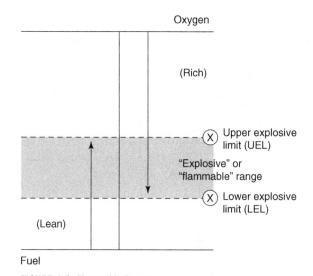

FIGURE 4-3 Flammable Ranges.

personnel prior to entry into a hazardous atmosphere. General precautions when operating around explosives; **flammable** and **combustible** gases, liquids, and solids; and oxidizers require elimination of as many sides of the "Fire Triangle" as possible, or the introduction of a substance to stop the interaction. These include the following:
- *Elimination of sources of ignition* (e.g., no smoking or open flames, stop "hot work," disconnect electrical sources, use only intrinsically safe equipment)
- *Elimination of sources of fuel* (e.g., stop the leak or release of fuels if it is safe to do so, remove excess fuels from the spill, release or work area)
- *Elimination of oxygen* (e.g., seal the area or container, close doors and windows)
- *Introduction of a substance to block the chemical reaction* (e.g., halogenated extinguishing agents)

2. **Toxic Atmospheres:** Releases of toxic atmospheres with the potential for introduction of toxic materials into the human body. Toxicity hazards include the hazardous environments of the U.S. Department of Transportation (DOT) "toxic" hazard classes:
   - Poisons
   - Corrosives
   - Radioactive materials
   - Other Regulated Materials (ORM) and hazardous items (e.g., magnetized materials, hot and cold materials, water-absorbing materials, or consumer quantities of other hazardous materials [i.e. hazardous substances])

These categories cover the 10 health hazard categories and 2 environmental hazard categories currently identified in the OSHA HAZCOM Standard (GHS). Four routes are possible for the introduction of **toxic materials** into the body. These are listed in **TABLE 4-1** in the order in which a toxic material may most rapidly enter an individual's bloodstream and be distributed through the body.

Typically, individuals may avoid these hazardous atmospheres by remaining away from areas where hazardous materials have been released (distance), avoiding areas of release for as long a period as possible (time), and using engineering and administrative controls and personal protective equipment or physical characteristics of the area of release to minimize contact (shielding).

Examples of ways to avoid hazardous atmospheres include the following:
- Keeping unnecessary people away from the hazard (time and distance)
- Minimizing contact with the material, such as not walking in the product (distance)
- Staying upwind or out of low areas if materials are heavier than air (distance)
- Remaining away from product areas for as long as possible (time, distance)
- Using protective equipment or barriers in order to remain separate from the hazardous materials (shielding)
- Prohibiting eating, drinking, or smoking in areas where hazardous materials have been released (distance, shielding)
- Washing hands and showering after potential contacts (time, distance, shielding)

U.S. Department of Transportation Hazardous Materials Classifications

Materials in transport (i.e., off fixed facilities) are governed by the rules of the U.S. DOT. Historically, the identification of hazardous shipments was initiated prior to the beginning of the 20th century in water and rail transportation (the two major transportation systems at the time). The rapid development and movement of explosive powders during and subsequent to the American Civil War propelled this need, evident through various 19th century disaster events. With the coming of industrialization, and in particular World War II, further identification was placed on chemical hazards in transport. Specifically, numbers (1–9) were used to identify hazard classes and were placed on bulk shipments identifying the severity of the hazard to both emergency and routine personnel in the area. Explosives were identified with the numeral 1, the most dangerous of the hazardous materials transport classes. Both maritime and industrial disasters, such as the Texas City, Texas explosion and fire that killed 581 and injured more than 3,000, necessitated

TABLE 4-1 Routes of Toxic Materials Introduction to the Body, Identified in Order of Concern

Routes	Comment
1. Injection	Introduction directly into the body and/or bloodstream
2. Inhalation	Introduction through the respiratory system
3. Ingestion	Introduction through the gastrointestinal system
4. Absorption	Introduction through skin contact with materials

this requirement. During this incident in April 1947, the ship *Grand Camp* exploded while carrying 2,300 tons of ammonium nitrate. After the initial response, a second vessel, the *High Flyer*, also carrying ammonium nitrate, exploded. Cargo quantity and hazards, as well as protective measures, were unknown to individuals at the Texas City disaster.

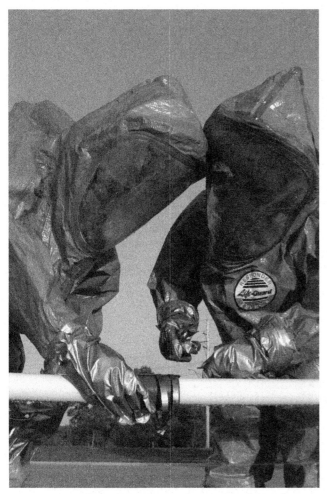

PHOTO 4-1 Hazmat Team–Minimizing the time hazardous materials are used is an administrative control, reducing potential contact time and subsequent toxicity. PPF and engineering controls additionally shield workers. Knowing the physical properties (e.g., heavier or lighter than air) of the hazardous material that individuals may be exposed to allows them to maintain maximum distances from the materials.

In 1976, with the organization of the Research and Special Programs Administration within the newly formed U.S. DOT (1966) (now identified as the Research and Innovative Technology Administration), hazardous materials recognition and identification standards were increased. Today, the nine hazard categories are identified by color, label, packaging material, and placard. Information concerning the nine U.S. DOT categories of hazardous materials is found in **TABLE 4-2**, as well as in the Appendices of this manual.

U.S. DOT Hazard Communication Requirements

The U.S. DOT additionally requires the communication of materials hazards to Hazardous Materials (Hazmat) employees (see Chapter 2). These communication requirements include triennial training for recognizing and identifying hazardous materials, using applicable U.S. DOT shipping tables (49 CFR 171–174), using transport manifests, using placards and labels as required in the tables, and additional safety requirements that are specified for specific transport modes. A **UN/NA identification number**, specific to a material, has been issued for more than 7,000 U.S. DOT hazardous materials. The adoption of a globally harmonized hazard classification and identification system is not expected to affect the communication of hazards in transport, or current regulatory requirements. Pictogram and labeling information is currently in place through DOT rules.

U.S. DOT Shipping Requirements

A page from the U.S. DOT shipping table is shown in **TABLE 4-3**. This table can be used to identify whether the materials are regulated by the U.S. DOT, and subsequently identify hazard class, identification numbers, packaging and labeling requirements, and special provisions for the transport of the material.

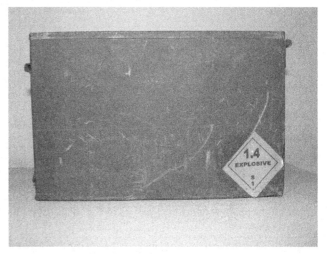

PHOTO 4-2 Explosives packaging and labels.

Hazardous Materials Manifest

A sample Hazardous Materials Shipping Paper is shown in **TABLE 4-4**. Major requirements include the following:
- The name(s) of the shipper
- An identification (usually an "X") that a hazardous material is found with the load
- The proper shipping description (from the U.S. DOT Table)

TABLE 4-2 UN/U.S. DOT Hazard Classes (Transportation)

Classes and Divisions Based on UN System	Examples of Materials by Classes and Divisions	General Hazard Properties (Not All Inclusive)
Class 1		
Division 1.1–Explosive with mass explosion hazard	Dynamite, TNT, black powder	Explosive; exposure to heat, shock, or contamination could result in thermal and mechanical hazards
Division 1.2–Explosive with projection hazard	ammunition, flares	
Division 1.3–Explosive with predominantly a fire hazard	propellant explosives, rocket motors, special fireworks	
Division 1.4–Explosive device with non-significant blast hazard	common fireworks, small arms ammunition	
Division 1.5–Very insensitive explosives, with a mass explosion hazard	ammonium nitrate–fuel oil mixtures	
Division 1.6–Extremely insensitive detonating articles		
Class 2		
Division 2.1–Flammable gas	propane, butadiene (inhibited) acetylene, methyl chloride	Under pressure; container may rupture violently (fire and nonfire); may be flammable, poisonous, a corrosive, an asphyxiant, and/or an oxidizer; may cause frost-bite
Division 2.2–Nonflammable, non-toxic gas	carbon dioxide anhydrous ammonia	
Division 2.3–Toxic gas	arsine, phosgene, chlorine methyl bromide	
Class 3		
Flammable Liquids (and combustible liquids in U.S.)	acetone amyl acetate, gasoline methyl alcohol, toluene	Flammable; container may rupture violently from heat/fire; may be corrosive, toxic, and/or thermally unstable
Combustible Liquid	**fuel oils**	
Class 4		
Division 4.1–Flammable solid	nitrocellulose, magnesium ribbon	Flammable, some spontaneously; may be water reactive, toxic, and/or corrosive; may be extremely difficult to extinguish
Division 4.2–Spontaneously combustible material	phosphorus, pyrophoric liquids and solids,	
Division 4.3–Dangerous When Wet material/water reactive substances	calcium carbide, potassium, sodium	
Class 5		
Division 5.1–Oxidizer	ammonium nitrate fertilizer	Supplies oxygen to support combustion; sensitive to heat, shock, friction, and/or contamination
Division 5.2–Organic peroxide	dibenzoyl peroxide, peroxyacetic acid, diacetal peroxide solution	

TABLE 4-2 (continued)

Classes and Divisions Based on UN System	Examples of Materials by Classes and Divisions	General Hazard Properties (Not All Inclusive)
Class 6		
Division 6.1–Toxic substance	aniline, arsenic tear gas carbon tetrachloride	Toxic by inhalation, ingestion, and skin and eye absorption; may be flammable
Division 6.2–Infectious substances	anthrax, botulism, rabies, tetanus	
Class 7		
Radioactive material	cobalt, uranium hexafluoride	May cause burns and biologic effects; energy and matter
Class 8		
Corrosive material	hydrochloric acid, sulfuric acid, sodium hydroxide, nitric acid hydrogen fluoride unslaked lime, metallic mercury	Disintegration of contacted tissues; may be fuming, water reactive
Class 9		
Miscellaneous hazardous material	dry ice, molten sulfur adipic acid, PCBs	
ORM-A	dry ice	
ORM-B	metallic mercury	
ORM-C	Oakum	
ORM-D	consumer quantities and substances	
ORM-E	hazardous substances, wastes	

Source: Federal Emergency Management Agency, U.S. Fire Administration, National Fire Academy. *Recognizing and Identifying Hazardous Materials, Participant Manual, 2nd ed.* Washington, D.C.: U.S. Government Printing Office, 1995.

- The emergency contact number for the owner of the hazardous materials
- The shipper's certification and signature

The proper shipping description of the material initiates with the marking (usually an "X") designating the Hazmat, and includes the proper shipping name from the U.S. DOT Table, the identification number, the hazard class, the packing group, the labeling and placard requirements, and other appropriate hazard or required safety information. A reportable quantity (RQ) designation may also be included. If the material is a waste, this notation will also be included. The RQ designation identifies that a notification must be made if the material is released into the environment.

Uniform Hazardous Waste Manifest

Shipments of hazardous wastes, because they are being transported to a licensed final location, are transported over the Environmental Protection Agency (EPA)–required Uniform Hazardous Waste Manifest. This manifest meets the requirements of the U.S. DOT for the shipment of hazardous materials and additionally includes the unique generator's identification number for the waste materials (a liability requirement of the Resource Conservation and Recovery Act [RCRA]). U.S. EPA waste numbers, not to be confused with U.S. DOT transport numbers, are also included. These identify the type of waste being shipped and disposed.

TABLE 4-5 includes a sample Uniform Hazardous Waste Manifest.

TABLE 4-6 identifies individuals designated by the U.S. DOT as responsible for Shipping Papers in the various modes of regulated transport.

U.S. DOT Placarding and Labeling

The last (i.e., outside) package holding any U.S. DOT hazardous material must be labeled or placarded. Marking of bulk transportation containers is with 14-inch by 14-inch color-coded identifiers of the hazardous material load carried (placards). Smaller labels are placed on individual packing containers.

TABLE 4-3 U.S. DOT Shipping Table Excerpt: §172.101: Hazardous Materials

Symbols (1)	Hazardous Materials Descriptions and Proper Shipping Names (2)	Hazard Class or Division (3)	Identification Numbers (4)	PG (5)	Label Codes (6)	Special Provisions (§ 172.102) (7)	Packaging (8) Exceptions (8a)	Packaging (8) Non-Bulk (8b)	Packaging (8) Bulk (8c)	Quantity Limitations (9) Passenger Aircraft/Rail (9a)	Quantity Limitations (9) Cargo Aircraft Only (9b)	Vessel Stowage (10) Location (10a)	Vessel Stowage (10) Other (10b)
	Phosgene	2.3	UN1076		2.3, 8	1, B7, B46	None	192	314	Forbidden	Forbidden	D	40

Special Provisions:

1	This material is poisonous by inhalation (see §171.8 of this subchapter) in Hazard Zone A (see §173.116(a) or §173.133(a) of this subchapter), and must be described as an inhalation hazard under the provisions of this subchapter.
B7	Safety relief devices are not authorized on multi-unit tank car tanks. Openings for safety relief devices on multi-unit tank car tanks shall be plugged or blank flanged.
B46	The detachable protective housing for the loading and unloading valve of multi-unit tank car tanks must withstand tank test pressure and must be approved by the Associate Administrator.

Packaging:

| 192 | Amount in kilograms of hazardous materials as a gross maximum. |
| 314 | Amount in kilograms of hazardous materials as a gross maximum. |

Vessel Stowage:

| D | Category "D" means the material must be stowed "on deck only" on a cargo vessel and on a passenger vessel carrying a number of passengers limited to not more than the larger of 25 passengers or one passenger per 3m of overall vessel length, but the material is prohibited on passenger vessels in which the limiting number of passengers is exceeded. |
| 40 | Stow "clear of living quarters". |

Source: Code of Federal Regulations, Title 49, Transportation, Part 172.101, Hazardous Materials Table, Washington, D.C.

TABLE 4-4 Hazardous Materials Shipping Paper

- "RQ" means that this is a reportable quantity
- Proper shipping name from Column 2 of the Hazardous Materials Table
- Hazard Class from Column 3 of the Table
- ID Number from Column 4 of the Hazardous Materials Table

SHIPPING PAPER Page 1 of 1
TO: Wafers R US FROM: Essex Corporation
88 Valley Street 5775 Dawson Avenue
Silicon Junction, CA Coleta, CA 93117

QTY	HM	DESCRIPTION	WEIGHT
1 cyl	RQ	Phosgene, 2.3, UN1076, Poison, Inhalation Hazard, Zone A	25 lbs.

This is to certify that the above-named materials are properly classified, described, packaged, marked, labeled, and placarded and are in proper condition for transportation according to the applicable regulations of the Department of Transportation.

Shipper: Essex Corp. Carrier: Knuckle Bros.
Per: Shultz Per:
Date: 6/17/88 Date:

SPECIAL INSTRUCTIONS:
24 hour Emergency Contact, Ed Shultz, 1-800-555-5555

PHOTO 4-3 Compressed gas cylinders. Well insulated cylinders, known as Dewars, contain substances at their thermal extremes.

PHOTO 4-4 A U.S. DOT labeled flammable liquid container.

TABLE 4-5 Uniform Hazardous Waste Manifest

State of New Jersey
Department of Environmental Protection
Division of Hazardous Waste Management
Manifest Section
CN 028, Trenton, NJ 08625

Type or print in block letters. (Form designed for use on elite (12-pitch) typewriter.)   Form Approved: OMB No. 2050-0039. Expires 9-30-88

UNIFORM HAZARDOUS WASTE MANIFEST	7. Generator's US EPA ID No.	Manifest Document No.	2. Page 1 of	Information in the shaded area is not required by Federal law.
3. Generator's Name and Mailing Address				A. State Manifest Document Number
4. Generator's Phone ( )				B. State Generator's ID
5. Transporter 1 Company Name	6.	US EPA ID Number		C. State Trans. ID
7. Transporter 2 Company Name	8.	US EPA ID Number		D. Transporter's Phone ( )
				E. State Trans. ID
9. Designated Facility Name and Site Address	10.	US EPA ID Number		F. Transporter's Phone ( )
				G. State Facility's ID
				H. Facility's Phone ( )

11. US DOT Description (*Including Proper Shipping Name, Hazard Class, and ID Number*) HM	12. Containers No. / Type	13. Total Quantity	14. Unit Wt/Vol	I. Waste No.
a.				
b.				
c.				
d.				

J. Additional Descriptions for Materials Listed Above	K. Handling Codes for Wastes Listed Above
a. c.	a. c.
b. d.	b. d.

15. Special Handling Instructions and Additional Information

16. GENERATOR'S CERTIFICATION: I hereby declare that the contents of this consignment are fully and accurately described above by proper shipping name and are classified, packed, marked, and labeled, and are in all respects in proper condition for transport by highway according to applicable international and national government regulations.
If I am a large quantity generator, I certify that I have a program in place to reduce the volume and toxicity of waste generated to the degree I have determined to be economically practicable and that I have selected the practicable method of treatment, storage or disposal currently available to me which minimizes the present or future threat to human health and the environment, OR, if I am a small quantity generator, I have made a good faith effort to minimize my waste generation and use the best waste management method that is available to me and that I can afford.

Printed/Typed Name	Signature	Month Day Year

17. Transporter 1 Acknowledgement of Receipt of Materials

Printed/Typed Name	Signature	Month Day Year

18. Transporter 2 Acknowledgement of Receipt of Materials

Printed/Typed Name	Signature	Month Day Year

19. Discrepancy Indication Space

20. Facility Owner or Operator. Certification of receipt of hazardous materials covered by this manifest except as noted in item 19.

Printed/Typed Name	Signature	Month Day Year

TABLE 4-6 Shipping Paper Responsibilities by Mode of Transportation			
Mode of Transportation	Title of Shipping Paper	Location of Shipping Papers	Responsible Person
Highway	Bill of Lading or Freight Bill	Cab of vehicle	Driver
Rail	Waybill and/or Consist	With member of train crew (conductor or engineer)	Conductor
Water	Dangerous Cargo	Wheelhouse or pipe-like container or barge	Captain or master
Air	Air Bill with Shippers Certification for Restricted Articles	Cockpit (may also be found attached to the outside of packages)	Pilot

Source: Recognizing and Identifying Hazardous Materials, Participant Manual, 2nd ed, Federal Emergency Management Agency. U.S. Fire Administration, National Fire Academy, U.S. Government Printing Office. Washington, D.C., 1995.

PHOTO 4-5 Flammable solid packaging and labeling.

PHOTO 4-7 A toxic material (fertilizer) container. Note: Labeling meets the requirements of the Federal Insecticide, Fungicide and Rodenticide Act (FIFRA).

PHOTO 4-6 Oxidizer: Liquid oxygen storage vessels.

PHOTO 4-8 A corrosive transport vehicle.

CHAPTER 4: Recognizing and Identifying Hazardous Materials

U.S. DOT labels are consistent with, although not exactly the same as those placed on packages by chemical manufacturers meeting globally harmonized labeling requirements. The U.S. Department of Labor and the Occupational Safety and Health Administration require that U.S. DOT labels be maintained on packages received at fixed facilities until they are empty. (Note: The definition of "empty" varies by product, container, and regulation. Some apparently empty containers may contain product and subsequently significant hazards. All containers should be assumed to contain hazardous quantities of materials unless identified/labeled as empty.)

Identification of Hazardous Materials at Fixed Facilities

National Fire Protection Association (NFPA) Code #704 System

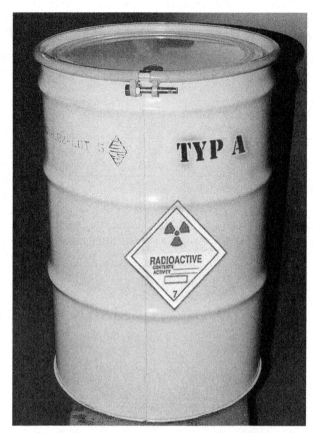

PHOTO 4-9 A bulk transport container for radioactive materials.

Hazardous materials and chemicals at fixed facilities may also be identified by systems such as the NFPA 704 "Diamond" System for the Identification of Fire Hazards of Materials. An explanation of the NFPA System is given in **TABLE 4-7**. Currently, NFPA has established groups to review differences between NFPA 704 (designed to identify the hazards of short term/acute exposures under conditions of fire, spill or similar emergency) and the globally harmonized OSHA HAZCOM 2012 standard. There is no immediate plan to change the system. While the NFPA 704 numbering is designed to identify related hazard, while HAZCOM 2012 globally harmonized numbering is used for the purpose of classifying hazards into categories for proper labeling and training.

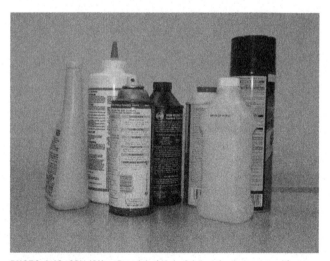

PHOTO 4-10 ORM (Other Regulated Materials) packaging are used for consumer quantities of hazardous materials and substances. These packages will have OSHA HAZCOM (GHS) labels.

Parts of placards and labels include the following:
- Unique pictograms
- Unique coloring
- Hazard class wording or four-digit identification number
- U.S. DOT hazard class numeration

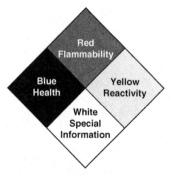

PHOTO 4-11 The NFPA 704 diamond. The health hazard is identified on a blue background, the fire hazard on a red background, and the reactivity hazard on a yellow background.

50 Operating Safely in Hazardous Environments: Second Edition

TABLE 4-7 NFPA 704 Marking System

The National Fire Protection Association 704 (Identification of the Hazards of Materials for Emergency Response) Marking System

The NFPA 704 marking system is commonly used within facilities to alert people to the type and degree of hazard. It may be found on nonbulk packages. However, it is not used in transportation. NFPA 704 is voluntary unless enacted by local law.

Description of NFPA 704 Marking

The NFPA 704 system employs a diamond-shaped symbol divided into four color-coded quadrants.
1. The blue in the left quadrant refers to the health hazard.
2. The red in the top quadrant indicates the flammability hazard.
3. The yellow in the right quadrant refers to the reactivity hazard.
4. The bottom quadrant carries special information—"OX" for oxidizers, SA for simple asphyxiant, or slash W̶ for water reactive materials, with an associated 2 (materials that react violently with water), or 3 (materials that react explosively with water).

Each colored quadrant contains a number from 0 to 4 that represents the relative degree of hazard.

NFPA 704 Marking System

Scale	Health Hazard	Flammability Hazard	Instability Hazard
4	Materials that, under emergency conditions, can be lethal.	Materials that rapidly or completely vaporize at atmospheric pressure and normal ambient temperature or that are readily dispersed in air and that will burn readily.	Materials that in themselves are readily capable of detonation or of explosive decomposition or reaction at normal temperatures and pressures.
3	Materials that, under emergency conditions, can cause serious or permanent injury.	Liquids and solids that can be ignited under almost all ambient temperature conditions.	Materials that in themselves are capable of detonation or explosive reaction but require strong initiating source or which must be heated under confinement before initiation.
2	Materials that, under emergency conditions, can cause temporary incapacitation or residual injury.	Materials that must be moderately heated or exposed to relatively high ambient temperatures before ignition can occur.	Materials that readily undergo violent chemical change at elevated temperatures and pressures.
1	Materials that, under emergency conditions, can cause significant irritation.	Materials that must be preheated before ignition can occur.	Materials that in themselves are normally stable, but which can become unstable at elevated temperature and pressures.
0	Materials that, under emergency conditions, offer no hazard beyond ordinary materials.	Materials that will not burn.	Materials that in themselves are normally stable, even under fire exposure conditions, and are not reactive with water.

Source: National Fire Protection Association. *NFPA Code 704, Standard System for the Identification of the Hazards of Materials for Emergency Response.* Batterymarch Park, Quincy, MA, 2007.

Globally Harmonized System of Classification and Labeling of Chemicals: HAZCOM 2012

OSHA has required employees to communicate the hazards of chemicals to employees since 1983, through its hazard communication standard. At the state or local level, Right-to-Know Legislation many times mimics the federal HAZCOM standard.

The standard was substantially refined in March of 2012, with the adoption of the Globally Harmonized System of Classification and Labeling of Chemicals, known as GHS or HAZCOM 2012, originally proffered by the International Labour Organization (ILO) to ensure international safe use of chemicals. GHS—HAZCOM 2012 now preempts state and local standards in this area, requires chemical manufacturers/importers to identify physical hazards, health hazards, environmental hazards (the environmental hazards section of the standard is not enforced by OSHA) and hazards not otherwise classified, and to list these on SDS and labels. Sixteen physical hazards, ten health hazards and two environmental hazard categories are currently defined.

Manufacturers, importers and distributors of hazardous chemicals are required to label containers with the following information:
- Product identifier
- Signal word (Danger or Warning)
- Hazard statement
- Pictogram
- Precautionary statement(s)
- Name, address and telephone number of the chemical manufacturer, importer or distributor

An exemplary product label is found at **TABLE 4-8**. The GHS HAZCOM required Pictograms are found at **TABLE 4-9**.

OSHA requires all workplace employees protected by HAZCOM 2012 receive training by December 1, 2013, and all provisions of the standard applicable to chemical manufacturers be in place by June 1, 2015.

Effective October 1, 2011, all employers may comply with the HAZCOM 2012.

Federal Insecticide, Fungicide, and Rodenticide Act Labeling Requirements

The Federal Insecticide, Fungicide, and Rodenticide Act (FIFRA) regulates labeling, production, and use of these materials. Such hazardous substances must be used in compliance with the specified label requirements. Labels

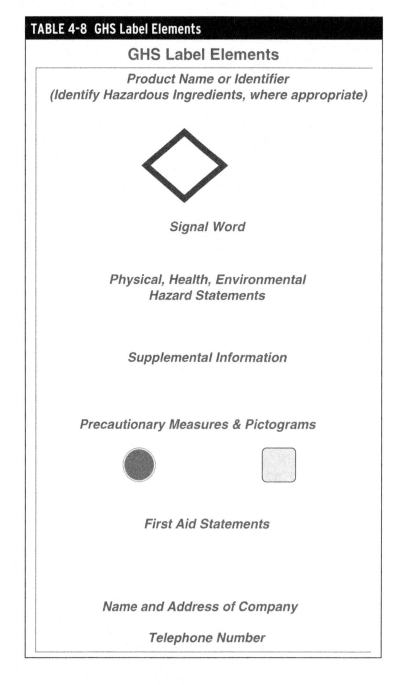

TABLE 4-8 GHS Label Elements

TABLE 4-9 GHS Pictograms and Hazard Classes

GHS Pictograms and Hazard Classes		
■ Oxidizers	■ Flammables ■ Self Reactives ■ Pyrophorics ■ Self-Heating ■ Emits Flammable Gas ■ Organic Peroxides	■ Explosives ■ Self Reactives ■ Organic Peroxides
■ Acute toxicity (severe)	■ Corrosives	■ Gases Under Pressure
■ Carcinogen ■ Respiratory Sensitizer ■ Reproductive Toxicity ■ Target Organ Toxicity ■ Mutagenicity ■ Aspiration Toxicity	■ Environmental Toxicity	■ Irritant ■ Dermal Sensitizer ■ Acute toxicity (harmful) ■ Narcotic Effects ■ Respiratory Tract ■ Irritation

of FIFRA-regulated products must be marked with an appropriate signal word: "DANGER" (severe hazards), "WARNING" (moderate hazards), or "CAUTION" (low hazards). **Photo 4-7** depicts a FIFRA compliant label.

Occupational Safety and Health Administration Color Coding

The U.S. Department of Labor and OSHA specify color-coding for safety-related items. This color-coding is listed in **TABLE 4-10**.

Material Safety Data Sheets/Safety Data Sheets

In addition, OSHA requires manufacturers of hazardous substances to produce material safety data sheets (MSDS), known as Safety Data Sheets (SDS) under HAZCOM 2012 for their materials. MSDS/SDS are produced when quantities of physical, health or environmentally hazardous constituents meet trigger amounts. Fixed facilities are required to maintain MSDS/SDS for all hazardous chemicals they use on site, train employees in their use, and make them continually available. The 16 required sections of an SDS, as described in HAZCOM 2012, are found in **TABLE 4-11**.

Initial Actions Where Hazardous Environments May Be Produced by Chemical Releases

At any unprotected intentional or unintentional release of a hazardous material or substance, protection for individuals and the environment should initiate with the recognition of the chemical or product from as great

TABLE 4-10 OSHA Color-Coding Requirements	
Color	Color Identification
Red	1. Fire Protection Equipment and Apparatus 2. "DANGER" Signs (Immediate Danger, Special Precautions Necessary) 3. Emergency "STOP" Bars and Buttons
Yellow	1. Physical Hazards 2. "CAUTION" Signs (Potential Hazards)
Green	1. Safety Instruction Signs
Dark Red/Fluorescent Yellow Orange Triangle	1. Slow Moving Vehicle Sign
Source: Hazardous Materials Transportation Act. Public Law 93-633, 49 USC 1801 et seq.	

a distance as possible. Safe action can then be taken. An exemplary order for recognition is listed below, from farthest to closest means of identification.

- Knowledge of the location or occupancy
- Visual identification of transportation container or container shape/size
- Identification by colors, markings, and other signage
- Identification by labels and site-specific identifiers
- Identification by shipping papers, MSDSs, or other product documentation
- Identification by sensory indicators or symptoms of exposure
- Incident documentation or post exposure medical surveillance

Subsequently, actions can occur to control potential hazards. This is particularly important for first responders to releases of hazardous materials, although in many cases, the best action may be "no action."

U.S. DOT Emergency Response Guidebook

The U.S. DOT *Emergency Response Guidebook* (*ERG-2012*) is primarily a guide to aide first responders in quickly identifying the specific or generic hazards of the materials involved in the incident and protecting themselves and the public during the initial response phase of a hazardous materials incident. User instructions are included.

Recommended initial distances (evacuations) for releases of U.S. DOT–listed hazardous materials can be found in *ERG-2012*. This reference book should be standard equipment for all individuals or groups that respond to releases or potential releases of Hazmats.

The guidebook lists all U.S. DOT Hazmats by proper shipping name and identification number. Its Response Guide lists initial evacuation distances for severe hazards, and gives response information for each material. In addition, the Response Guide provides information by general U.S. DOT category and by transport vehicle container style. This is particularly helpful in an emergency when a specific product name or identification number cannot be acquired.

Once initial safety information has been reviewed, response or work objectives can be identified, as well as a strategy to accomplish safe activities. Again, engineering and administrative work plans should be used (and are usually required by OSHA) whenever possible to safeguard workers. That is, it is better to eliminate the hazardous environment prior to personnel entry than to protect the employee (such as with personal protective equipment) while the hazard remains.

If a RQ of a hazardous material is released, a report must be made to the National Response Center by calling 1-800-424-8802. Manned 24 hours a day, this interagency reporting center acts as the notification point for federally reportable incidents. Reporting to the National Response Center does not eliminate responsibilities to notify local emergency organizations (such as through the 911 system) for immediate response needs.

CHEMTREC/Chemical Manufacturers Association

Additional chemical response and safety information can be acquired from the Chemical Manufacturers Association (CMA). CMA operates a 24-hour-a-day emergency response center (CHEMTREC) with telecommunicators who can produce additional information and recommendations concerning releases of hazardous materials. In addition, CHEMTREC communicators can reach product manufacturers for first-hand information. The CHEMTREC emergency number is 1-800-424-9300.

The CMA also operates a nonemergency referral service and a lending library of audiovisuals. The CMA nonemergency number is 1-800-262-8200.

TABLE 4-11 Minimum Information for a Safety Data Sheet (SDS)		
1.	Identification of the substance or mixture and of the supplier	• GHS product identifier. • Other means of identification. • Recommended use of the chemical and restrictions on use. • Supplier's details (including name, address, phone number, etc.). • Emergency phone number.
2.	Hazards identification	• GHS classification of the substance/mixture and any national or regional information. • GHS label elements, including precautionary statements. (Hazard symbols may be provided as a graphical reproduction of the symbols in black and white or the name of the symbol, e.g., flame, skull and crossbones.) • Other hazards which do not result in classification (e.g., dust explosion hazard) or are not covered by the GHS.
3.	Composition/information on ingredients	Substance • Chemical identity. • Common name, synonyms, etc. • CAS number, EC number, etc. • Impurities and stabilizing additives which are themselves classified and which contribute to the classification of the substance. Mixture • The chemical identity and concentration or concentration ranges of all ingredients which are hazardous within the meaning of the GHS and are present above their cutoff levels. NOTE: For information on ingredients, the competent authority rules for CBI take priority over the rules for product identification.
4.	First aid measures	• Description of necessary measures, subdivided according to the different routes of exposure, i.e., inhalation, skin and eye contact, and ingestion. • Most important symptoms/effects, acute and delayed. • Indication of immediate medical attention and special treatment needed, if necessary.
5.	Firefighting measures	• Suitable (and unsuitable) extinguishing media. • Specific hazards arising from the chemical (e.g., nature of any hazardous combustion products). • Special protective equipment and precautions for firefighters.
6.	Accidental release measures	• Personal precautions, protective equipment and emergency procedures. • Environmental precautions. • Methods and materials for containment and cleaning up.
7.	Handling and storage	• Precautions for safe handling. • Conditions for safe storage, including any incompatibilities.
8.	Exposure controls/personal protection.	• Control parameters, e.g., occupational exposure limit values or biological limit values. • Appropriate engineering controls. • Individual protection measures, such as personal protective equipment.
9.	Physical and chemical properties	• Appearance (physical state, color, etc.). • Odor. • Odor threshold. • pH. • melting point/freezing point. • initial boiling point and boiling range. • flash point. • evaporation rate. • flammability (solid, gas). • upper/lower flammability or explosive limits. • vapor pressure. • vapor density. • relative density. • solubility(ies). • partition coefficient: n-octanol/water. • autoignition temperature. • decomposition temperature.

(continues)

	TABLE 4-11 (continued)	
10.	Stability and reactivity	• Chemical stability. • Possibility of hazardous reactions. • Conditions to avoid (e.g., static discharge, shock or vibration). • Incompatible materials. • Hazardous decomposition products.
11.	Toxicological information	Concise but complete and comprehensible description of the various toxicological (health) effects and the available data used to identify those effects, including: • information on the likely routes of exposure (inhalation, ingestion, skin and eye contact); • Symptoms related to the physical, chemical and toxicological characteristics; • Delayed and immediate effects and also chronic effects from short- and long-term exposure; • Numerical measures of toxicity (such as acute toxicity estimates).
12.	Ecological information	• Ecotoxicity (aquatic and terrestrial, where available). • Persistence and degradability. • Bioaccumulative potential. • Mobility in soil. • Other adverse effects.
13.	Disposal considerations	• Description of waste residues and information on their safe handling and methods of disposal, including the disposal of any contaminated packaging.
14.	Transport information	• UN Number. • UN Proper shipping name. • Transport Hazard class(es). • Packing group, if applicable. • Marine pollutant (Yes/No). • Special precautions which a user needs to be aware of or needs to comply with in connection with transport or conveyance either within or outside their premises.
15.	Regulatory information	• Safety, health and environmental regulations specific for the product in question.
16.	Other information including information on preparation and revision of the SDS	

Additional Sources for Response Information

The National Library of Medicine Tox Line and Tox Net, provided by the National Institutes of Health, provide toxic substance safety information. The service can be accessed online (www.nlm.nih.gov). The National Institute for Occupational Safety and Health (www.cdc.gov/niosh) provides information on workplace safety and health hazards. Additional hazardous materials response and release information can be found in the following sources.

- *Emergency Handling of Hazardous Materials in Surface Transport and Emergency Action Guides*, published by the American Association of Railroads, 50 "F" Street, NW, Washington, D.C. 2001, 2010.
- *Pocket Guide to Chemical Hazards*, published by the National Institute for Occupational Safety and Health (2007), which can be acquired from the U.S. Government Printing Office, Washington, D.C. 20402.

Chapter Summary

Recognition and identification of hazardous materials and substances allow individuals to plan for timely protection. Key to this process is a knowledge of the various classification systems (U.S. DOT, NFPA), the general behaviors of materials (fire [physical hazard] or toxicity [health or environmental hazard]), and mechanism for acquiring immediate response (ERG) and specific substance data (shipping papers, MSDSs).

Terms

Combustible: Capable of supporting the process of combustion. Combustible liquids and gases generally have a **flash point** between 100°F and 200°F. Materials with a flash point above 200°F may combust but generally require heating to increase the rate of off-gassing.

Flammable: Readily supplying fuel to support the process of combustion. Flammable liquids and gases generally have a flash point below 100°F.

Flammable Range: The proportion of O_2 to fuel necessary to support combustion.

Flash Point: The point at which a material volatilizes in sufficient quality to cause a fire, in the presence of O_2 and a source of ignition.

Hazardous Chemical: Any substance or mixture which is classified under the OSHA HAZCOM (GHS) Standard as a physical or health hazard, a simple asphyxiant, combustible dust, pyrophoric gas or hazard not otherwise classified.

Hazardous Material: Anything that can cause harm to people or the environment. There are roughly 7000 hazardous materials identified by the U.S. Department of Transportation. Hazardous materials include hazardous chemicals and wastes.

Hazardous Waste: A hazardous material or substance that is prepared for being returned to the environment. Hazardous wastes are defined by the U.S. EPA in the Resource Conservation and Recovery Act of 1976.

Toxic Materials: Substances that cause harm or illness when injected, inhaled, ingested, or absorbed through the skin of the human body.

UN/NA Identification Number: A unique four-digit identification number assigned by the U.S. DOT to each hazardous material. This number is required to be displayed on shipping papers and packages of hazardous materials, including portable tanks, cargo tanks, and tank car shipments. It may be displayed on an orange panel adjacent to placards on vehicles or in the middle of required placards.

References

Clean Air Act, Public Law 91-604, 42 U.S.C. (December 31, 1970) 7401 et seq.

Clean Water Act, Public Law 92-500, 33 U.S.C. (October 18, 1972) 1251 et seq.

Comprehensive Environmental Response, Compensation and Liability Act, Public Law 96-510 (December 11, 1980), amended at Public Law 99-499 (amended 1982, 1986).

Federal Emergency Management Agency, United States Fire Administration, National Fire Academy. *Recognizing and Identifying Hazardous Materials, Participant Manual*, 2nd ed., Washington, D.C.: Government Printing Office, 1995.

Federal Insecticide, Fungicide and Rodenticide Act, Public Law 92-516, 7 U.S.C. (October 21, 1972) 136 et seq.

Hazardous Materials Transportation Act, Public Law 93-633, 49 U.S.C. (1976) 1801 et seq.

National Fire Protection Association. *NFPA Code #704: Standard System for the Identification of the Hazards of Materials for Emergency Response.* Quincy, MA: NFPA, 2012.

Resource Conservation and Recovery Act, Public Law 94-580, 42 U.S.C. (October 21, 1976) 6901 et seq.

Safe Drinking Water Act, Public Law 93-523, 42 U.S.C. (December 16, 1974) 300f et seq.

Toxic Substances Control Act, 15 U.S.C., (September 28, 1976) 2601 et seq.

U.S. Department of Labor, Occupational Safety and Health Administration: Hazard Communication: Final Rule: Docket # OSHA-H0221C-2006-0062; Washington, DC. March 26, 2012.

U.S. Department of Transportation, Research and Special Programs Administration. *Emergency Response Guidebook: A Guidebook for First Responders During the Initial Place of a Dangerous Goods/Hazardous Materials Incident*, Neenah, WI: J.J. Keller, 2012.

The Williams-Steiger Occupational Safety and Health Act, Public Law 91-596, 29 U.S.C. et seq., (December 29, 1970).

Exercise 4

Identify the primary hazards associated with each identifier of a hazardous substance:

1. _____

2. _____

3. _____

4. _____

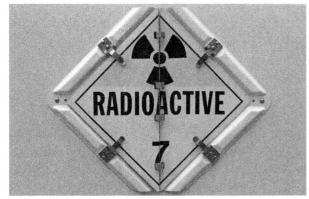

5. _____

6. Radioactive materials are typically found in _____ containers.

7. Flammable liquids have a flashpoint of _____.

8. Combustible liquids have a flashpoint of _____.

9–13. Five items are required to be found on a U.S. DOT compliant Hazardous Material Shipping Paper. These are:

9. _____

10. _____

11. _____

12. _____

13. _____

14. Pesticides, such as those regulated under FIFRA, with severe toxicity hazards are identified by the signal word _____.

15–18. Identify the hazards described on this NFPA 704 "Diamond":

15. Health Hazard _____

16. Flammability Hazard _____

17. Reactivity Hazard _____

18. Other Hazard _____

19. Using *ERG-2012*, identify the Hazardous Material with a Chemical ID #1017 _____.

CHAPTER 4: Recognizing and Identifying Hazardous Materials

20. What is the initial evacuation distance for a small release of this material? _____.

21–22. Identify the hazard(s) associated with these OSHA HAZCOM: GHS pictograms:

 _____    _____

5 Quantifying Hazardous Environments

Key Concepts

- Immediately Dangerous to Life or Health (IDLH) Atmospheres
- Permissible Exposure Limit (PEL)
- Threshold Limit Value (TLV)
- Recommended Exposure Limit (REL)
- Oxygen Testing
- Combustible Vapor Testing
- Ionizing Radiation Testing
- Toxic Materials Testing

Quantification of Hazardous Atmospheres

The quantification of hazardous atmospheres allows for appropriate health and safety decision making. These decisions may be not to enter the atmosphere, to use engineering controls, to administrate safe solutions to entry, to enter partially, or to use personal protective equipment for entry. Subsequently, levels of control may be selected.

The decision-making process of quantification initiates with the recognition of a hazardous atmosphere or environment. Subsequently, appropriate test equipment or methodologies are chosen. Standards for comparison should be chosen prior to any testing to facilitate the decision-making process and ensure limits of quantification on tests are appropriate. These standards may be the **permissible exposure limit (PEL)** enforced by the Occupational Safety and Health Administration (OSHA), the **threshold limit value (TLV)** recommended by the American Conference of Governmental Industrial Hygienists (ACGIH), or the **recommended exposure limit (REL)** set by the National Institute for Occupational Safety and Health (NIOSH) for occupational exposures (OSHA, ACGIH, and NIOSH deal with the occupational setting). The American Industrial Hygiene Association additionally published workplace employee exposure limits (WEEL) and biological employee exposure limits (BEEL), which were occupational exposure recommendations for substances typically not addressed by OSHA, ACGIH, or NIOSH. Funding for the WEEL/BEEL project was discontinued in 2012, although older documents remain in existence, and an alternate sponsor for this project is anticipated. **Action levels**—typically a level approaching the occupational exposure limit requiring some preliminary safety action—and **short-term exposure limits (STELs)**—fifteen to thirty minute exposures above the occupational exposure limit that should not be exceeded at any time during the workday—may also be used for comparison purposes. The ACGIH identifies STEL values (excursions above the PEL for short periods during the workday) are allowable 4 times during a work shift, with a minimum of 1 hour between STEL periods). **Ceiling concentrations,** which are short-term exposure limits or levels, are not to be exceeded for any period of time during the workday.

In the public health setting, the **No Observed Adverse Effects Level (NOAEL)** or the **Lowest Observed Adverse Effects Level (LOAEL)** are used for comparison purposes, as well as minimal risk levels published by the Department of Health and Human Services (HHS), Agency for Toxic Substances and Disease Registry (ATSDR). The United States Environmental Protection Agency (EPA) publishes **Risk-Based Concentrations (RBC)** for nearly 600 hazardous substances, which can be extrapolated to identify levels of exposure concerns for the general public. It is important to separate occupational (work time) from public (lifetime) exposure limits, because the exposure time/dose varies greatly between these groups.

Once atmospheres are identified and characterized, recharacterization should occur if new work areas are entered, new processes are introduced, or work changes occur. Entities such as the American Industrial Hygiene Association (Emergency Response Planning Guidelines) and the U.S. EPA (Acute Exposure Guideline Levels) publish data and exposure guidelines for once-in-a-lifetime or rare exposures to chemicals, typically used for decision making during emergent events effecting the public (e.g., evacuations).

Quantification of exposures may occur at the site or hazardous atmosphere itself, or off-site, in simulated work environments. Individuals operating safely in hazardous environments should be knowledgeable of appropriate on-site testing methodologies, in particular if environments change or are capable of changing during the course of work.

Off-site quantification (laboratory or materials testing analysis) is performed when on-site quantification in particular limits of detection necessary for comparison are not sufficient for decision-making. This may be because of economic, equipment, or technological concerns.

The Process of Quantification

In general, the process for on-site quantification of hazardous atmospheres includes the following steps:
1. Determining the toxin to be tested, and the appropriate exposure limits to which the test results will be compared;
2. Determining the equipment to be used for quantifying toxins within the necessary limits of detection;
3. Calibrating or testing equipment to ensure operability within the ranges selected (generally according to manufacturers' instructions);
4. Determining interferences to the tests and eliminating the interferences to quantification (or consideration of same);
5. Performing quantification testing;

TABLE 5-1 Instrument Calibration Log

INSTRUMENT CALIBRATION LOG

DATE	INSTRUMENT	CHALLENGE AGENT	KNOWN CONCENTRATION	READING	ADJUSTED?		SIGNATURE
					YES	NO	
					YES	NO	
					YES	NO	
					YES	NO	
					YES	NO	
					YES	NO	
					YES	NO	
					YES	NO	
					YES	NO	

6. Performing post-test calibration or checks to ensure equipment has operated appropriately during the course of the test;
7. Comparing data and making decisions.

Testing or collecting samples for off-site analysis may follow the same protocol, with the exception of on-site quantification of data. Chain-of-custody logs, or documentation of sample collection procedures, are essential for both on- and off-site testing, whereas calibration logs and documentation are required for on-site test equipment. An example of calibration, quantification, and on-site sample collection logs (documentation) can be found in **TABLES 5-1** and **5-2**. **TABLE 5-3** contains a typical log used for the documentation of on-site sample results. A typical laboratory chain-of-custody form, documenting custody of samples while being processed off site for quantification, is found in **TABLE 5-4**.

Information in **TABLE 5-5** further describes environmental monitoring devices used for on-site quantification of hazardous atmospheres.

Testing for Physical Hazards

The environmental hazards of noise, insufficient lighting, and temperature extremes may be monitored on site. Sound pressure levels should be monitored in employees' personal hearing zones by use of personal dosimetry equipment. In addition, on-site estimations of personal exposure may be made with direct reading sound level meters. As some sound pressure levels vary greatly through a work regimen, personal dosimetry is the preferred method. Calibration of equipment should be performed before any site readings and calculations are made.

Light meters are also available for on-site quantification of work site illumination levels. These generally provide illumination quantification in a measurement comparable to a standard (such as foot-candles).

In addition to temperature and relative humidity on site, metering devices are available to estimate employee-exposure potentials to temperature extremes. These include monitoring devices (e.g., wet bulb globe

TABLE 5-2 On-Site Sample Collection Log

PROJECT NAME: _____
PROJECT NUMBER: _____
DATE: _____

ON-SITE SAMPLE COLLECTION LOG

TIME (START/STOP, IF APPROPRIATE)	SAMPLE NUMBER	COLLECTION PROCEDURE	COLLECTION DEVICE	CALIBRATION	LOCATION	ANALYZED BY AND DATE

TABLE 5-3 Results of On-Site Monitoring

PROJECT NAME: _____
PROJECT NUMBER: _____
DATE OF SAMPLING: _____

RESULTS OF ON-SITE MONITORING

| TIME | MONITORING DEVICE | CALIBRATION | O_2/CGI | | RADIATION MONITOR READING | OTHER | LOCATION | PURPOSE | INITIALS |
			% LEL	% O_2	mR/hr				

TABLE 5-4 Typical Laboratory Chain of Custody

REPORT TO:		INVOICE TO:	
Company		Company	
Contact		Contact	
Address		Address	
City		City	
State		State	
Telephone		Telephone	
Fax		Fax	

PROJECT INFORMATION

Project Name _____

Quote # _____

P.O. # _____

Public Water Supplier # _____

CHECK ANALYSES REQUESTED

LAB USE ONLY

Lab Sample I.D.	Temp. C	Preserv.	Client Sample Identification	Sample Type					# of bottles	Date Sampled	Time Sampled	Sampled By
				Solid	Liquid	Other	Comp.	Grab				

Relinquished By:		Date	Time	Accepted By:	
Relinquished By:		Date	Time	Accepted at Lab By:	

*Any sample delivered after 5:00 p.m. will not be entered into our system until 8:00 a.m. the next business day. Page ___ of ___

CHAPTER 5: Quantifying Hazardous Environments

TABLE 5-5 On-Site Environmental Monitoring Device Characteristics

Hazard Monitored	Instrument	Application	Detection Method	Notes
Combustible Gas/Vapor	Combustible Gas Indicator	Measures the concentration of a combustible gas or vapor	A filament is heated by burning the combustible gas/vapor; the increase in heat is measured	Calibrated before use
Oxygen Level	Oxygen Meter	Measures the percentage of oxygen in air	Uses an electrochemical sensor to measure the partial pressure of oxygen in air	Calibrated before each use in normal air
Ionizing Radiation	Geiger-Muller (G-M) Counter Scintillation Tube	Environmental radiation monitor; some monitors can distinguish among the types of ionizing radiation	G-M: ionizing radiation reacts with inert gas producing electric current proportional to ionizing radiation. Scintillator: ionizing radiation produces photons of light within a tube. Crystals are specific to types of radiation (e.g., Sodium iodide crystal for gamma radiation).	Must be calibrated annually at a specialized facility
Organics	Colorimetric Tube	Measures concentration of specific gases and vapors	Substance reacts with the indicator chemical, producing a stain whose length in the tube is proportional to the concentration of the substance.	Conduct leak test before use. Check flow rate and volume periodically. Check shelf life before use.
	Flame Ionizing Detector (FID) with Gas Chromatograph (GC) Option	Measures total concentration of organics in survey mode; identifies and measures specific compounds in GC mode	Gases and vapors are ionized in a flame; a current is produced in proportion to the number of carbon atoms present.	Requires experience to operate. Fuel source is hydrogen.
	Photoionizing Detector	Measures total concentrations of substance(s); some identification of compounds is possible if more than one probe is used.	Ultraviolet radiation ionizes molecules, produces ions proportional to concentration.	Does not detect methane. Compounds have different ionization potentials. The proper strength bulb must be used to "photo ionize" substances or substances may be missed.
	Portable Infrared Spectrophotometer	Designed to quantify one or two component mixtures.	Infrared radiation is passed through a sample; each compound will absorb infrared radiation at a specific frequency; amount of absorption is proportional to concentration.	Requires knowledge of infrared radiation frequencies for chemicals. Short battery life; needs to be operated on a stable surface (tabletop).
	Catalytic Combustion Meter (Super Sensitive Combustible Gas Indicator [CGIS])	Measures substances capable of undergoing combustion	Oxidation takes place on the surface of a heated catalytic bed element; oxidation is proportional to concentration.	Similar to CGI, but used for ppm measurements.
Inorganics (Volatile)	Colorimetric Tube	Measures concentration of specific inorganic gases and vapors		See previous note.

TABLE 5-5 (continued)				
Hazard Monitored	Instrument	Application	Detection Method	Notes
	Photoionizing Detector	Measures total concentration of some inorganics		See previous note.
Inorganics (Volatile)– (cont'd)	Portable Infrared Spectrophotometer	Designed to quantify one or two component mixtures; will detect oxides of nitrogen, ammonia, hydrogen cyanide, hydrogen fluoride, and sulfur dioxide	See previous description.	See previous note.
	Specific Chemical Monitor	Measures concentration of specific gases and vapors	See previous description.	Limited number of chemicals can be detected; even though specific, there can be interferences.
Aerosols/[1] Particulate	Direct Reading Instrument for Analyzing Airborne Particulates	Measures and sizes the concentration of aerosols in air	Operates on one of four basic techniques: 1. Optical 2. Electrical 3. Piezoelectric 4. Beta attenuation	Individual instruments have specific notes. Instruments are available to measure fibers.

Source: U.S. Department of Health and Human Services, Public Health Service, Centers for Disease Control, National Institute for Occupational Safety and Health. *Occupational Safety and Health Guidelines Manual for Hazardous Waste Site Activities–DHHS Publication 85-115.* Washington, D.C.: Government Printing Office, October 1985.

[1]These direct reading instruments will read out total or respirable aerosol matter, not the composition of the aerosols. The content, such as lead, pesticides, dust, fume, mist, fog, spray, or smoke, must be separately analyzed.

PHOTO 5-1 Noise Dosimetry Equipment integrates specific sound pressure levels [employee dose] in an employee-hearing zone.

PHOTO 5-2 A Typical Light Meter.

temperature integrating devices) to estimate employees' stress from environmental heat. These devices integrate ambient temperatures, solar load, and humidity. Refer to Chapter 3 for further information concerning the application of these devices for use in stressful environments.

In addition, devices integrating measurements for cold temperature extremes, such as temperature and wind chill, are available.

Immediately Dangerous to Life or Health Atmospheres

There are generally considered to be four quantifiable **immediately dangerous to life or health (IDLH)** atmospheres. IDLH atmospheres must be tested and identified on-site. IDLH atmospheres include:

1. Oxygen deficiency;
2. Increased fire or explosion hazard due to flammable vapors or excess oxygen;
3. High levels of toxic materials;
4. High levels of ionizing radiation.

TABLE 5-6 describes these IDLH Atmospheres.

Additionally, NIOSH defines an IDLH atmosphere as one that produces **NO** irreversible health effects, or certain transient effects (eye or respiratory irritation, disorientation, incoordination), which could prevent escape, as a result of 30 minutes of exposure. OSHA utilizes the term **imminent danger** to characterize situations that place an employee in danger that could reasonably be expected to cause death or serious physical harm immediately or before effective enforcement procedures. While this term includes unprotected work in IDLH atmospheres, it additionally covers physical hazards that may produce these detrimental effects.

Testing for Oxygen in Atmospheres

Oxygen constitutes 20.9% of the air we breathe and is essential for human respiration. In addition, it is necessary in sufficient quantities to test for (and to have) flammable atmospheres. Consequently, oxygen testing is generally performed first in any suspect oxygen deficient, flammable, or toxic atmosphere.

Although oxygen testing may be performed with length of stain, direct reading indicators (see discussion that follows), most testing occurs with oxygen-metering equipment. Most oxygen-metering equipment measures the partial pressure of oxygen in air. Once calibrated in a known atmosphere and checked with a calibration source, metering equipment is taken to

TABLE 5-6 Immediately Dangerous to Life or Health Atmospheres		
Atmospheres	**Quantification**	**Comment**
Oxygen Deficient Atmosphere	Less than 19.5% O_2	Test with calibrated O_2 meter.
Flammable Atmosphere	Greater than 23% O_2	Test with calibrated O_2 meter.
	Greater than 20% LEL (10% LEL in a permit required confined space, 25% LEL in certain maritime situations).	Test with calibrated combustible gas analyzer.
	Flammable Dusts	Less than 5 feet in visibility.
Toxic Atmosphere	Quantify toxin with appropriate sampling and analysis procedure.	Identify IDLH level of toxin and compare results. High levels of toxins may also displace O_2 or produce flammable atmospheres.
Ionizing Radiation	Use of energy quantification device, such as Geiger-Mueller tube; a level of 5 rem (0.05 Seiverts) (not to be exceeded in a year with no more than 1.25 rem (0.0125 Sv) per calendar quarter) is recommended as a protection guideline for workers; the NRC public exposure limit is 0.5 rem (0.005 Sv)/year; the Emergency Worker Guidelines are: • 5 rem (0.05 Sv): Allowable dose • 10 rem (0.1 Sv): Valuable property protection, with consent • 25 rem (0.25 Sv): Lifesaving operations only, with consent These levels identify irradiated individuals. Contaminated items/individuals are: • > 0.5 mR/Hr (0.005 Gray/hr) (Surfaces) • > 0.1 mR/Hr (0.001 Gray/hr) (Thyroid)	Quantification may be difficult, unless the isotope and its energy level are known; therefore, all levels above background that are monitored with survey meters should be considered hazardous.
Source: Sample, Direct Reading Length of Stain Tube Information: Carbon Dioxide. Cite: Mine Safety Appliance Company – Detector Tube and Dosimeter Handbook, Pittsburgh, PA 1988.		

the atmosphere to be tested. Whenever possible, testing for any immediately dangerous situations should be performed from outside the environment. If this is not possible, appropriate protection should be afforded to employees during the initial testing and entry, assuming that IDLH atmospheres exist.

Once oxygen levels are recorded, appropriate decision-making can occur. In general, oxygen below 19.5% is considered immediately dangerous to life or health (see Chapter 3 for discussion on asphyxiant atmospheres).

Conversely, oxygen concentrations above 23% are considered flammability hazards because they may increase burning temperatures and characteristic burning rates of materials.

Combustible Gas Testing

Once oxygen testing has been performed, tests for flammable or combustible vapors in atmospheres can be performed. Because of the nature of the oxidation/reduction reaction, metering equipment should be used to ensure oxygen atmospheres of at least 16% are present. Below this level, flammable vapor readings may be skewed. Flammable vapor readings may also be inaccurate if the material is above its **upper explosive limit (UEL)**, because most metering equipment does not provide information in this range. Appropriate flammability testing can be performed once oxygen levels have been determined. Manufacturers' specifications for specific flammable vapor testing equipment should be

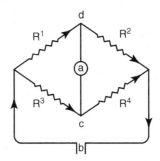

a. Galvanometer.
b. Battery.
c, d. Bridge.
R^1, R^2 Resistances to be compared.
R^3, R^4 Known resistances which can be varied. When galvanometer shows no current. $R^1R^2 = R^3R^4$.

FIGURE 5-1 The Wheatstone bridge is an instrument for the measurement of differential resistance in an electrical circuit.

consulted before testing. The flammable vapor test is generic; that is, it tests for the potential flammability of the total atmosphere without identifying specific flammable vapors within the concentration of air (oxygen)/vapor to be tested.

Flammable vapor testing identifies, in many cases, the percentage of flammable vapors in air relative to the **lower explosive limit (LEL)** of a known substance (calibration gas). A discussion concerning explosive ranges is found in Chapter 4.

Many combustible/explosive vapor meters operate on the principle of a Wheatstone bridge (**FIGURE 5-1**). Information concerning explosive vapor/oxygen mixtures for common flammable substances is found in **TABLE 5-7**. The Wheatstone bridge is an electrical circuit that measures the known resistance passing over a wire.

As the atmosphere to be tested is drawn through or passed over the Wheatstone bridge and heated by the electrical circuitry, the resistance within the bridge declines in a calibrated ratio, similar in response to the atmosphere against which the meter was calibrated. The device then gives an indication of the percentage of the LEL (i.e., the minimum amount of flammable vapor with sufficient oxygen concentration in air necessary for a fire or explosion) that has been achieved.

It is important to understand the concept of flammable ranges when metering and monitoring for flammable gases and vapors because atmospheres below, within, and above flammable ranges may occur, all of which may be considered hazardous.

In general, 10% to 25% of the LEL of any flammable vapor is considered a hazardous atmosphere for

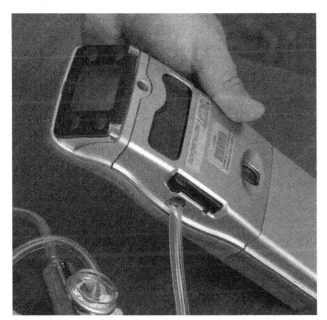

PHOTO 5-3 O2/CGI Metering Device.

TABLE 5-7 Flash Points and Flammable (Explosive) Ranges for Common Flammable Substances		
Substance	Flash Point	Explosive Range
Gasoline	-45°F (-48°C)	1.4-7.6%
Ammonia	-108°F (-78°C)	15-28%
Benzene	12°F (-11°C)	1.2-7.8%
Butyl Alcohol	84°F (29°C)	1.4-11.2%
Carbon Monoxide	-312°F (-191°C)	12.5-74%
Ethylene Oxide	-20°F (-29°C)	3-100%
Propane	-44°F (-42°C)	2.1-9.5%
Toluene	40°F (4°C)	1.1-7.1%

Source: U.S. Public Health Service, Centers for Disease Control and Prevention, National Institute for Occupational Safety and Health Manual of Analytical Methods, 4th Edition, Diversion of Physical Sciences and Engineering, Cincinnati, OH, August 1994.

employee work because of the potential for atmospheres to increase rapidly in flammability, or to exist within other areas of the workspace not tested. Many general industry safety regulations prohibit work without specialized safety protocol in atmospheres greater than 20% of the LEL of any vapor in air. OSHA has published three flammable vapor levels considered IDLH. They are 10% of the LEL of any flammable vapor in air in a permit-required confined space, 20% of the LEL of any flammable vapor in air for work at hazardous waste sites or for emergency responses to releases of hazardous substances, and 25% of the LEL of any flammable vapor in air in a maritime application (with certain exceptions allowing for no more than 10% LEL). In controlled situations, 25% of the LEL is often used as a safety guideline, although it is not considered IDLH. Explosive Ranges for various flammable substances are found in Table 5-7.

As with oxygen testing, combustible gas analyzers should be calibrated prior to and after use and maintained in compliance with manufacturers' specifications. In addition to the Wheatstone bridge principle, other technologies are capable of detecting and providing information on flammable atmospheres in work places.

Toxic Materials Testing

Unlike oxygen or flammable vapor testing, toxic material quantification is generally restricted to the specific toxins identified. Consequently, it is essential to identify all toxic materials to be quantified or that may interfere with tests of atmospheres. Once this has been accomplished, a specific test mechanism can be developed and deployed.

Length of stain direct reading indicator tubes is one method of determining concentrations of toxic materials on site. In general, a known quantity of atmosphere is drawn through a tube, series of tubes, or filter and calibrated tube for the purpose of producing a known reaction. This known reaction is then measured and an estimate of a toxic material in the atmosphere is given. Direct reading length of stain or colormetric tubes provide information concerning different quantities of hazardous atmospheres based on their ability to react with the atmospheres. Therefore, it is essential that testers identify appropriate levels for comparison, review the comparison units, and ensure that these units are used or are convertible to the manufactured detector tube quantifiers. A typical example of manufacturer's information for using a length of stain detector tube is found in **TABLE 5-8**.

A second means of detecting toxic materials is through the use of direct-reading instrumentation. Many manufacturers now use sensor-rich technology for direct quantification and reporting of toxic materials. As with length of stain detection and reaction tubes, these devices are specific to toxic materials to be monitored and, in many cases, have potential interferences.

Consequently, when using equipment such as carbon monoxide monitors, hydrogen sulfide monitors, and sulfur dioxide monitors, care should be taken to ensure interferences to detection are either not present or are accounted for.

Interferences to detection may also exist in the form of temperature or humidity extremes. Therefore, known temperature monitoring devices and humidity monitoring devices (such as a sling psychrometer) should be available to ensure the atmosphere to be tested is within appropriate environmental ranges for accurate detection and precision.

Ionizing Radiation

Ionizing radiation exists in low levels in ambient atmospheres, and high levels of ionizing radiation may be detected with a variety of equipment. Typically, Geiger-Muller sensing tubes are used for this purpose. These tubes allow entry and detection of radioactive particles or energy into the tube, and subsequently convert the energy detected to an electrical impulse and a face reading (deflection) on the meter. Once calibrated and checked to a known source, ionizing radiation higher

TABLE 5-8 Sample, Direct Reading Length of Stain Tube Information: Carbon Dioxide

Detector Tube	MSA AUER

CO_2-100

Part No.: 497606

Instructions for Use

1. Application
   Detection of carbon dioxide (CO_2) in air or technical gases.
2. Detector Tube Sampling Pump
   MSA AUER Gas-Tester® IIH, Kwik-Draw™ Pump, Gas-Tester® I / ThumbPump™-Sampler, Toximeter®II, or other suitable detector tube pumps. Observe respective instructions for use.
3. Measuring Range
   100 ppm ... 3000 ppm carbon dioxide at n = 10 (10 strokes).
4. Chemical Reaction and Color Change
   Reaction of carbon dioxide with hydrazine. Change of pH-value causes color change of an acid-base-indicator.
   Color change: white → blue
5. Sampling Procedure
   - Check detector tube pump for leakage.
   - Break off both tube tips.
   - Insert detector tube tightly into pump.
   Gas-Tester, Kwik-Draw pump, ThumpPump Sampler: Arrow on tube must point toward the pump.
   Toximeter II: insert tube into inlet side (white arrow). Arrow on tube must point toward the pump.
   Factor: see package.
   - Perform 10 pump strokes.
   - Read concentration at end of color zone within 2 minutes after sampling.
   - Used detector tubes cannot be used repeatedly.
   - Duration of one pump stroke: 30 ... 40 seconds.
6. Ambient Conditions During Sampling
   - Detector tubes can be used without compensation of the reading between 0 °C and 40 °C (32 °F and 104 °F) and between 10% rh [0.5 g/m^3 at 0 °C (32 °F)] and 90% rh [46 g/m3 at 40 °C (104 °F)].
   - Pressure compensation: multiply reading (in ppm) with compensation factor F.

   $$F = \frac{1013 \text{ (mbar)}}{\text{actual atm. pressure (mbar)}} = \frac{760 \text{ (mm Hg)}}{\text{actual atm. pressure (mm Hg)}}$$

7. Interferences and Cross Sensitivities
   No interference from:
   - hydrogen, methane, ethane, propane, butanes, carbon dioxide.
   - higher saturated hydrocarbons (e.g. hexanes, octanes), olefinic hydrocarbons (e.g. ethylene), aromatic hydrocarbons (e.g. benzene) up to 1% vol.
   - sulfur dioxide, hydrogen sulfide, nitrogen dioxide, hydrogen chloride and other acidic gases up to 100 ppm.
   - ammonia up to 100 ppm
8. Overall Uncertainty
   Up to ±15% in the range above 1000 ppm.
   Up to ±25% in the range 500 ppm ... 1000 ppm.
   (Expressed as relative standard deviation).
9. Storage and Transport
   Up to 25 °C (77 °F) and protected from light. Expiration date: see back of package.
10. Safety Advice/Disposal
    For tubes contents the following indications of danger apply: R 21/22-34.
    Safety advice S: 2-24/25-26-28 (water).
    Tubes must be kept away from unauthorized persons. For disposal as waste observe the legal regulations applicable in the individual country of use.

Manufactured by AUERGESELLSCHAFT GMBH, Germany

Source: Mine Safety Appliance Company. Sample, Direct Reading Length of Stain Tube Information: Carbon Dioxide. *Detector Tube and Dosimeter Handbook.* Pittsburgh, PA, 1988.

than background levels can be identified. As with other quantification procedures, the specific isotope should be identified wherever possible to ensure accurate comparisons based on the calibration source and the meter. Consequently, the use of standard survey meters is many times qualitative as opposed to quantitative when identifying ionizing radiation.

Specific metering equipment and calibration devices for radioisotopes are manufactured and may be used at sites where the isotope is identified. A discussion of on-site radiological testing equipment is given in Table 5-5.

Collection of Samples for Off-Site Analysis

Field staff may provide collection of samples for off-site analysis. Samples generally are personal breathing zone or exposure samples, or environmental air samples. Personal breathing zone (PBZ) sampling is performed by collection of a known quantity of breathing zone air on or through a known media. Pump collection flow rates, volumes, and collection media are predetermined and validated by such groups as OSHA, NIOSH, and a variety of state public health agencies.

It is important to remain within the parameters of a validated testing protocol for quality control and assurance purposes when collecting samples for off-site analysis. Once samples have been collected, timely transportation must be provided according to the analysis procedure. The analysis lab should be contacted before collection to ensure all collection and transport procedures are known.

Collection information for two typical laboratory analysis procedures published in the *Manual of Analytical Methods* of the Department of Health and Human Services (HHS), Centers for Disease Control and Prevention, National Institute for Occupational Safety and Health is presented in **TABLE 5-9** (Asbestos and other Fibers) and **TABLE 5-10** (Lead).

Although a personal sample need not be performed on all employees in hazardous areas, it is important to identify the most hazardously exposed or lead employee and place monitoring equipment on or around this person to determine all employees' maximum exposure. Then additional employees' characterizations may be based on these results, or a determination can be made whether additional sampling is necessary.

Environmental sampling may be used to characterize worst-case employee exposure or to provide information on general work environments. Environmental samples are often collected at the conclusion of remedial activities to ensure the space is acceptable for non-protected occupancy. As opposed to PBZ sampling, environmental sampling is performed in a known location (possibly a typical location or a worst-case scenario). Once again, validated procedures are followed and a known volume of air is collected on appropriate media through a calibrated pumping device. When deviating at all from published sampling criteria, industrial hygiene personnel must be contacted to ensure the validity of the test regimen.

Analysis of known sample levels is important to ensure quality control during off-site analysis of field samples. Field blanks, transportation blanks, laboratory blanks, or spiked samples are many times used to ensure quality control in the transport-and-analysis process. Field blanks (opened media) over which an atmosphere has not passed are used to ensure quality of the sampling media. Transportation blanks (opened sampling media over which an environment has not been passed and closed during the transportation process) are used to ensure that contamination or cross-contamination during transport does not occur.

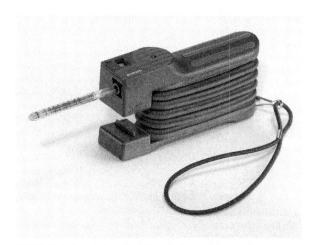

PHOTO 5-4 Sample Length of Stain Detector Tube and Pump.

Laboratory blanks (known as clean media) are used to ensure that contamination of analytical equipment is not present. Spiked samples (known quality control samples) are used to ensure that calibration of laboratory equipment and analysis procedures are appropriate. Most validated sample procedures require that field blanks be provided with each set of sample media collected or submitted for analysis.

TABLE 5-9 Asbestos and Other Fibers by PCM

ASBESTOS and OTHER FIBERS by PCM 7400

| FORMULA: Various | MW: Various | CAS: see Synonyms | RTECS: Various |

METHOD: 7400, Issue 2   **EVALUATION:** FULL   Issue 1: Rev. 3 on 15 May 1989
Issue 2: 15 August 1994

OSHA: 0.1 asbestos fiber (> 5 µm long)/cc; 1 f/cc, 30 min excursion; carcinogen
MSHA: 2 asbestos fibers/cc
NIOSH: 0.1 f/cc (fibers > 5 µm long), 400 L; carcinogen
ACGIH: 0.2 f/cc crocidolite; 0.5 f/cc amosite; 2 f/cc chrysotile and other asbestos; carcinogen

PROPERTIES: solid, fibrous, crystalline, anisotropic

SYNONYMS [CAS #]: actinolite [77536-66-4] or ferroactinolite [15669-07-5]; amosite [12172-73-5]; anthophyllite [77536-67-5]; chrysotile [12001-29-5]; serpentine [18786-24-8]; crocidolite [12001-28-4]; tremolite [77536-68-6]; amphibole asbestos [1332-21-4]; refractory ceramic fibers [142844-00-6]; fibrous glass

SAMPLING	MEASUREMENT
SAMPLER: FILTER (0.45- to 1.2-µm cellulose ester membrane, 25-mm; conductive cowl on cassette)	**TECHNIQUE:** LIGHT MICROSCOPY, PHASE CONTRAST
	ANALYTE: fibers (manual count)
FLOW RATE*: 0.5 to 16 L/min	**SAMPLE PREPARATION:** acetone - collapse/triacetin - immersion method [2]
VOL-MIN*: 400 L @ 0.1 fiber/cc **-MAX*:** (step 4, sampling)	**COUNTING RULES:** described in previous version of this method as "A" rules [1,3]
*Adjust to give 100 to 1300 fiber/mm^2	
SHIPMENT: routine (pack to reduce shock)	**EQUIPMENT:** 1. positive phase-contrast microscope 2. Walton-Beckett graticule (100-µm field of view) Type G-22 3. phase-shift test slide (HSE/NPL)
SAMPLE STABILITY: stable	
BLANKS: 2 to 10 field blanks per set	**CALIBRATION:** HSE/NPL test slide
ACCURACY	**RANGE:** 100 to 1300 fibers/mm^2 filter area
RANGE STUDIED: 80 to 100 fibers counted	**ESTIMATED LOD:** 7 fibers/mm^2 filter area
BIAS: see EVALUATION OF METHOD	**PRECISION ($\bar{S}_r$):** 0.10 to 0.12 [1]; see EVALUATION OF METHOD
OVERALL PRECISION ($\hat{S}_{rT}$): 0.115 to 0.13 [1]	
ACCURACY: see EVALUATION OF METHOD	

APPLICABILITY: The quantitative working range is 0.04 to 0.5 fiber/cc for a 1000-L air sample. The LOD depends on sample volume and quantity of interfering dust, and is <0.01 fiber/cc for atmospheres free of interferences. The method gives an index of airborne fibers. It is primarily used for estimating asbestos concentrations, though PCM does not differentiate between asbestos and other fibers. Use this method in conjunction with electron microscopy (e.g., Method 7402) for assistance in identification of fibers. Fibers < ca. 0.25 µm diameter will not be detected by this method [4]. This method may be used for other materials such as fibrous glass by using alternate counting rules (see Appendix C).

INTERFERENCES: If the method is used to detect a specific type of fiber, any other airborne fiber may interfere since all particles meeting the counting criteria are counted. Chain-like particles may appear fibrous. High levels of non-fibrous dust particles may obscure fibers in the field of view and increase the detection limit.

OTHER METHODS: This revision replaces Method 7400, Revision #3 (dated 5/15/89).

(continues)

TABLE 5-9 (continued)

REAGENTS:

1. Acetone,* reagent grade.
2. Triacetin (glycerol triacetate), reagent grade.

   *See SPECIAL PRECAUTIONS.

EQUIPMENT:

1. Sampler: field monitor, 25-mm, three-piece cassette with ca. 50-mm electrically conductive extension cowl and cellulose ester filter, 0.45- to 1.2-µm pore size, and backup pad.
   NOTE 1: Analyze representative filters for fiber background before use to check for clarity and background. Discard the filter lot if mean is ≥ 5 fibers per 100 graticule fields. These are defined as laboratory blanks. Manufacturer-provided quality assurance checks on filter blanks are normally adequate as long as field blanks are analyzed as described below.
   NOTE 2: The electrically conductive extension cowl reduces electrostatic effects. Ground the cowl when possible during sampling.
   NOTE 3: Use 0.8-µm pore size filters for personal sampling. The 0.45-µm filters are recommended for sampling when performing TEM analysis on the same samples. However, their higher pressure drop precludes their use with personal sampling pumps.
   NOTE 4: Other cassettes have been proposed that exhibit improved uniformity of fiber deposit on the filter surface, e.g., bellmouthed sampler (Envirometrics, Charleston, SC). These may be used if shown to give measured concentrations equivalent to sampler indicated above for the application.
2. Personal sampling pump, battery or line-powered vacuum, of sufficient capacity to meet flow-rate requirements (see step 4 for flow rate), with flexible connecting tubing.
3. Wire, multi-stranded, 22-gauge; 1" hose clamp to attach wire to cassette.
4. Tape, shrink- or adhesive-.
5. Slides, glass, frosted-end, pre-cleaned, 25- × 75-mm.
6. Cover slips, 22- × 22-mm, No. 1½, unless otherwise specified by microscope manufacturer.
7. Lacquer or nail polish.
8. Knife, #10 surgical steel, curved blade.
9. Tweezers.

TABLE 5-9 (continued)

EQUIPMENT (continued):

10. Acetone flash vaporization system for clearing filters on glass slides (see ref. [5] for specifications or see manufacturer's instructions for equivalent devices).
11. Micropipets or syringes, 5-µL and 100- to 500-µL.
12. Microscope, positive phase (dark) contrast, with green or blue filter, adjustable field iris, 8 to 10× eyepiece, and 40 to 45× phase objective (total magnification ca. 400×); numerical aperture = 0.65 to 0.75.
13. Graticule, Walton-Beckett type with 100-µm diameter circular field (area = 0.00785 mm^2) at the specimen plane (Type G-22). Available from Optometrics USA, P.O. Box 699, Ayer, MA 01432 [phone (508)-772-1700], and McCrone Accessories and Components, 850 Pasquinelli Drive, Westmont, IL 60559 [phone (312) 887-7100].
    NOTE: The graticule is custom-made for each microscope. (see APPENDIX A for the custom-ordering procedure).
14. HSE/NPL phase contrast test slide, Mark II. Available from Optometrics USA (address above).
15. Telescope, ocular phase-ring centering.
16. Stage micrometer (0.01-mm divisions).

SPECIAL PRECAUTIONS: Acetone is extremely flammable. Take precautions not to ignite it. Heating of acetone in volumes greater than 1 mL must be done in a ventilated laboratory fume hood using a flameless, spark-free heat source.

SAMPLING:

1. Calibrate each personal sampling pump with a representative sampler in line.
2. To reduce contamination and to hold the cassette tightly together, seal the crease between the cassette base and the cowl with a shrink band or light colored adhesive tape. For personal sampling, fasten the (uncapped) open-face cassette to the worker's lapel. The open face should be oriented downward.
   NOTE: The cowl should be electrically grounded during area sampling, especially under conditions of low relative humidity. Use a hose clamp to secure one end of the wire (Equipment, Item 3) to the monitor's cowl. Connect the other end to an earth ground (i.e., cold water pipe).
3. Submit at least two field blanks (or 10% of the total samples, whichever is greater) for each set of samples. Handle field blanks in a manner representative of actual handling of associated samples in the set. Open field blank cassettes at the same time as other cassettes just prior to sampling. Store top covers and cassettes in a clean area (e.g., a closed bag or box) with the top covers from the sampling cassettes during the sampling period.
4. Sample at 0.5 L/min or greater [6]. Adjust sampling flow rate, Q (L/min), and time, t (min), to produce a fiber density, E, of 100 to 1300 fibers/mm^2 (3.85×10^4 to 5×10^5 fibers per 25-mm filter with effective

(continues)

TABLE 5-9 (continued)

collection area $A_c = 385$ mm²) for optimum accuracy. These variables are related to the action level (one-half the current standard), L (fibers/cc), of the fibrous aerosol being sampled by:

$$t = \frac{A_c \times E}{Q \times L \times 10^3}.$$

- NOTE 1: The purpose of adjusting sampling times is to obtain optimum fiber loading on the filter. The collection efficiency does not appear to be a function of flow rate in the range of 0.5 to 16 L/min for asbestos fibers [7]. Relatively large diameter fibers (>3 µm) may exhibit significant aspiration loss and inlet deposition. A sampling rate of 1 to 4 L/min for 8 h is appropriate in atmospheres containing ca. 0.1 fiber/cc in the absence of significant amounts of non-asbestos dust. Dusty atmospheres require smaller sample volumes (≤400 L) to obtain countable samples. In such cases take short, consecutive samples and average the results over the total collection time. For documenting episodic exposures, use high flow rates (7 to 16 L/min) over shorter sampling times. In relatively clean atmospheres, where targeted fiber concentrations are much less than 0.1 fiber/cc, use larger sample volumes (3000 to 10000 L) to achieve quantifiable loadings. Take care, however, not to overload the filter with background dust. If ≥50% of the filter surface is covered with particles, the filter may be too overloaded to count and will bias the measured fiber concentration.
- NOTE 2: OSHA regulations specify a minimum sampling volume of 48 L for an excursion measurement, and a maximum sampling rate of 2.5 L/min [3].

5. At the end of sampling, replace top cover and end plugs.
6. Ship samples with conductive cowl attached in a rigid container with packing material to prevent jostling or damage.
    - NOTE: Do not use untreated polystyrene foam in shipping container because electrostatic forces may cause fiber loss from sample filter.

SAMPLE PREPARATION:

- NOTE 1: The object is to produce samples with a smooth (non-grainy) background in a medium with refractive index ≤ 1.46. This method collapses the filter for easier focusing and produces permanent (1–10 years) mounts which are useful for quality control and interlaboratory comparison. The aluminum "hot block" or similar flash vaporization techniques may be used outside the laboratory [2]. Other mounting techniques meeting the above criteria may also be used (e.g., the laboratory fume hood procedure for generating acetone vapor as described in Method 7400—revision of 5/15/85, or the non-permanent field mounting technique used in P&CAM 239 [3,7–9]). Unless the effective filtration area is known, determine the area and record the information referenced against the sample ID number [1,9–11].
- NOTE 2: Excessive water in the acetone may slow the clearing of the filter, causing material to be washed off the surface of the filter. Also, filters that have been exposed to high humidities prior to clearing may have a grainy background.

7. Ensure that the glass slides and cover slips are free of dust and fibers.
8. Adjust the rheostat to heat the "hot block" to ca. 70 °C [2].
    - NOTE: If the "hot block" is not used in a fume hood, it must rest on a ceramic plate and be isolated from any surface susceptible to heat damage.
9. Mount a wedge cut from the sample filter on a clean glass slide.
    a. Cut wedges of ca. 25% of the filter area with a curved-blade surgical steel knife using a rocking motion to prevent tearing. Place wedge, dust side up, on slide.
       NOTE: Static electricity will usually keep the wedge on the slide.
    b. Insert slide with wedge into the receiving slot at base of "hot block". Immediately place tip of a micropipet containing ca. 250 µL acetone (use the minimum volume needed to consistently clear the filter sections) into the inlet port of the PTFE cap on top of the "hot block" and inject the

TABLE 5-9 (continued)

   acetone into the vaporization chamber with a slow, steady pressure on the plunger button while holding pipet firmly in place. After waiting 3 to 5 s for the filter to clear, remove pipet and slide from their ports.
   CAUTION: Although the volume of acetone used is small, use safety precautions. Work in a well-ventilated area (e.g., laboratory fume hood). Take care not to ignite the acetone. Continuous use of this device in an unventilated space may produce explosive acetone vapor concentrations.
   c. Using the 5-µL micropipet, immediately place 3.0 to 3.5 µL triacetin on the wedge. Gently lower a clean cover slip onto the wedge at a slight angle to reduce bubble formation. Avoid excess pressure and movement of the cover glass.
   NOTE: If too many bubbles form or the amount of triacetin is insufficient, the cover slip may become detached within a few hours. If excessive triacetin remains at the edge of the filter under the cover slip, fiber migration may occur.
   d. Mark the outline of the filter segment with a glass marking pen to aid in microscopic evaluation.
   e. Glue the edges of the cover slip to the slide using lacquer or nail polish [12]. Counting may proceed immediately after clearing and mounting are completed.
   NOTE: If clearing is slow, warm the slide on a hotplate (surface temperature 50 °C) for up to 15 min to hasten clearing. Heat carefully to prevent gas bubble formation.

CALIBRATION AND QUALITY CONTROL:

10. Microscope adjustments. Follow the manufacturer's instructions. At least once daily use the telescope ocular (or Bertrand lens, for some microscopes) supplied by the manufacturer to ensure that the phase rings (annular diaphragm and phase-shifting elements) are concentric. With each microscope, keep a logbook in which to record the dates of microscope cleanings and major servicing.
    a. Each time a sample is examined, do the following:
        (1) Adjust the light source for even illumination across the field of view at the condenser iris. Use Kohler illumination, if available. With some microscopes, the illumination may have to be set up with bright field optics rather than phase contract optics.
        (2) Focus on the particulate material to be examined.
        (3) Make sure that the field iris is in focus, centered on the sample, and open only enough to fully illuminate the field of view.
    b. Check the phase-shift detection limit of the microscope periodically for each analyst/microscope combination:
        (1) Center the HSE/NPL phase-contrast test slide under the phase objective.
        (2) Bring the blocks of grooved lines into focus in the graticule area.
            NOTE: The slide contains seven blocks of grooves (ca. 20 grooves per block) in descending order of visibility. For asbestos counting, the microscope optics must completely resolve the grooved lines in block 3 although they may appear somewhat faint, and the grooved lines in blocks 6 and 7 must be invisible when centered in the graticule area. Blocks 4 and 5 must be at least partially visible but may vary slightly in visibility between microscopes. A microscope which fails to meet these requirements has resolution either too low or too high for fiber counting.
        (3) If image quality deteriorates, clean the microscope optics. If the problem persists, consult the microscope manufacturer.
11. Document the laboratory's precision for each counter for replicate fiber counts.
    a. Maintain as part of the laboratory quality assurance program a set of reference slides to be used on a daily basis [13]. These slides should consist of filter preparations including a range of loadings and background dust levels from a variety of sources including both field and reference samples (e.g., PAT, AAR, commercial samples). The Quality Assurance Officer should maintain custody of the reference slides and should supply each counter with a minimum of one reference

(continues)

TABLE 5-9 (continued)

   slide per workday. Change the labels on the reference slides periodically so that the counter does not become familiar with the samples.
   b. From blind repeat counts on reference slides, estimate the laboratory intra- and intercounter precision. Obtain separate values of relative standard deviation (S_r) for each sample matrix analyzed in each of the following ranges: 5 to 20 fibers in 100 graticule fields, >20 to 50 fibers in 100 graticule fields, and >50 to 100 fibers in 100 graticule fields. Maintain control charts for each of these data files.
      NOTE: Certain sample matrices (e.g., asbestos cement) have been shown to give poor precision [9].
12. Prepare and count field blanks along with the field samples. Report counts on each field blank.
    NOTE 1: The identity of blank filters should be unknown to the counter until all counts have been completed.
    NOTE 2: If a field blank yields greater than 7 fibers per 100 graticule fields, report possible contamination of the samples.
13. Perform blind recounts by the same counter on 10% of filters counted (slides relabeled by a person other than the counter). Use the following test to determine whether a pair of counts by the same counter on the same filter should be rejected because of possible bias: Discard the sample if the absolute value of the difference between the square roots of the two counts (in fiber/mm^2) exceeds $2.77XS_r'$ where X = average of the square roots of the two fiber counts (in fiber/mm^2) and $S_r' = S_r / 2$ where S_r is the intracounter relative standard deviation for the appropriate count range (in fibers) determined in step 11. For more complete discussions see reference [13].
    NOTE 1: Since fiber counting is the measurement of randomly placed fibers which may be described by a Poisson distribution, a square root transformation of the fiber count data will result in approximately normally distributed data [13].
    NOTE 2: If a pair of counts is rejected by this test, recount the remaining samples in the set and test the new counts against the first counts. Discard all rejected paired counts. It is not necessary to use this statistic on blank counts.
14. The analyst is a critical part of this analytical procedure. Care must be taken to provide a non-stressful and comfortable environment for fiber counting. An ergonomically designed chair should be used, with the microscope eyepiece situated at a comfortable height for viewing. External lighting should be set at a level similar to the illumination level in the microscope to reduce eye fatigue. In addition, counters should take 10- to 20-minute breaks from the microscope every one or two hours to limit fatigue [14]. During these breaks, both eye and upper back/neck exercises should be performed to relieve strain.
15. All laboratories engaged in asbestos counting should participate in a proficiency testing program such as the AIHA-NIOSH Proficiency Analytical Testing (PAT) Program for asbestos and routinely exchange field samples with other laboratories to compare performance of counters.

MEASUREMENT:

16. Center the slide on the stage of the calibrated microscope under the objective lens. Focus the microscope on the plane of the filter.
17. Adjust the microscope (Step 10).
    NOTE: Calibration with the HSE/NPL test slide determines the minimum detectable fiber diameter (ca. 0.25 µm) [4].
18. Counting rules: (same as P&CAM 239 rules [1,10,11]: see examples in APPENDIX B).
    a. Count any fiber longer than 5 µm which lies entirely within the graticule area.
       (1) Count only fibers longer than 5 µm. Measure length of curved fibers along the curve.
       (2) Count only fibers with a length-to-width ratio equal to or greater than 3:1.
    b. For fibers which cross the boundary of the graticule field:
       (1) Count as ½ fiber any fiber with only one end lying within the graticule area, provided that the fiber meets the criteria of rule a above.

TABLE 5-9 (continued)

   (2) Do not count any fiber which crosses the graticule boundary more than once.
   (3) Reject and do not count all other fibers.
   c. Count bundles of fibers as one fiber unless individual fibers can be identified by observing both ends of a fiber.
   d. Count enough graticule fields to yield 100 fibers. Count a minimum of 20 fields. Stop at 100 graticule fields regardless of count.
19. Start counting from the tip of the filter wedge and progress along a radial line to the outer edge. Shift up or down on the filter, and continue in the reverse direction. Select graticule fields randomly by looking away from the eyepiece briefly while advancing the mechanical stage. Ensure that, as a minimum, each analysis covers one radial line from the filter center to the outer edge of the filter. When an agglomerate or bubble covers ca. 1/6 or more of the graticule field, reject the graticule field and select another. Do not report rejected graticule fields in the total number counted.
   NOTE 1: When counting a graticule field, continuously scan a range of focal planes by moving the fine focus knob to detect very fine fibers which have become embedded in the filter. The small-diameter fibers will be very faint but are an important contribution to the total count. A minimum counting time of 15 s per field is appropriate for accurate counting.
   NOTE 2: This method does not allow for differentiation of fibers based on morphology. Although some experienced counters are capable of selectively counting only fibers which appear to be asbestiform, there is presently no accepted method for ensuring uniformity of judgment between laboratories. It is, therefore, incumbent upon all laboratories using this method to report total fiber counts. If serious contamination from non-asbestos fibers occurs in samples, other techniques such as transmission electron microscopy must be used to identify the asbestos fiber fraction present in the sample (see NIOSH Method 7402). In some cases (i.e., for fibers with diameters >1 µm), polarized light microscopy (as in NIOSH Method 7403) may be used to identify and eliminate interfering non-crystalline fibers [15].
   NOTE 3: Do not count at edges where filter was cut. Move in at least 1 mm from the edge.
   NOTE 4: Under certain conditions, electrostatic charge may affect the sampling of fibers. These electrostatic effects are most likely to occur when the relative humidity is low (below 20%), and when sampling is performed near the source of aerosol. The result is that deposition of fibers on the filter is reduced, especially near the edge of the filter. If such a pattern is noted during fiber counting, choose fields as close to the center of the filter as possible [5].
   NOTE 5: Counts are to be recorded on a data sheet that provides, as a minimum, spaces on which to record the counts for each field, filter identification number, analyst's name, date, total fibers counted, total fields counted, average count, fiber density, and commentary. Average count is calculated by dividing the total fiber count by the number of fields observed. Fiber density (fibers/mm^2) is defined as the average count (fibers/field) divided by the field (graticule) area (mm^2/field).

CALCULATIONS AND REPORTING OF RESULTS

20. Calculate and report fiber density on the filter, E (fibers/mm^2), by dividing the average fiber count per graticule field, F/n_f, minus the mean field blank count per graticule field, B/n_b, by the graticule field area, A_f (approx. 0.00785 mm^2):

$$E = \frac{(F/n_f - B/n_b)}{A_f}, \text{ fibers/mm}^2.$$

   NOTE: Fiber counts above 1300 fibers/mm^2 and fiber counts from samples with >50% of filter area covered with particulate should be reported as "uncountable" or "probably biased." Other fiber counts outside the 100–1300 fiber/mm^2 range should be reported as having "greater than optimal variability" and as being "probably biased."

21. Calculate and report the concentration, C (fibers/cc), of fibers in the air volume sampled, V (L), using the effective collection area of the filter, A_c (approx. 385 mm^2 for a 25-mm filter):

(continues)

TABLE 5-9 (continued)

$$C = \frac{EA_c}{V \times 10^3}.$$

NOTE: Periodically check and adjust the value of A_c, if necessary.

22. Report intralaboratory and interlaboratory relative standard deviations (from Step 11) with each set of results.

    NOTE: Precision depends on the total number of fibers counted [1,16]. Relative standard deviation is documented in references [1,15–17] for fiber counts up to 100 fibers in 100 graticule fields. Comparability of interlaboratory results is discussed below. As a first approximation, use 213% above and 49% below the count as the upper and lower confidence limits for fiber counts greater than 20 (Figure 1).

EVALUATION OF METHOD:

Method Revisions:

This method is a revision of P&CAM 239 [10]. A summary of the revisions is as follows:

1. Sampling:
   The change from a 37-mm to a 25-mm filter improves sensitivity for similar air volumes. The change in flow rates allows for 2-m³ full-shift samples to be taken, providing that the filter is not overloaded with non-fibrous particulates. The collection efficiency of the sampler is not a function of flow rate in the range 0.5 to 16 L/min [10].
2. Sample preparation technique:
   The acetone vapor-triacetin preparation technique is a faster, more permanent mounting technique than the dimethyl phthalate/diethyl oxalate method of P&CAM 239 [2,4,10]. The aluminum "hot block" technique minimizes the amount of acetone needed to prepare each sample.
3. Measurement:
   a. The Walton-Beckett graticule standardizes the area observed [14,18,19].
   b. The HSE/NPL test slide standardizes microscope optics for sensitivity to fiber diameter [4,14].
   c. Because of past inaccuracies associated with low fiber counts, the minimum recommended loading has been increased to 100 fibers/mm² filter area (a total of 78.5 fibers counted in 100 fields, each with field area = 0.00785 mm².) Lower levels generally result in an overestimate of the fiber count when compared to results in the recommended analytical range [20]. The recommended loadings should yield intracounter S_r in the range of 0.10 to 0.17 [21–23].

Interlaboratory Comparability:

An international collaborative study involved 16 laboratories using prepared slides from the asbestos cement, milling, mining, textile, and friction material industries [9]. The relative standard deviations (S_r) varied with sample type and laboratory. The ranges were:

Rules	Intralaboratory S_r	Interlaboratory S_r	Overall S_r
AIA (NIOSH A Rules)*	0.12 to 0.40	0.27 to 0.85	0.46
Modified CRS (NIOSH B Rules)†	0.11 to 0.29	0.20 to 0.35	0.25

*Under AIA rules, only fibers having a diameter less than 3 µm are counted and fibers attached to particles larger than 3 µm are not counted. NIOSH A Rules are otherwise similar to the AIA rules.
†See Appendix C.

A NIOSH study conducted using field samples of asbestos gave intralaboratory S_r in the range 0.17 to 0.25 and an interlaboratory S_r of 0.45 [21]. This agrees well with other recent studies [9,14,16].

TABLE 5-9 (continued)

At this time, there is no independent means for assessing the overall accuracy of this method. One measure of reliability is to estimate how well the count for a single sample agrees with the mean count from a large number of laboratories. The following discussion indicates how this estimation can be carried out based on measurements of the interlaboratory variability, as well as showing how the results of this method relate to the theoretically attainable counting precision and to measured intra- and interlaboratory S_r. (NOTE: The following discussion does not include bias estimates and should not be taken to indicate that lightly loaded samples are as accurate as properly loaded ones).

Theoretically, the process of counting randomly (Poisson) distributed fibers on a filter surface will give an S_r that depends on the number, N, of fibers counted:

$$S_r = 1/N^{1/2}.$$

Thus S_r is 0.1 for 100 fibers and 0.32 for 10 fibers counted. The actual S_r found in a number of studies is greater than these theoretical numbers [17,19–21].

An additional component of variability comes primarily from subjective interlaboratory differences. In a study of ten counters in a continuing sample exchange program, Ogden [15] found this subjective component of intralaboratory S_r to be approximately 0.2 and estimated the overall S_r by the term:

$$\frac{[N+(0.2\times N)^2]^{1/2}}{N}.$$

Ogden found that the 90% confidence interval of the individual intralaboratory counts in relation to the means were $+2\,S_r$ and $-1.5\,S_r$. In this program, one sample out of ten was a quality control sample. For laboratories not engaged in an intensive quality assurance program, the subjective component of variability can be higher.

In a study of field sample results in 46 laboratories, the Asbestos Information Association also found that the variability had both a constant component and one that depended on the fiber count [14]. These results gave a subjective interlaboratory component of S_r (on the same basis as Ogden's) for field samples of ca. 0.45. A similar value was obtained for 12 laboratories analyzing a set of 24 field samples [21]. This value falls slightly above the range of S_r (0.25 to 0.42 for 1984–85) found for 80 reference laboratories in the NIOSH PAT program for laboratory-generated samples [17].

A number of factors influence S_r for a given laboratory, such as that laboratory's actual counting performance and the type of samples being analyzed. In the absence of other information, such as from an interlaboratory quality assurance program using field samples, the value for the subjective component of variability is chosen as 0.45. It is hoped that the laboratories will carry out the recommended interlaboratory quality assurance programs to improve their performance and thus reduce the S_r.

The above relative standard deviations apply when the population mean has been determined. It is more useful, however, for laboratories to estimate the 90% confidence interval on the mean count from a single sample fiber count (Figure 1). These curves assume similar shapes of the count distribution for interlaboratory and intralaboratory results [16].

For example, if a sample yields a count of 24 fibers, Figure 1 indicates that the mean interlaboratory count will fall within the range of 227% above and 52% below that value 90% of the time. We can apply these percentages directly to the air concentrations as well. If, for instance, this sample (24 fibers counted) represented a 500-L volume, then the measured concentration is 0.02 fibers/mL (assuming 100 fields counted, 25-mm filter, 0.00785 mm² counting field area). If this same sample were counted by

(continues)

TABLE 5-9 (continued)

a group of laboratories, there is a 90% probability that the mean would fall between 0.01 and 0.08 fiber/mL. These limits should be reported in any comparison of results between laboratories.

Note that the S_r of 0.45 used to derive Figure 1 is used as an estimate for a random group of laboratories. If several laboratories belonging to a quality assurance group can show that their interlaboratory S_r is smaller, then it is more correct to use that smaller S_r. However, the estimated S_r of 0.45 is to be used in the absence of such information. Note also that it has been found that S_r can be higher for certain types of samples, such as asbestos cement [9].

Quite often the estimated airborne concentration from an asbestos analysis is used to compare to a regulatory standard. For instance, if one is trying to show compliance with an 0.5 fiber/mL standard using a single sample on which 100 fibers have been counted, then Figure 1 indicates that the 0.5 fiber/mL standard must be 213% higher than the measured air concentration. This indicates that if one measures a fiber concentration of 0.16 fiber/mL (100 fibers counted), then the mean fiber count by a group of laboratories (of which the compliance laboratory might be one) has a 95% chance of being less than 0.5 fibers/mL; i.e., $0.16 + 2.13 \times 0.16 = 0.5$.

It can be seen from Figure 1 that the Poisson component of the variability is not very important unless the number of fibers counted is small. Therefore, a further approximation is to simply use +213% and −49% as the upper and lower confidence values of the mean for a 100-fiber count.

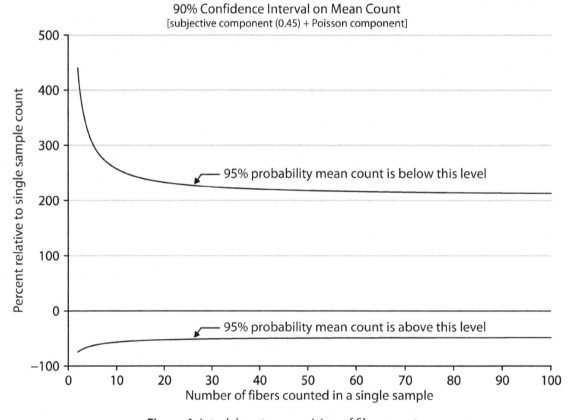

Figure 1. Interlaboratory precision of fiber counts.

TABLE 5-9 (continued)

The curves in Figure 1 are defined by the following equations:

$$U_{CL} = \frac{2X + 2.25 + [(2.25 + 2X)^2 - 4(1 - 2.25S_r^2)X^2]^{1/2}}{2(1 - 2.25S_r^2)} \text{ and}$$

$$L_{CL} = \frac{2X + 4 - [(4 + 2X)^2 - 4(1 - 4S_r^2)X^2]^{1/2}}{2(1 - 4S_r^2)},$$

where S_r = subjective interlaboratory relative standard deviation, which is close to the total interlaboratory S_r when approximately 100 fibers are counted,
 X = total fibers counted on sample,
 L_{CL} = lower 95% confidence limit, and
 U_{CL} = upper 95% confidence limit.

Note that the range between these two limits represents 90% of the total range.

REFERENCES:

[1] Leidel, N. A., S. G. Bayer, R. D. Zumwalde, and K. A. Busch. USPHS/NIOSH Membrane Filter Method for Evaluating Airborne Asbestos Fibers, U.S. Department of Health, Education, and Welfare, Publ. (NIOSH) 79-127 (1979).

[2] Baron, P. A. and G. C. Pickford. "An Asbestos Sample Filter Clearing Procedure," *Appl. Ind. Hyg., 1*, 169–171, 199 (1986).

[3] Occupational Safety and Health Administration, U.S. Department of Labor, Occupational Exposure to Asbestos, Tremolite, Anthophyllite, and Actinolite Asbestos; Final Rules, 29 CFR Part 1910.1001 Amended June 20, 1986.

[4] Rooker, S. J., N. P. Vaughn, and J. M. LeGuen. "On the Visibility of Fibers by Phase Contrast Microscopy," *Amer. Ind. Hyg. Assoc. J., 43*, 505–515 (1982).

[5] Baron, P. and G. Deye, "Electrostatic Effects in Asbestos Sampling," Parts I and II, *Amer. Ind. Hyg. Assoc. J., 51*, 51–69 (1990).

[6] Johnston, A. M., A. D. Jones, and J. H. Vincent. "The Influence of External Aerodynamic Factors on the Measurement of the Airborne Concentration of Asbestos Fibers by the Membrane Filter Method," *Ann. Occup. Hyg., 25*, 309–316 (1982).

[7] Beckett, S.T., "The Effects of Sampling Practice on the Measured Concentration of Airborne Asbestos," *Ann. Occup. Hyg., 21*, 259–272 (1980).

[8] Jankovic, J. T., W. Jones, and J. Clere. "Field Techniques for Clearing Cellulose Ester Filters Used in Asbestos Sampling," *Appl. Ind. Hyg., 1*, 145–147 (1986).

[9] Crawford, N. P., H. L. Thorpe, and W. Alexander. "A Comparison of the Effects of Different Counting Rules and Aspect Ratios on the Level and Reproducibility of Asbestos Fiber Counts," Part I: Effects on Level (Report No. TM/82/23), Part II: Effects on Reproducibility (Report No. TM/82/24), Institute of Occupational Medicine, Edinburgh, Scotland (December, 1982).

[10] NIOSH Manual of Analytical Methods, 2nd ed., Vol. 1., P&CAM 239, U.S. Department of Health, Education, and Welfare, Publ. (NIOSH) 77-157-A (1977).

[11] Revised Recommended Asbestos Standard, U.S. Department of Health, Education, and Welfare, Publ. (NIOSH) 77-169 (1976); as amended in NIOSH statement at OSHA Public Hearing, June 21, 1984.

[12] Asbestos International Association, AIA Health and Safety Recommended Technical Method #1 (RTMI). "Airborne Asbestos Fiber Concentrations at Workplaces by Light Microscopy" (Membrane Filter Method), London (1979).

[13] Abell, M., S. Shulman and P. Baron. "The Quality of Fiber Count Data," *Appl. Ind. Hyg., 4*, 273–285 (1989).

[14] "A Study of the Empirical Precision of Airborne Asbestos Concentration Measurements in the Workplace by the Membrane Filter Method," Asbestos Information Association, Air Monitoring Committee Report, Arlington, VA (June, 1983).

(continues)

TABLE 5-9 (continued)

[15] McCrone, W., L. McCrone and J. Delly, "Polarized Light Microscopy," Ann Arbor Science (1978).
[16] Ogden, T. L. "The Reproducibility of Fiber Counts," Health and Safety Executive Research Paper 18 (1982).
[17] Schlecht, P. C. and S. A. Schulman. "Performance of Asbestos Fiber Counting Laboratories in the NIOSH Proficiency Analytical Testing (PAT) Program," *Am. Ind. Hyg. Assoc. J., 47*, 259–266 (1986).
[18] Chatfield, E. J. Measurement of Asbestos Fiber Concentrations in Workplace Atmospheres, Royal Commission on Matters of Health and Safety Arising from the Use of Asbestos in Ontario, Study No. 9, 180 Dundas Street West, 22nd Floor, Toronto, Ontario, CANADA M5G 1Z8.
[19] Walton, W. H. "The Nature, Hazards, and Assessment of Occupational Exposure to Airborne Asbestos Dust: A Review," *Ann. Occup. Hyg., 25*, 115–247 (1982).
[20] Cherrie, J., A.D. Jones, and A.M. Johnston. "The Influence of Fiber Density on the Assessment of Fiber Concentration Using the membrane filter Method." *Am. Ind. Hyg. Assoc. J., 47*(8), 465–74 (1986).
[21] Baron, P. A. and S. Shulman. "Evaluation of the Magiscan Image Analyzer for Asbestos Fiber Counting." *Am. Ind. Hyg. Assoc. J.*, (in press).
[22] Taylor, D. G., P. A. Baron, S. A. Shulman and J. W. Carter. "Identification and Counting of Asbestos Fibers," *Am. Ind. Hyg. Assoc. J. 45*(2), 84–88 (1984).
[23] "Potential Health Hazards of Video Display Terminals," NIOSH Research Report, June 1981.
[24] "Reference Methods for Measuring Airborne Man-Made Mineral Fibers (MMMF)," WHO/EURO Technical Committee for Monitoring an Evaluating Airborne MMMF, World Health Organization, Copenhagen (1985).
[25] Criteria for a Recommended Standard…Occupational Exposure to Fibrous Glass, U.S. Department of Health, Education, and Welfare, Publ. (NIOSH) 77-152 (1977).

METHOD WRITTEN BY:

Paul A. Baron, Ph.D., NIOSH/DPSE.

APPENDIX A. CALIBRATION OF THE WALTON-BECKETT GRATICULE

Before ordering the Walton-Beckett graticule, the following calibration must be done to obtain a counting area (D) 100 µm in diameter at the image plane. The diameter, d_c (mm), of the circular counting area and the disc diameter must be specified when ordering the graticule.

1. Insert any available graticule into the eyepiece and focus so that the graticule lines are sharp and clear.
2. Set the appropriate interpupillary distance and, if applicable, reset the binocular head adjustment so that the magnification remains constant.
3. Install the 40 to 45× phase objective.
4. Place a stage micrometer on the microscope object stage and focus the microscope on the graduated lines.
5. Measure the magnified grid length of the graticule, L_o (µm), using the stage micrometer.
6. Remove the graticule from the microscope and measure its actual grid length, L_a (mm). This can best be accomplished by using a stage fitted with verniers.
7. Calculate the circle diameter, d_c (mm), for the Walton-Beckett graticule:

$$d_c = \frac{L_a}{L_o} \times D.$$

   Example: If L_o = 112 µm, L_a = 4.5 mm, and D = 100 µm, then d_c = 4.02 mm.
8. Check the field diameter, D (acceptable range 100 µm ± 2 µm) with a stage micrometer upon receipt of the graticule from the manufacturer. Determine field area (acceptable range 0.00754 mm^2 to 0.00817 mm^2).

TABLE 5-9 (continued)

APPENDIX B. COMPARISON OF COUNTING RULES

Figure 2 shows a Walton-Beckett graticule as seen through the microscope. The rules will be discussed as they apply to the labeled objects in the figure.

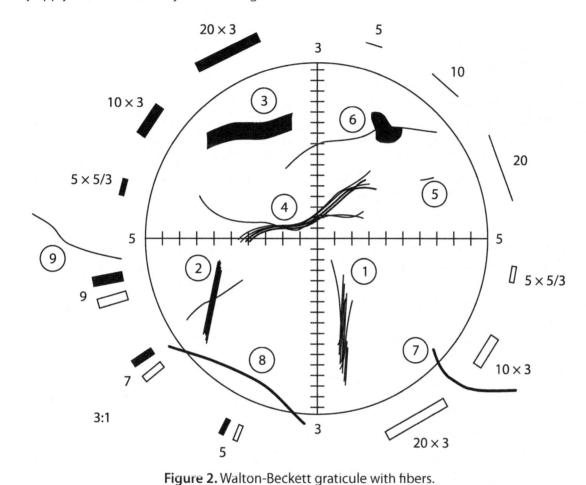

Figure 2. Walton-Beckett graticule with fibers.

TABLE 5-9 (continued)

These rules are sometimes referred to as the "A" rules:

Object	Count	Discussion
1	1 fiber	Optically observable asbestos fibers are actually bundles of fine fibrils. If the fibrils seem to be from the same bundle, the object is counted as a single fiber. Note, however, that all objects meeting length and aspect ratio criteria are counted whether or not they appear to be asbestos.
2	2 fibers	If fibers meeting the length and aspect ratio criteria (length >5 µm and length-to-width ratio > 3 to 1) overlap, but do not seem to be part of the same bundle, they are counted as separate fibers.
3	1 fiber	Although the object has a relatively large diameter (>3 µm), it is counted as fiber under the rules. There is no upper limit on the fiber diameter in the counting rules. Note that fiber width is measured at the widest compact section of the object.
4	1 fiber	Although long fine fibrils may extend from the body of a fiber, these fibrils are considered part of the fiber if they seem to have originally been part of the bundle.
5	Do not count	If the object is ≤ 5 µm long, it is not counted.
6	1 fiber	A fiber partially obscured by a particle is counted as one fiber. If the fiber ends emanating from a particle do not seem to be from the same fiber and each end meets the length and aspect ratio criteria, they are counted as separate fibers.
7	½ fiber	A fiber which crosses into the graticule area one time is counted as ½ fiber.
8	Do not count	Ignore fibers that cross the graticulate boundary more than once.
9	Do not count	Ignore fibers that lie outside the graticule boundary.

APPENDIX C. ALTERNATE COUNTING RULES FOR NON-ASBESTOS FIBERS

Other counting rules may be more appropriate for measurement of specific non-asbestos fiber types, such as fibrous glass. These include the "B" rules given below (from NIOSH Method 7400, Revision #2, dated 8/15/87), the World Health Organization reference method for man-made mineral fiber [24], and the NIOSH fibrous glass criteria document method [25]. The upper diameter limit in these methods prevents measurements of non-thoracic fibers. It is important to note that the aspect ratio limits included in these methods vary. NIOSH recommends the use of the 3:1 aspect ratio in counting fibers.

It is emphasized that hybridization of different sets of counting rules is not permitted. Report specifically which set of counting rules are used with the analytical results.

"B" Counting Rules

1. Count only *ends* of fibers. Each fiber must be longer than 5 µm and less than 3 µm diameter.
2. Count only ends of fibers with a length-to-width ratio equal to or greater than 5:1.
3. Count each fiber end which falls within the graticule area as one end, provided that the fiber meets rules 1 and 2 above. Add split ends to the count as appropriate if the split fiber segment also meets the criteria of rules 1 and 2 above.
4. Count visibly free ends which meet rules 1 and 2 above when the fiber appears to be attached to another particle, regardless of the size of the other particle. Count the end of a fiber obscured by another particle if the particle covering the fiber end is less than 3 µm in diameter.

TABLE 5-9 (continued)

5. Count free ends of fibers emanating from large clumps and bundles up to a maximum of 10 ends (5 fibers), provided that each segment meets rules 1 and 2 above.
6. Count enough graticule fields to yield 200 ends. Count a minimum of 20 graticule fields. Stop at 100 graticule fields, regardless of count.
7. Divide total end count by 2 to yield fiber count.

APPENDIX D. EQUIVALENT LIMITS OF DETECTION AND QUANTITATION

	Fiber density on filter*		Fiber concentration in air, f/cc	
	Fibers per 100 fields	Fibers/mm²	400-L air sample	1000-L air sample
	200	255	0.25	0.10
	100	127	0.125	0.05
LOQ	80.0	102	0.10	0.04
	50	64	0.0625	0.025
	25	32	0.03	0.0125
	20	25	0.025	0.010
	10	12.7	0.0125	0.005
	8	10.2	0.010	0.004
LOD	5.5	7	0.00675	0.0027

*Assumes 385 mm² effective filter collection area, and field area = 0.00785 mm², for relatively "clean" (little particulate aside from fibers) filters.

Source: U.S. Public Health Service, Centers for Disease Control and Prevention, National Institute for Occupational Safety and Health Manual of Analytical Methods, 4th Edition, Diversion of Physical Sciences and Engineering, Cincinnati, OH, August 1994.

TABLE 5-10 NIOSH Procedure #7300: Elements by ICP

ELEMENTS by ICP 7300
(Nitric/Perchloric Acid Ashing)

MW: Table 1   CAS: Table 2   RTECS: Table 2

METHOD: 7300, Issue 3	EVALUATION: PARTIAL	Issue 1: 15 August 1990
		Issue 3: 15 March 2003

OSHA: Table 2
NIOSH: Table 2
ACGIH: Table 2

PROPERTIES: Table 1

ELEMENTS:
aluminum*	calcium	lanthanum	nickel	strontium	tungsten*
antimony*	chromium*	lithium*	potassium	tellurium	vanadium*
arsenic	cobalt*	magnesium	phosphorus	tin	yittrium
barium	copper	manganese*	selenium	thallium	zinc
beryllium*	iron	molybdenum*	silver	titanium	zirconium*
cadmium	lead*				

*Some compounds of these elements require special sample treatment.

SAMPLING | MEASUREMENT

SAMPLER: FILTER (0.8-µm, cellulose ester membrane, or 5.0-µm, polyvinyl chloride membrane)

TECHNIQUE: INDUCTIVELY COUPLED ARGON PLASMA, ATOMIC EMISSION SPECTROSCOPY (ICP-AES)

FLOWRATE: 1 to 4 L/min

ANALYTE: elements above

VOL-MIN: Table 1
-MAX: Table 1

ASHING REAGENTS: conc. HNO_3/ conc. $HClO_4$ (4:1), 5 mL; 2mL increments added as needed

SHIPMENT: routine

CONDITIONS: room temperature, 30 min; 150 °C to near dryness

SAMPLE STABILITY: stable

FINAL SOLUTION: 4% HNO_3, 1% $HClO_4$, 25 mL

BLANKS: 2 to 10 field blanks per set

WAVELENGTH: depends upon element; Table 3

ACCURACY

BACKGROUND CORRECTION: spectral wavelength shift

RANGE STUDIED: not determined

CALIBRATION: elements in 4% HNO_3, 1% $HClO_4$

BIAS: not determined

RANGE: varies with element [1]

OVERALL PRECISION ($\hat{S}_{rT}$): not determined

ESTIMATED LOD: Tables 3 and 4

ACCURACY: not determined

PRECISION ($\hat{S}$): Tables 3 and 4

APPLICABILITY: The working range of this method is 0.005 to 2.0 mg/m^3 for each element in a 500-L air sample. This is simultaneous elemental analysis, not compound specific. Verify that the types of compounds in the samples are soluble with the ashing procedure selected.

INTERFERENCES: Spectral interferences are the primary interferences encountered in ICP-AES analysis. These are minimized by judicious wavelength selection, interelement correction factors and background correction [1-4].

OTHER METHODS: This issue updates issues 1 and 2 of Method 7300, which replaced P&CAM 351 [3] for trace elements. Flame atomic absorption spectroscopy (e.g., Methods 70XX) is an alternate analytical technique for many of these elements. Graphite furnace AAS (e.g., 7102 for Be, 7105 for Pb) is more sensitive.

TABLE 5-10 (continued)

REAGENTS:

1. Nitric acid (HNO_3), conc., ultra pure.
2. Perchloric acid ($HClO_4$), conc., ultra pure.*
3. Ashing acid: 4:1 (v/v) HNO_3:$HClO_4$. Mix 4 volumes conc. HNO_3 with 1 volume conc. $HClO_4$.
4. Calibration stock solutions, 1000 µg/mL. Commercially available, or prepared per instrument manufacturer's recommendation (see step 12).
5. Dilution acid, 4% HNO_3, 1% $HClO_4$. Add 50 mL ashing acid to 600 mL water; dilute to 1 L.
6. Argon.
7. Distilled, deionized water.

* See SPECIAL PRECAUTIONS.

EQUIPMENT:

1. Sampler: cellulose ester membrane filter, 0.8-µm pore size; or polyvinyl chloride membrane, 5.0-µm pore size; 37-mm diameter, in cassette filter holder.
2. Personal sampling pump, 1 to 4 L/min, with flexible connecting tubing.
3. Inductively coupled plasma-atomic emission spectrometer, equipped as specified by the manufacturer for analysis of elements of interest.
4. Regulator, two-stage, for argon.
5. Beakers, Phillips, 125-mL, or Griffin, 50-mL, with watchglass covers.**
6. Volumetric flasks, 10-, 25-,100-mL., and 1-L**
7. Assorted volumetric pipets as needed.**
8. Hotplate, surface temperature 150 °C.

** Clean all glassware with conc. nitric acid and rinse thoroughly in distilled water before use.

SPECIAL PRECAUTIONS: All perchloric acid digestions are required to be done in a perchloric acid hood. When working with concentrated acids, wear protective clothing and gloves.

SAMPLING:

1. Calibrate each personal sampling pump with a representative sampler in line.
2. Sample at an accurately known flow rate between 1 and 4 L/min for a total sample size of 200 to 2000 L (see Table 1) for TWA measurements. Do not exceed a filter loading of approximately 2 mg total dust.

SAMPLE PREPARATION:

3. Open the cassette filter holders and transfer the samples and blanks to clean beakers.
4. Add 5 mL ashing acid. Cover with a watchglass. Let stand 30 min at room temperature.
   NOTE: Start a reagent blank at this step.
5. Heat on hotplate (120 °C) until ca. 0.5 mL remains
   NOTE 1: Recovery of lead from some paint matrices may require other digestion techniques. See Method 7082 (Lead by Flame AAS) for an alternative hotplate digestion procedure or Method 7302 for a microwave digestion procedure.
   NOTE 2: Some species of Al, Be, Co, Cr, Li, Mn, Mo, V, and Zr will not be completely solubilized by this procedure. Alternative solubilization techniques for most of these elements can be found elsewhere [5-10]. For example, aqua regia may be needed for Mn [6,12].
6. Add 2 mL ashing acid and repeat step 5. Repeat this step until the solution is clear.
7. Remove watchglass and rinse into the beaker with distilled water.
8. Increase the temperature to 150 °C and take the sample to near dryness (ca. 0.5 mL).
9. Dissolve the residue in 2 to 3 mL dilution acid.
10. Transfer the solutions quantitatively to 25-mL volumetric flasks.
11. Dilute to volume with dilution acid.
    NOTE: If more sensitivity is required, the final sample volume may be held to 10 mL.

(continues)

TABLE 5-10 (continued)

CALIBRATION AND QUALITY CONTROL:

12. Calibrate the spectrometer according to the manufacturers recommendations.
    NOTE: Typically, an acid blank and 1.0 µg/mL multielement working standards are used. The following multielement combinations are chemically compatible in 4% HNO_3/1% $HClO_4$:
    a. Al, As, Ba, Be, Ca, Cd, Co, Cr, Cu, Fe, La, In, Na
    b. Ag, K, Li, Mg, Mn, Ni, P, Pb, Se, Sr, Tl, V, Y, Zn, Sc
    c. Mo, Sb, Sn, Te, Ti, W, Zr
    d. Acid blank
13. Analyze a standard for every ten samples.
14. Check recoveries with at least two spiked blank filters per ten samples.

MEASUREMENT:

15. Set spectrometer to conditions specified by manufacturer.
16. Analyze standards and samples.
    NOTE: If the values for the samples are above the range of the standards, dilute the solutions with dilution acid, reanalyze and apply the appropriate dilution factor in the calculations.

CALCULATIONS:

17. Obtain the solution concentrations for the sample, C_s (µg/mL), and the average media blank, C_b (µg/mL), from the instrument.
18. Using the solution volumes of sample, V_s (mL), and media blank, V_b (mL), calculate the concentration, C (mg/m^3), of each element in the air volume sampled, V (L):

$$C = \frac{C_s V_s - C_b V_b}{V}, mg/m^3$$

NOTE: µg/L ≡ mg/m^3

EVALUATION OF METHOD:

Issues 1 and 2
Method, 7300 was originally evaluated in 1981 [2,3]. The precision and recovery data were determined at 2.5 and 1000 µg of each element per sample on spiked filters. The measurements used for the method evaluation in Issues 1 and 2 were determined with a Jarrell-Ash Model 1160 Inductively Coupled Plasma Spectrometer operated according to manufacturer's instructions.

Issue 3
In this update of NIOSH Method 7300, the precision and recovery data were determined at approximately 3x and 10x the instrumental detection limits on commercially prepared spiked filters [12] using 25.0 mL as the final sample volume. Tables 3 and 4 list the precision and recovery data, instrumental detection limits, and analytical wavelengths for mixed cellulose ester (MCE) and polyvinyl chloride (PVC) filters. PVC Filters which can be used for total dust measurements and then digested for metals measurements were tested and found to give good results. The values in Tables 3 and 4 were determined with a Spectro Analytical Instruments Model End On Plasma (EOP)(axial) operated according to manufacturer's instructions.

TABLE 5-10 (continued)

REFERENCES:

[1] Millson M, Andrews R [2002]. Backup data report, Method 7300, unpublished report, NIOSH/DART.
[2] Hull RD [1981]. Multielement Analysis of Industrial Hygiene Samples, NIOSH Internal Report, presented at the American Industrial Hygiene Conference, Portland, Oregon.
[3] NIOSH [1982]. NIOSH Manual of Analytical Methods, 2nd ed., V. 7, P&CAM 351 (Elements by ICP), U.S. Department of Health and Human Services, Publ. (NIOSH) 82-100.
[4] NIOSH [1994]. Elements by ICP: Method 7300, Issue 2. In: Eller PM, Cassinelli ME, eds., NIOSH Manual of Analytical Methods, 4th ed. Cincinnati, OH: U.S. Department of Health and Human Services, Centers for Disease Control and Prevention, National Institute for Occupational Safety and Health, DHHS (NIOSH) Publication No. 94-113.
[5] NIOSH [1994]. Lead by FAAS: Method 7082. In: Eller PM, Cassinelli ME, eds., NIOSH Manual of Analytical Methods, 4th ed. Cincinnati, OH: U.S. Department of Health and Human Services, Centers for Disease Control and Prevention, National Institute for Occupational Safety and Health, DHHS (NIOSH) Publication No. 94-113.
[6] NIOSH [1977]. NIOSH Manual of Analytical Methods, 2nd ed., V. 2, S5 (Manganese), U.S. Department of Health, Education, and Welfare, Publ. (NIOSH) 77-157-B.
[7] NIOSH [1994]. Tungsten, soluble/insoluble: Method 7074. In: Eller PM, Cassinelli ME, eds., NIOSH Manual of Analytical Methods, 4th ed. Cincinnati, OH: U.S. Department of Health and Human Services, Centers for Disease Control and Prevention, National Institute for Occupational Safety and Health, DHHS (NIOSH) Publication No. 94-113.
[8] NIOSH [1979]. NIOSH Manual of Analytical Methods, 2nd ed., V. 5, P&CAM 173 (Metals by Atomic Absorption), U.S. Department of Health, Education, and Welfare, Publ. (NIOSH) 79-141.
[9] NIOSH [1977]. NIOSH Manual of Analytical Methods, 2nd ed., V. 3, S183 (Tin), S185 (Zirconium), and S376 (Molybdenum), U.S. Department of Health, Education, and Welfare, Publ. (NIOSH) 77-157-C.
[10] ISO [2001]. Workplace air - Determination of metals and metalloids in airborne particulate matter by inductively coupled plasma atomic emission spectrometry - Part 2: Sample preparation. International Organization for Standardization. ISO 15202-2:2001(E).
[11] ASTM [1985]. 1985 Annual Book of ASTM Standards, Vol. 11.01; Standard Specification for Reagent Water; ASTM, Philadelphia, PA, D1193-77 (1985).
[12] Certification Inorganic Ventures for spikes.

METHOD REVISED BY:

Mark Millson and Ronnee Andrews, NIOSH/DART.

Method originally written by Mark Millson, NIOSH/DART, and R. DeLon Hull, Ph.D., NIOSH/DSHEFS, James B. Perkins, David L. Wheeler, and Keith Nicholson, DataChem Labortories, Salt Lake City, UT.

(continues)

TABLE 5-10 (continued)

TABLE 1. PROPERTIES AND SAMPLING VOLUMES

Element (Symbol)	Properties		Air Volume, L @ OSHA PEL	
	Atomic Weight	MP, °C	MIN	MAX
Silver (Ag)	107.87	961	250	2000
Aluminum (Al)	26.98	660	5	100
Arsenic (As)	74.92	817	5	2000
Barium (Ba)	137.34	710	50	2000
Beryllium (Be)	9.01	1278	1250	2000
Calcium (Ca)	40.08	842	5	200
Cadmium (Cd)	112.40	321	13	2000
Cobalt (Co)	58.93	1495	25	2000
Chromium (Cr)	52.00	1890	5	1000
Copper (Cu)	63.54	1083	5	1000
Iron (Fe)	55.85	1535	5	100
Potassium (K)	39.10	63.65	5	1000
Lanthanum	138.91	920	5	1000
Lithium (Li)	6.94	179	100	2000
Magnesium (Mg)	24.31	651	5	67
Manganese (Mn)	54.94	1244	5	200
Molybdenum (Mo)	95.94	651	5	67
Nickel (Ni)	58.71	1453	5	1000
Phosphorus (P)	30.97	44	25	2000
Lead (Pb)	207.19	328	50	2000
Antimony (Sb)	121.75	630.5	50	2000
Selenium (Se)	78.96	217	13	2000
Tin (Sn)	118.69	231.9	5	1000
Strontium (Sr)	87.62	769	10	1000
Tellurium (Te)	127.60	450	25	2000
Titanium (Ti)	47.90	1675	5	100
Thallium (Tl)	204.37	304	25	2000
Vanadium (V)	50.94	1890	5	2000
Tungsten (W)	183.85	3410	5	1000
Yttrium (Y)	88.91	1495	5	1000
Zinc (Zn)	65.37	419	5	200
Zirconium (Zr)	91.22	1852	5	200

TABLE 5-10 (continued)

TABLE 2. EXPOSURE LIMITS, CAS #, RTECS

Element (Symbol)	CAS #	RTECS	Exposure Limits, mg/m³ (Ca = carcinogen)		
			OSHA	NIOSH	ACGIH
Silver (Ag)	7440-22-4	VW3500000	0.01 (dust, fume, metal)	0.01 (metal, soluble)	0.1 (metal) 0.01 (soluble)
Aluminum (Al)	7429-90-5	BD0330000	15 (total dust) 5 (respirable)	10 (total dust) 5 (respirable fume) 2 (salts, alkyls)	10 (dust) 5 (powders, fume) 2 (salts, alkyls)
Arsenic (As)	7440-38-2	CG0525000	varies	C 0.002, Ca	0.01, Ca
Barium (Ba)	7440-39-3	CQ8370000	0.5	0.5	0.5
Beryllium (Be)	7440-41-7	DS1750000	0.002, C 0.005	0.0005, Ca	0.002, Ca
Calcium (Ca)	7440-70-2	--	varies	varies	varies
Cadmium (Cd)	7440-43-9	EU9800000	0.005	lowest feasible, Ca	0.01 (total), Ca 0.002 (respir.), Ca
Cobalt (Co)	7440-48-4	GF8750000	0.1	0.05 (dust, fume)	0.02 (dust, fume)
Chromium (Cr)	7440-47-3	GB4200000	0.5	0.5	0.5
Copper (Cu)	7440-50-8	GL5325000	1 (dust, mists) 0.1 (fume)	1 (dust) 0.1 (fume)	1 (dust, mists) 0.2 (fume)
Iron (Fe)	7439-89-6	NO4565500	10 (dust, fume)	5 (dust, fume)	5 (fume)
Potassium (K)	7440-09-7	TS6460000	--	--	--
Lanthanum	7439-91-0	--	--	--	--
Lithium (Li)	7439-93-2	--	--	--	--
Magnesium (Mg)	7439-95-4	OM2100000	15 (dust) as oxide 5 (respirable)	10 (fume) as oxide	10 (fume) as oxide
Manganese (Mn)	7439-96-5	OO9275000	C 5	1; STEL 3	5 (dust) 1; STEL 3 (fume)
Molybdenum (Mo)	7439-98-7	QA4680000	5 (soluble) 15 (total insoluble)	5 (soluble) 10 (insoluble)	5 (soluble) 10 (insoluble)
Nickel (Ni)	7440-02-0	QR5950000	1	0.015, Ca	0.1 (soluble) 1 (insoluble, metal)
Phosphorus (P)	7723-14-0	TH3500000	0.1	0.1	0.1
Lead (Pb)	7439-92-1	OF7525000	0.05	0.05	0.05
Antimony (Sb)	7440-36-0	CC4025000	0.5	0.5	0.5
Selenium (Se)	7782-49-2	VS7700000	0.2	0.2	0.2
Tin (Sn)	7440-31-5	XP7320000	2	2	2
Strontium (Sr)	7440-24-6	--	--	--	--
Tellurium (Te)	13494-80-9	WY2625000	0.1	0.1	0.1
Titanium (Ti)	7440-32-6	XR1700000	--	--	--
Thallium (Tl)	7440-28-0	XG3425000	0.1 (skin) (soluble)	0.1 (skin) (soluble)	0.1 (skin)
Vanadium (V)	7440-62-2	YW240000	--	C 0.05	--
Tungsten	7440-33-7	--	5	5 10 (STEL)	5 10 (STEL)
Yttrium (Y)	7440-65-5	ZG2980000	1	N/A	1
Zinc (Zn)	7440-66-6	ZG8600000	--	--	--
Zirconium (Zr)	7440-67-7	ZH7070000	5	5, STEL 10	5, STEL 10

(continues)

TABLE 5-10 (continued)

TABLE 3. MEASUREMENT PROCEDURES AND DATA [1].
Mixed Cellulose Ester Filters (0.45 μm)

Element (a)	Wavelength (nm)	Est. LOD μg/Filter	LOD ng/mL	Certified 3x LOD (b)	% Recovery (c)	Percent RSD (N=25)	Certified 10x LOD (b)	% Recovery (c)	Percent RSD (N=25)
Ag	328	0.042	1.7	0.77	102.9	2.64	3.21	98.3	1.53
Al	167	0.115	4.6	1.54	105.4	11.5	6.40	101.5	1.98
As	189	0.140	5.6	3.08	94.9	2.28	12.9	93.9	1.30
Ba	455	0.005	0.2	0.31	101.8	1.72	1.29	97.7	0.69
Be	313	0.005	0.2	0.31	100.0	1.44	1.29	98.4	0.75
Ca	317	0.908	36.3	15.4	98.7	6.65	64.0	100.2	1.30
Cd	226	0.0075	0.3	0.31	99.8	1.99	1.29	97.5	0.88
Co	228	0.012	0.5	0.31	100.8	1.97	1.29	98.4	0.90
Cr	267	0.020	0.8	0.31	93.4	16.3	1.29	101.2	2.79
Cu	324	0.068	2.7	1.54	102.8	1.47	6.40	100.6	0.92
Fe	259	0.095	3.8	1.54	103.3	5.46	6.40	98.0	0.95
K	766	1.73	69.3	23.0	90.8	1.51	96.4	97.6	0.80
La	408	0.048	1.9	0.77	102.8	2.23	3.21	100.1	0.92
Li	670	0.010	0.4	0.31	110.0	1.91	1.29	97.7	0.81
Mg	279	0.098	3.9	1.54	101.1	8.35	6.40	98.0	1.53
Mn	257	0.005	0.2	0.31	101.0	1.77	1.29	94.7	0.73
Mo	202	0.020	0.8	0.31	105.3	2.47	1.29	98.6	1.09
Ni	231	0.020	0.8	0.31	109.6	3.54	1.29	101.2	1.38
P	178	0.092	3.7	1.54	84.4	6.19	6.40	82.5	4.75
Pb	168	0.062	2.5	1.54	109.4	2.41	6.40	101.7	0.88
Sb	206	0.192	7.7	3.08	90.2	11.4	12.9	**41.3**	32.58
Se	196	0.135	5.4	2.3	87.6	11.6	9.64	84.9	4.78
Sn	189	0.040	1.6	0.77	90.2	18.0	3.21	**49**	21.79
Sr	407	0.005	0.2	0.31	101.0	1.55	1.29	97.3	0.65
Te	214	0.078	3.1	1.54	102.0	2.67	6.40	97.4	1.24
Ti	334	0.050	2.0	0.77	98.4	2.04	3.21	93.4	1.08
Tl	190	0.092	3.7	1.54	100.9	2.48	6.40	99.1	0.80
V	292	0.028	1.1	0.77	103.2	1.92	3.21	98.3	0.84
W	207	0.075	3.0	1.54	**72.2**	10.1	6.40	**57.6**	14.72
Y	371	0.012	0.5	0.31	100.5	1.80	1.29	97.4	0.75
Zn	213	0.310	12.4	4.60	102.2	1.87	19.3	95.3	0.90
Zr	339	0.022	0.9	0.31	88.0	19.4	1.29	**25**	57.87

(a) Bold values are qualitative only because of low recovery.
(b) Values are certified by Inorganic Ventures INC. at 3x and 10x the approximate instrumental LOD.
(c) Values reported were obtained with a Spectro Analytical Instruments EOP ICP; performance may vary with instrument and should be independently verified.

TABLE 5-10 (continued)

TABLE 4. MEASUREMENT PROCEDURES AND DATA [1].
Polyvinyl Chloride Filter (5.0 µm)

Element (c)	Wavelength (nm)	Est. LOD µg per filter	LOD ng/mL	Certified 3x LOD (b)	% Recovery (a)	Percent RSD (N=25)	Certified[17] 10x LOD (b)	% Recovery (a)	Percent RSD (N=25)
Ag	328	0.042	1.7	0.78	104.2	8.20	3.18	81.8	18.9
Al	167	0.115	4.6	1.56	77.4	115.24	6.40	92.9	20.9
As	189	0.140	5.6	3.10	100.7	5.13	12.70	96.9	3.2
Ba	455	0.005	0.2	0.31	102.4	3.89	1.270	99.8	2.0
Be	313	0.005	0.2	0.31	106.8	3.53	1.270	102.8	2.1
Ca	317	0.908	36.3	15.6	**68.1**	12.66	64.00	96.8	5.3
Cd	226	0.0075	0.3	0.31	105.2	5.57	1.27	101.9	2.8
Co	228	0.012	0.5	0.31	109.3	4.67	1.27	102.8	2.8
Cr	267	0.020	0.8	0.31	109.4	5.31	1.27	103.4	4.1
Cu	324	0.068	2.7	1.56	104.9	5.18	6.40	101.8	2.4
Fe	259	0.095	3.8	1.56	88.7	46.82	6.40	99.1	9.7
K	766	1.73	69.3	23.4	96.4	4.70	95.00	99.2	2.2
La	408	0.048	1.9	0.78	**45.5**	4.19	3.18	98.8	2.6
Li	670	0.010	0.4	0.31	107.7	4.80	1.27	110.4	2.7
Mg	279	0.098	3.9	1.56	**54.8**	20.59	6.40	**64.5**	5.7
Mn	257	0.005	0.2	0.31	101.9	4.18	1.27	99.3	2.4
Mo	202	0.020	0.8	0.31	106.6	5.82	1.27	98.1	3.8
Ni	231	0.020	0.8	0.31	111.0	5.89	1.27	103.6	3.2
P	178	0.092	3.7	1.56	101.9	17.82	6.40	86.5	10.4
Pb	168	0.062	2.5	1.56	109.6	6.12	6.40	103.2	2.9
Sb	206	0.192	7.7	3.10	**64.6**	22.54	12.70	**38.1**	30.5
Se	196	0.135	5.4	2.30	83.1	26.23	9.50	76.0	17.2
Sn	189	0.040	1.6	0.78	85.7	27.29	3.18	**52.0**	29.4
Sr	407	0.005	0.2	0.31	**71.8**	4.09	1.27	81.2	2.7
Te	214	0.078	3.1	1.56	109.6	7.49	6.40	97.3	3.8
Ti	334	0.050	2.0	0.78	101.0	9.46	3.18	92.4	5.5
Tl	190	0.092	3.7	1.56	110.3	4.04	6.40	101.9	2.0
V	292	0.028	1.1	0.78	108.3	3.94	3.18	102.5	2.6
W	207	0.075	3.0	1.56	**74.9**	15.79	6.40	**44.7**	19.6
Y	371	0.012	0.5	0.31	101.5	3.63	1.27	101.4	2.5
Zn	213	0.310	12.4	4.70	91.0	68.69	19.1	101.0	9.6
Zr	339	0.022	0.9	0.31	**70.7**	54.20	1.27	**40.4**	42.1

(a) Values reported were obtained with a Spectro Analytical Instruments EOP ICP; performance may vary with instrument and should be independently verified.
(b) Values are certified by Inorganic Ventures INC. at 3x and 10x the approximate instrumental LOD [12].
(c) Bold values are qualitative only because of low recovery. Other digestion techniques may be more appropriate for these elements and their compounds.

NIOSH Manual of Analytical Methods (NMAM), Fourth Edition

PHOTO 5-5 Geiger-Mueller Ionizing Radiation Detection Equipment.

Resampling Time Frames

Hazardous atmosphere quantification should be performed before or during initial job regimens. Resampling should occur if job conditions change, environmental conditions change, or at routine intervals, such as annually. Samples identifying particularly toxic materials (e.g., lead) or high levels of toxins may necessitate repetitive sampling at more frequent intervals.[1]

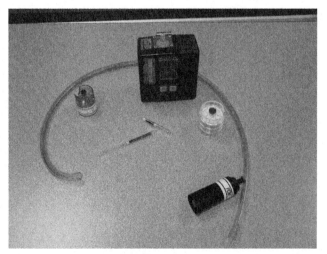

PHOTO 5-6 Personal Air Collection Device.

---

[1] The OSHA Lead Standard 29 CFR 1926.62(d) requires sampling at 6 month intervals if the exposure characterization is above the action level, and quarterly if the exposure characterization is above the Permissible Exposure Limit.

As mentioned earlier, collection and analysis are only the initial steps in the quantification of a hazardous atmosphere. Comparison to known exposure criteria completes the quantification process.

Chapter Summary

Once hazardous atmospheres have been identified, quantification may occur on site or through collection of the atmosphere for off-site analysis. If validated procedures are followed, comparison of field or lab data can be made to published and validated exposure levels. Subsequently, safe actions and options can be reviewed and implemented.

Terms

Action Level: The level of a hazardous atmosphere above which actions should occur to be protective of employees. This level is below (e.g., 1/2) the PEL, REL, or TLV.

Ceiling Concentration: The quantity of a material above which individuals should never be exposed for any duration during a workday. The ceiling concentration is considered a short-term exposure limit.

Immediately Dangerous to Life or Health (IDLH): The level of an atmosphere below which irreversible health effects are not anticipated or certain transient effects would not preclude an evacuation/escape (severe eye or respiratory irritation, disorientation, uncoordination) with 30 minutes of exposure. Concentrations below these levels are protective of workers who may experience failure of respiratory protection equipment in these atmospheres and subsequently need to evacuate the space.

Imminent Danger: Any condition or practice that creates a danger that could reasonably be expected to cause death or serious harm immediately or before the danger can be eliminated by OSHA enforcement procedures. Employees have the right to refuse work, without retribution, if an imminent danger exists, subsequent to employer notification. If identified by an OSHA inspector, the inspector must notify both employers and employees of the hazard, and procedures will be immediately initiated through the federal court system to eliminate the situation.

Lower Explosive Limit (LEL): The minimum concentration of fuel in air at which a fire will occur in the presence of a source of ignition and oxygen.

Concentrations of fuel below this level are considered too lean to burn.

Lowest Observed Adverse Effects Level (LOAEL): The level at which adverse health effects have been recorded. LOAELs are published for acute exposures (0–14 days), intermediate exposures (14–365 days), and long-term/chronic exposures (> 365 days) by the U.S. Department of Health and Human Services, Agency for Toxic Substances and Disease Registry and are considered public health standards.

No Observed Adverse Effects Level (NOAEL): The level below which no adverse effects have been seen in the general population. NOAELs are published for acute exposures (0–14 days), intermediate exposures (14–365 days), and long-term/chronic exposures (> 365 days), by the U.S. Department of Health and Human Services, Agency for Toxic Substances and Disease Registry, and are considered public health standards.

Permissible Exposure Limit (PEL): The permissible exposure limit, the limit of a hazardous atmosphere, as published by OSHA, to which an employee may be exposed. Exposures above this level are cited as a workplace employer violation by OSHA.

Recommended Exposure Limit (REL): The recommended exposure limit of a hazardous atmosphere, as published by NIOSH. Exposures above this level may produce symptoms of exposure, according to the risk assessment methodology used by NIOSH. REL are typically published for a 10-hour workday, up to 40 hours per week.

Risk Based Concentration: Chemical concentrations corresponding to fixed levels of risk (i.e., a Hazard Quotient (HQ) of 1, or lifetime cancer risk of 1E-6, whichever occurs at a lower concentration) in water, air, fish tissue, and soil. RBC tables contain Reference Doses (RfDs) and Cancer Slope Factors (CSFs) for chemicals. These toxicity factors have been combined with "standard" exposure scenarios to calculate RBC. RBC are developed by the U.S. Environmental Protection Agency.

Short-Term Exposure Limit (STEL): The short-term (time-specified, typically 15 minutes) exposure limit of a hazardous atmosphere. Individuals may be exposed above the full shift exposure limit if the STEL is not exceeded. The ACGIH identifies that the STEL may be applied no more than 4 times in a work shift, with a 1-hour timeframe between exposures for excursions above the full shift exposure.

Threshold Limit Value (TLV): The threshold limit value for a hazardous atmosphere, as published by the American Conference of Governmental Industrial Hygienists. The TLV levels are designed to assist in the control of health hazards of employees, when applied to work situations by qualified personnel.

Upper Explosive Limit (UEL): The upper explosive limit or maximum concentration of fuel in air at which a fire will occur in the presence of a source of ignition and oxygen. Concentrations of fuel above the UEL are considered too rich to burn.

References

American Conference of Governmental Industrial Hygienists. *2011 TLVs and BEIs: Threshold Limit Values for Chemical Substances and Physical Agents, Biological Exposure Indices.* Cincinnati: ACGIH, 2011.

American Industrial Hygiene Association (AIHA). *Emergency Response Planning Guidelines and Workplace Environmental Exposure Levels, Handbook.* Fairfax, VA: AIHA Foundation, 2011.

Detector Tube Handbook: Air Investigations and Technical Gas Analysis with Draegen Tubes. Lubeck, Germany: Kurt Leichmife, 1999.

Lionel Electronic Laboratories. *Instruction and Operation Manual: Radiological Survey Meter OCDM.* New York: Lionel Electronic Laboratories, 2000.

———. *Instruction Manual: Microguard Portable Alarm.* Pittsburgh: Mine Safety Appliance Company, 1987.

MSA-Aver: Instructions for Detector Tube #497606: Carbon Dioxide, 2011.

National Institute for Occupational Safety and Health, Occupational Safety and Health Administration, USCG/Environmental Protection Agency. *Occupational Safety and Health Guidelines Manual for Hazardous Waste Site Activities—DHHS Publication 85-115.* Washington, D.C.: Government Printing Office, October 1985.

Perper, J.B., and Dauson, B.J. *Direct Reading Colormetric Indicator Tube Manual,* 2nd ed. Fairfax, VA: AIHA Publications, 1993.

Radiation Alert: Inspector, User Manual. Summer Team, TN: SE International, 1996.

SKC, Inc. *The Essential Reference for Air Sampling, Comprehensive Catalogue and Air Sampling Guide.* Eighty-Four, PA: SKC, 2011.

U.S. Department of Health and Human Services, Centers for Disease Control and Prevention, National Institute for Occupational Safety and Health. *NIOSH Pocket Guide to Hazardous Chemicals—*

DHHS-97-140. Washington, D.C.: Government Printing Office, November 18, 2010.

U.S. Department of Labor. *General Industry Safety Regulations 29 CFR 1910*. Washington, D.C.: Government Printing Office, Current.

U.S. Environmental Protection Agency, Office of Pesticides and Toxic Substances. *Guidance for Controlling Asbestos-Containing Materials in Buildings—EPA 560/5-85-084*. Washington, D.C.: Government Printing Office, June 1985.

U.S. Public Health Service, Centers for Disease Control and Prevention. *National Institute for Occupational Safety and Health, Manual of Analytical Methods*, 5th ed. Cincinnati: NIOSH—Division of Physical Sciences and Engineering, August 2011.

Exercise 5

1–4. Identify and quantify four Immediately Dangerous to Life or Health Environments; and identify monitoring equipment from Table 5-5 that may be used to perform a site measurement for the IDLH environment.

Environment	IDLH Level	Monitoring Equipment
1.		
2.		
3.		
4.		

5–8. Using Table 5-5 as a reference, identify the appropriate site measurement device(s) most helpful in the following situations:

   5. Ambient dust levels at a demolition site: _____

   6. Chlorine (CL_2), gas released from a storage container: _____

   7. Oxygen levels in a sanitary sewer system: _____

   8. Organic vapors, found at a landfill: _____

9–10. Identify two styles of devices which may be used to quantify personal sound pressure (noise) levels_____

11–16. Using Table 5-8 as a guide; perform a CO_2 test in the room you currently occupy:

   11. What is the range of the test device? _____

   12. How much air (or how many "strokes") must be pulled through the test device for a compliant test?

13. What color change must you observe? _____

14. What interferences to the test are possible? _____

15. Can they be reasonably anticipated in the room you currently occupy? _____

16. What is the appropriate and safe disposal practice for any testing equipment used during this sampling regimen? _____
_____

17. Identify the proper height for employee breathing zone sampling: _____

18–19. Using Table 5-9, from the *NIOSH Manual of Analytical Methods*, Method 7400, identify the flow rate and volume necessary to collect a proper sample for this asbestos collection for laboratory analyses:

Flow Rate: _____ Volume: _____

20–21. Using Table 5-10, from the *NIOSH Manual of Analytical Methods*, Method 7300, what are the minimum and maximum air volumes that should be collected for a lead in air sample?

Minimal Volume: _____ Maximum Volume: _____

22–25. Using Table 5-10, identify the Permissible Exposure Limits; the Recommended Exposure Limit, and the Threshold Limit Value for the element Nickel (Ni). Is this a carcinogenic element?

22. OSHA—PEL: _____

23. NIOSH—REL: _____

24. ACGIH—TLV: _____

25. Carcinogenicity: _____

6 Protective Planning

Key Concepts

- Emergency Planning Zone (EPZ)
- Health and Safety Plan (HASP)
- Confined Space Entry Permit
- Ready for Hot Work Permit
- Asbestos Abatement Permit/Notification
- Lead-Based-Paint Abatement Permit/Notification

Introduction

Once hazardous environments have been identified, plans for protecting individuals and the public must be completed before hazardous activities or emergency response work begins. In general, public protection is initiated through appropriate planning for removal of individuals from hazardous areas (i.e., evacuations), although response from agencies of public safety (Hazmat teams, fire services) is also included. Full evacuation procedures include warning individuals of approaching hazards, transporting individuals to safe areas, broadcasting informational programs concerning shelter, and initiating programs for housing individuals once they are removed from hazardous areas.

Employee protection in hazardous areas is derived from two styles of **health and safety plans (HASPs)**. General health and safety plans identify characteristics and programs required of employers to protect employees. Areas such as general training, medical surveillance programs, protective equipment, and preventative maintenance programs for emergency equipment are found in these styles of HASPs.

Second, site- or job-specific HASPs identify procedures used to protect individuals in specific hazardous environments. These include the appropriate site-specific monitoring, metering, and protective equipment and procedures, as well as engineering and administrative controls, to be used at these sites. Specific equipment safeguarding procedures may also be included. A component of site-specific HASPs, emergency action and emergency response plans identify appropriate procedures for employees to follow in the event that evacuation from work areas with hazardous environments is necessary, or a response is necessary, into a hazardous environment to mitigate emergencies.

Information concerning emergency action and response plans is covered in Chapter 9.

Finally, environmental protection plans include preplans for the protection of air, water, and soil resources from hazardous environments.

Protecting the Public

There are various requirements and subsequent plans for public protection, including those listed below.

Nuclear Power Plant Emergency Planning Zones

The Nuclear Regulatory Commission (NRC) requires, and the U.S. Department of Homeland Security—Federal Emergency Management Agency (FEMA) has implemented, emergency planning directives for the **emergency planning zones (EPZs)** that exist in a 10-mile area surrounding each of the nation's 104 commercial nuclear reactors, found at 65 active nuclear power plants. There are currently 4 reactors and 2 plants under construction. Three million people reside in these EPZs. This requirement for emergent evacuation and medical services due to a nuclear power emergency occurred in part in response to the identified lack of emergency planning and preparedness at the Three Mile Island nuclear disaster, Harrisburg, Pennsylvania, in March 1979.

PHOTO 6-1 Emergency planning zones are established in the 10-mile area around nuclear power plants. Specialized evacuation procedures are developed and publicized, as well as warning and communication systems.

Chemical Emergency Planning Zones

Title III of the Superfund Amendments and Reauthorization Act (SARA) of 1986 (the Emergency Planning and Community Right-to-Know Act) requires individuals or facilities that possess any of 477 extremely hazardous substances on-site, or 10,000 pounds of other hazardous substances, to report site information to local emergency planning commissions (LEPCs) or state emergency response agencies for the purpose of preplanning public protection.

This federally required program was an extension of the voluntary Chemical Awareness and Emergency Response (CAER) Program, promulgated by the Chemical Manufacturers Association, prior to the Bhopal, India methyl isocyanate chemical release. The Bhopal chemical release was a prime factor in the Congressional SARA directive in 1986. In addition to public evacuations, the SARA directives facilitate the planning of emergency response procedures to chemical releases.

Hazardous Area Emergency Planning

In addition, certain hazardous areas within the country, such as the following, require additional awareness and public protection preplanning.

1. Areas adjacent to the chemical weapons stockpile programs of the U.S. Department of Defense. (There are currently 7 large chemical weapons stockpile sites.)
2. Areas adjacent to uncontrolled hazardous waste (Superfund) sites, listed under the Comprehensive Environmental Response Compensation and Liability Act of 1980, regulated by the U.S. Environmental Protection Agency. (There are currently 1280 active sites, with 62 under consideration for listing. 347 sites have undergone remediation and have been de-listed.)
3. Areas with known threats from natural hazards or man-made structures, such as sites downstream from U.S. Army Corps of Engineers–regulated dams.
4. Areas adjacent to regulated Spill Prevention, Control and Countermeasure (SPCC) sites. These SPCC sites contain oils or hazardous substances for which downstream notification of drinking water users is required in the event of a release.

PHOTO 6-3 SPCC sites contain oil or hazardous substances requiring specialized protection and planning.

PHOTO 6-4 DMAT deployment occurred subsequent to the 2010 Haiti earthquakes. DMAT, as well as USAR Teams, may be deployed both nationally and internationally as a component of Federal response plans.

PHOTO 6-2 Areas around chemical facilities require emergency planning in compliance with Title III of the Superfund Amendments and Reauthorization Act of 1986. These areas may also require US EPA compliant Risk Management Plans.

PHOTO 6-5 PCB-Containing electrical equipment is required by the EPA to be labeled with the PCB mark. Many sites must be registered with the U.S. EPA, and local fire response agencies.

5. Urban areas subject to natural or man-made disasters including those that are target areas for terrorist threats. These areas are protected under the Urban Area Security Initiative (UASI) Program, promulgated by the Department of Homeland Security (DHS). Note: FEMA has organized 27 Urban Search and Rescue (USAR) Teams, for response purposes to these areas subsequent to emergent events. The U.S. Department of Health and Human Services maintains 106 Disaster

Medical Assistance Teams (DMAT), trained to augment local health services during large scale disasters. Both USAR and DMAT teams may also respond internationally.

6. Public risk management planning areas (RMPs) providing protection to the public in and around areas where extremely hazardous chemicals or explosives are used or stored. These plans are approved by the U.S. Environmental Protection Agency (U.S. EPA). Sites where specific hazardous substances are used, such as polychlorinated biphenyl–containing electrical equipment must also be listed with the U.S. EPA and local emergency response agencies, to ensure an appropriate response to a fire or release occurs.

7. Occupant protection plans, required in target housing and child-occupied facilities, to ensure occupants are not exposed to lead dusts during lead-based–paint removal. Required elements of occupant protection plans are found in **TABLE 6-1**.

The Public Protection Process

In general, community awareness and emergency planning initiates with a report of a hazardous condition or a potential for hazardous conditions at a site.

As a second step, emergency planners identify EPZs and the actions necessary to protect or evacuate individuals in these areas. In 1972, President Nixon identified the National Oceanic and Atmospheric Administration's Weather Radio System as the mechanism for alerting the public to national disasters. In part, this was in response to the potential threat of nuclear disasters at the time. Since implemented, emergency broadcast systems have been integrated into radio, television, and other mechanisms of communications for early warning purposes. Currently, "reverse 911" systems use all forms of communication technologies to alert citizens to shelter-in-place, evacuate, or take other protective actions. Messages are delivered to landlines or cell phones and to computerized communication devices. In certain areas, early warning devices, such as siren alert systems, remain in place to warn the public of potentially hazardous conditions.

Emergency planners have also developed evacuation routes, evacuation mechanisms (where required), and staging areas for evacuees that have been identified as capable of sheltering known numbers of individuals.

It is generally considered to be the responsibility of government agencies to design and implement public evacuations. In most states, a declaration of emergency or a specific order of a state official (e.g., the governor) is required to activate emergency public evacuation

TABLE 6-1 Required Protections During Lead-Based-Paint Removal	
Element	**Comment**
Occupant Protection Plan and written description of the measures to protect occupants from lead-based-paint removal hazards	Must be completed by an EPA/state-licensed Lead Supervisor or Project Designer
Occupant Access Safety: 1. Relocate for duration of project 2. Relocate during working hours 3. No relocation; remain outside of work areas	For options 2 & 3, a lead-safe passageway must be provided for bathrooms and exit/entry
Work area setup and engineering controls	Barriers designed to protect materials inside the work area and to minimize the spread of lead dust outside of the work area must remain in place until clearance
Prohibited Activities: 1. Open flame burning 2. Machine grinding or sanding, unless HEPA filtered 3. Dry scraping 4. Heat gun operation > 1100F	Dry scraping permitted only in conjunction with heat guns or around electrical outlets or when treating defective paint spots totaling no more than 2 SF in any one room, or no more than 20 SF on exterior surfaces
Waste storage and disposal protocols	Safe/secure management protocols for packaging, labeling, onsite storage, and appropriate waste disposal offsite
Project completion and clearance requirements	Clearance conducted by an EPA/state-licensed Lead Inspector or Risk Assessor

areas and plans. However, certain preplanned situations (nuclear disasters) may have preapproval for early warning of the public through contact or warning systems.

The Employee Protection Process

Employer Health and Safety Plans (HASPs)

General employer HASPs are required of employers of individuals who enter hazardous areas. HASPs may take the form of hazardous waste operations HASPs (29 CFR 1910.120/29 CFR 1926.65/40 CFR 311), process safety management plans (29 CFR 1910.119), confined space entry plans or permits (29 CFR 1910.146), bloodborne pathogen exposure control plans (29 CFR 1910.1300), chemical hygiene plans (29 CFR 1910.1450), tunnel rescue plans (29 CFR 1926), hot work plans (NFPA 251B), and toxic material exposure control programs (e.g., Asbestos 29 CFR 1910.1001 and Lead 29 CFR 1926.62).

Site-Specific Health and Safety Planning Requirement

In addition to employer HASPs, site-specific planning for hazards—including specific metering, monitoring, administrative controls, engineering controls, and site-specific protective equipment—is required. Typically, these plans are developed for site activities after initial characterization of hazardous work. **FIGURE 6-1** provides an example of a site-specific HASP that is typically used for hazardous waste sites. **FIGURE 6-2** provides an entry plan for Permit Required Confined Space entry. **FIGURE 6-3** gives an example of a Ready for Hot Work Permit. Although company-wide programs address general requirements, site-specific HASPs identify specific procedures to be followed to ensure employee safety in hazardous work places.

Emergency Action and Response Plans

Emergency action plans (EAPs) are required for hazardous work sites. EAPs for hazardous environment work sites should be written, although the Occupational Safety and Health Administration allows verbal communication of EAPs in work sites with fewer than 10 employees. EAPs identify safe areas of refuge, evacuation routes, emergency decontamination procedures, and similar defensive actions for individuals working in hazardous environments. Conversely, emergency response plans identify offensive actions to be taken in hazardous environments in the event an emergency occurs. EAPs require that employees who identify or witness emergencies must evacuate to safe locations. Emergency response plans specify responsibilities and training provided to select employees who may implement emergency rescue, containment or control of hazardous atmospheres, firefighting, or first aid.

Further information concerning EAPs and emergency response plans is found in Chapter 9.

Environmental Protection

Protection of the air, water, and soil from hazardous atmospheres arises through appropriate handling of materials and appropriate waste disposal. Any emissions to the air must be permitted under the Clean Air Act. Potentially hazardous sites are registered and inspected by federal, state, or municipal air quality personnel. Similar inspections occur at water discharge or solid waste disposal sites.

FIGURE 6-4 identifies a permit for work with asbestos-containing materials required by the National Emission Standards for Hazardous Air Pollutants of the Clean Air Act, and enforced by the U.S. EPA or state environmental agencies.

FIGURE 6-5 identifies a similar permit for work with lead-based paint in target housing and child-occupied facilities, a required action referenced by both the U.S. EPA and the U.S. Department of Housing and Urban Development (HUD).

Any emission to water sources must be permitted under the National Pollution Discharge Elimination System (NPDES) of the Clean Water Act, with appropriate protective plans in place.

Any solid waste disposal or sanitary sewage must be processed through a licensed facility. A variety of solid waste exists, including those materials discussed in the Resource Conservation and Recovery Act section of Chapter 2.

In general, transportation of hazardous substances is regulated by the U.S. Department of Transportation (DOT) with requirements found in the DOT shipping table 49 CFR Part 172. An example of a standard shipping table identifying the material, appropriate packaging labels, transportation requirements such as containerization, and any exemptions or authorizations for the transportation requirements is found in Chapter 5.

PLAN/PERMIT # _____

SITE SAFETY AND HEALTH PLAN

A. SITE DESCRIPTION:

Site Name: _____ Effective Date: _____ Expires: _____

Location: _____

Prepared by: _____ Signature: _____

Version: [ ] Original   [ ] Revision

Original plan dated: _____

SITE DESCRIPTION AND CONTAMINATION CHARACTERIZATION:

SITE TYPE: (Check as many as applicable)

[ ] Active   [ ] Landfill   [ ] Residential   [ ] Recreational
[ ] Inactive   [ ] Agriculture   [ ] Industrial   [ ] Natural Area
[ ] Secure   [ ] Commercial   [ ] Military   [ ] Unknown
[ ] Unsecured   [ ] Other (specify) _____

SURROUNDING POPULATION:

[ ] Residential   [ ] Industrial   [ ] Urban
[ ] Rural   [ ] Other (specify)

TOPOGRAPHY: _____

DESCRIPTION OF ON-SITE ACTIVITIES:

[ ] Preliminary Assessment   [ ] Remedial Action
[ ] Site Inspection   [ ] Remedial Design
[ ] Remedial Investigation and Feasibility Study   [ ] Remedial Action Study/Pilot
[ ] Emergency Action   [ ] Other (specify) _____
                                              _____
                                              _____

B. OBJECTIVES: _____
_____
_____
_____

C. ON-SITE ORGANIZATION AND COORDINATION:

The following personnel are designated to carry out job functions on-site; as stated in this safety plan [specify name; company; contact number].

Project Team Leader: _____

Project Safety Officer: _____

Site Safety Officer: _____

Field Team Leader: _____

Field Team Members: _____
_____
_____
_____
_____

Federal Agency Reps: _____
_____

State Agency Reps: _____
_____

Local Agency Reps: _____
_____

Contractors: _____
_____
_____
_____
_____

FIGURE 6-1 Site-specific HASP.

All personnel arriving or departing the site will log in and out with the Site Safety Officer. All activities on site must be cleared through the project Team Leader. The Site Safety Officer will maintain a job log.

D. ON-SITE CONTROL:

_____ has been designated to coordinate access control and security on site. A safe perimeter has been established at _____

No unauthorized person should be within this area.

The on-site control point and staging area have been established at _____

Control boundaries have been established, and the Exclusion Zone (the contaminated area), Contamination Reduction Zone (CRZ) and Support Zone (clean area) have been identified and designated as follows:

Support Zone: _____

CRZ: _____

Exclusion Zone: _____

These boundaries are identified by: _____

E. HAZARD ASSESSMENT AND RISK ANALYSIS:

HAZARD EVALUATION: (Check all that apply)

[ ] Heat Stress   [ ] Cold Stress   [ ] Work at Heights
[ ] Oxygen Deficiency [ ] Radiological   [ ] Noise
[ ] Toxic Materials   [ ] Other Chemical Hazards   [ ] Biological
[ ] Explosion/Flammable [ ] Danger - Wildlife   [ ] Excavation/Trenching
[ ] Surface/Terrain   [ ] Confined Space   [ ] Electrical
[ ] Other (specify): _____

The following substance(s) are known or suspected to be on-site. The primary hazards of each are identified.

Substances Involved	Concentrations	Primary Hazards

OVERALL HAZARD EVALUATION:

[ ] High   [ ] Medium   [ ] Low   [ ] Unknown

This site Safety and Health Plan addresses all hazards identified in Section E through the following program elements:

_____

FIGURE 6-1 (continued)

F. PERSONAL PROTECTIVE EQUIPMENT:

NOTE: Completion of this section certifies hazard assessment, the effective date of this plan, in compliance with the requirements of 29CFR1910.132.

<u>EMERGENCY RESPONSE OPERATIONS</u>

Based on evaluation of potential hazards, the following levels of personal protection have been designated for the applicable work areas or tasks:

Location	Job Function	Level of Protection
Exclusion Zone	_____	A B C D
	_____	A B C D
Contamination Reduction Zone	_____	A B C D
	_____	A B C D

Specific protective equipment for each level of protection is a s follows: [A safety factor of (2) has been assigned to all skin protection devices.]

[Respirators selected meet the requirements of certification and protection factors issued by the National Institute for Occupational Safety and Health of the Centers for Disease Control and Prevention: U.S. Public Health Service]. [All respirators will be used in compliance with a written respirator program].

Level A _____   Level C _____
         _____   _____
         _____   _____
         _____   _____
         _____   _____

Level B _____   Level D _____
         _____   _____
         _____   _____
         _____   _____
         _____   _____

Other: _____
         _____

<u>NON-EMERGENCY OPERATIONS</u>

Site Characterization And Inspection: _____
Remedial Operations: _____
Post Remedial Operations: _____

NOTE: Completion of this section certifies hazard assessment, the effective date of this plan, in compliance with the requirements of 29CFR1910.132.

Based on evaluation of potential hazards, the following levels of personal protection have been designated for the applicable work areas or tasks:

Location	Job Function	Level of Protection
Exclusion Zone	_____	A B C D
	_____	A B C D
	_____	A B C D
	_____	A B C D

FIGURE 6-1 (continued)

PERSONAL PROTECTIVE EQUIPMENT: (cont'd)

Contamination
Reduction Zone _____   A B C D

                              _____   A B C D

                              _____   A B C D

                              _____   A B C D

Specific protective equipment for each level of protection is as follows: [A safety factor of (2) has been assigned to all skin protection devices.]

Level A _____   Level C _____

Level B _____   Level D _____

Other: _____

NOTE: Completion of this section certifies hazard assessment, the effective date of this plan, in compliance with the requirements of 29CFR1910.132.

G. COMMUNICATION PROCEDURES:

The following standard hand signals will be used in case of failure of radio communications:

- Hand gripping throat--------Out of air, can't breathe
- Grip partner's wrist--------Leave area immediately
- Hands on top of head-----Need assistance
- Thumbs up-----OK, I'm all right, I understand
- Thumbs down-----No, negative

Telephone communication to the control point should be established as soon as practicable. The control point phone number is:

_____

Channel _____ has been designated as the radio frequency for personnel in the Exclusion Zone. All other on-site communications will use channel _____ .

H. DECONTAMINATION PROCEDURES:

Personnel Decontamination: [ ] Not needed
The following procedure shall be utilized in decontamination of personnel:

_____
_____
_____
_____

Sampling Equipment Decontamination: [ ] Not needed
The following stations shall be utilized in decontamination of sampling equipment:

_____
_____
_____
_____

FIGURE 6-1 (continued)

Heavy Equipment Decontamination: [ ] Not needed
The following stations shall be utilized in decontamination of heavy equipment:
_____
_____
_____
_____
_____

Emergency Decontamination will include the following stations:
_____
_____
_____
_____
_____

The following decontamination equipment is required:
_____
_____
_____
_____
_____

I. SITE SAFETY AND HEALTH PROCEDURES:

1. _____ is the designated Site Safety Officer and is directly responsible to the Project Team Leader for safety recommendations on site.
2. Emergency Medical Care
   _____ are the certified First Aid/CPR/Emergency Medical Technicians/Paramedic/Occupational Health Nurse/Physician on-site. [Circle as Appropriate] and can be contacted on site by _____ .

These individuals have been trained in the requirements of the Bloodborne Pathogen Standard 29CFR1910.130. First Aid medical procedures are performed under the protection of an Exposure Control Plan.

The name of the Closest Acceptable Medical Facility:
_____
_____

Directions to Medical Facility: _____
_____
_____
_____

Phone: _____ Other Contact: _____

Minutes from site: _____

The following person(s) was contacted and briefed on the situation, the potential hazards, and the substances involved:

Name: _____ Medical information on-site is maintained
at: _____

Date: _____

The municipal ambulance service is: _____

Phone: _____
Response Time: _____
Safe Meeting Point: _____

The following person(s) was contacted and briefed on the situation, and the potential hazards:
Name: _____
Date: _____

FIGURE 6-1 (continued)

First-Aid equipment is available on-site at the following locations:
- First-aid kit _____
- Emergency eye wash _____
- Emergency shower _____
- O_2 _____

Emergency medical information for items identified in Section E: Hazard Assessment

Substance	Exposure Symptoms	First-aid Instructions

An accident report is required to be filed with the site safety officer and project manager for all incidents requiring first aid; all incidents resulting in property damage, or exposure/potential exposures to individual not covered by this safety plan.

3. Fire & Rescue Equipment, as listed below, is available: (Trained individuals only will utilize fire extinguishers).
   Fire Extinguishers:
   Location: _____   Rating: _____
   _____   _____
   _____   _____
   _____   _____
   _____   _____

Other emergency response equipment: _____
_____
_____

A fire brigade/rescue team is/is not available for response: _____

Alerting Procedures: _____
Authorized Actions: _____
_____

HOT WORK PROCEDURES:

As a minimum, this requires compliance with all rules as stated in section 5; Site Safety Rules; and the following:

a) A fire watch will be maintained during & 30 minutes after hot work activities occur. The fire watch will be trained in the area of fire suppression equipment.
b) All combustibles, will be cleared a minimum of 30 ft. from hot work.
c) No oxidizers or combustible/flammable vapors will be present while hot work occurs. (Less than 10% LEL); or inerted atmospheres.
d) Additional requirements: _____
_____

The following fire safety/rescue/emergency official was contacted and briefed on the situation, the potential hazard and the substances involved:

NAME: _____ DATE: _____
ORGANIZATION: _____
RESPONSE TIME: _____ MEETING POINT: _____

FIGURE 6-1 (continued)

4. List of emergency phone numbers:

Agency/Facility	Phone	Contact
Police		
Fire/Rescue		
Hospital		
Ambulance		
Public Health Advisor		

5. Environmental Monitoring

The following environmental monitoring instruments shall be used on site at the specified intervals.

Combustible Gas Indicator	[ ] continuous	[ ] hourly	[ ] periodic
Oxygen Monitor	[ ] continuous	[ ] hourly	[ ] periodic
Colormetric Tubes	[ ] continuous	[ ] hourly	[ ] periodic
_____	_____		
_____	_____		
_____	_____		
HNU/OVA	[ ] continuous	[ ] hourly	[ ] periodic
Other_____	[ ] continuous	[ ] hourly	[ ] periodic
_____	[ ] continuous	[ ] hourly	[ ] periodic

A log of environmental test dates, equipment calibrations and results will be maintained by site safety officer, and is found at Appendix A.

6. Emergency Procedures and Emergency Action Plan

The following standard emergency procedures will be used by on-site personnel. The Site Safety Officer shall be notified of any on-site emergencies and be responsible for ensuring that the appropriate procedures are followed.

Emergency Alarm Signal: _____

Secondary Emergency Alarm Signal: _____

Designated Meeting Location: _____

Emergency Rescue/Firefighting/First Aid will be provided on site as designated in Section I.3.

Personnel Injury in the Exclusion Zone: In the event of an injury in the Exclusion Zone, the emergency signal shall be sounded. Injured employees shall be removed through decon or emergency decon. All site personnel shall assemble at the designated meeting location. No persons shall reenter the Exclusion Zone until the cause of the injury or symptoms is determined and work continuation is authorized by the site safety officer and project manager.

Upon notification of an injury in the contamination reduction or Support Zone, the Project Team Leader and Site Safety Officer will assess the nature of the injury and rescue/emergency services will be provided as designated in I.3. If the cause of the injury or loss of the injured person does not affect the performance of site personnel, operations will continue. If the injury increases the risk to others, the designated emergency signal shall be sounded and all site personnel shall evacuate for further instructions. Activities on-site will stop until approval to continue by the project manager and site safety officer.

Fire/Explosion: In the event of a fire or explosion on-site, the designated emergency signal shall be sounded and all site personnel shall evacuated through standard decon to the designated location. The fire response/rescue team/fire department shall be alerted.

Equipment Failure: If any site worker experiences a failure or alteration of protective equipment that affects the protection afforded by the equipment, that person and his/her buddy shall immediately evacuate to support areas through standard decon. Reentry shall not be permitted until the equipment has been repaired or replaced. The team approved for re-entry by the site safety officer. If any other equipment on-site fails to operate properly, the Project Team Leader and site Safety Officer shall by notified and determine the effect of this failure on continuing operations on-site. If the failure affects the safety of personnel or

FIGURE 6-1 (continued)

prevents completion of the Work Plan tasks, all personnel shall leave the Exclusion Zone until the situation is evaluated and appropriate actions taken.

The following emergency (secondary) escape routes are designated for use in those situations where egress from the Exclusion Zone cannot occur through the decontamination line:

_____
_____
_____
_____

In all situations, when an on-site emergency results in evacuation of the Exclusion Zone, personnel shall not reenter until:
1. The conditions resulting in the emergency have been corrected.
2. The hazards have been reassessed.
3. The Site Safety Plan has been reviewed.
4. Site personnel have been briefed on any changes in the Site Safety Plan.

7. Site Safety Rules
    A. No smoking is permitted on-site, or within controlled areas (exclusion and contamination reduction zones).
    B. Eating, drinking or the application of cosmetics may only be permitted in areas approved by the site safety officer.
    C. In the event air monitoring shows atmospheres in excess of 10% L.E.L. on site, no sources of ignition will be allowed on-site, until the hazard is reduced to levels below 10% LEL. This occurrence shall be indicated by a designated sign, and communicated by the site safety officer.
    D. In the event I.D.L.H. atmospheres are present, trained and equipped standby personnel will be available on-site (mandatory rescue team of two people, minimum). This includes work in atmospheres in excess of 10% L.E.L., <19.5% Oxygen or in excess of published IDLH (NIOSH) levels. Work with shock sensitive or reactive materials shall be considered IDLH. This includes crystallized laboratory chemicals and overpressurized or stressed cylinders, drums or containers. In these situations only necessary employees will be permitted in exclusion zones; remote operations shall be used; where possible and a signal will be used to designate the initiation of these handling activities.
    E. 15 feet wide driveways will be maintained on-site at all times.
    F. No combustible materials will be stored within 10 feet of any structure.
    G. No oxidizers will be stored/used on grounds without amendment to this site safety plan.
    H. All hot work must first be approved by the site safety and health officer.
    I. Sanitary facilities:
       Hand washing facilities located at:_____
       Shower facilities are located at: _____
       Change/locker facilities are located at: _____
       Toilets are located at:_____
       Drinking water is located at:_____
    J. All individual working on this project designated as hazardous waste workers or technicians will be medically pre-approved to work with hazardous wastes, toxic materials, or specialized personal protective equipment. This includes site safety personnel. Any individual utilizing respiratory protection equipment shall be medically pre-approved to use a respirator. Each contractor will be responsible for maintaining this approved documentation. An approved fit test will be provided for all individuals using both negative and positive pressure respirators, within six months of use. The site safety officer shall audit this program and may request this information.

8. Personal Monitoring
   Engineering controls; administrative procedures and personal protective equipment will be upgraded/downgraded at the following action levels: _____
   _____
   _____
   _____
   _____

FIGURE 6-1 (continued)

The following personal monitoring will be in effect on-site:

Personal exposure sampling (air):

EMPLOYEE CLASS	TYPE	DATE/SHIFT

Posting of monitoring results shall be at: _____
Maintenance of monitoring results shall be at: _____

Medical Monitoring (biological exposure indices): _____

Heat stress monitoring is/is not required; the following procedures shall be followed [anticipated air temperature _____]:

_____
_____
_____
_____

A Heat Alert Program is/is not required on site. The following procedures shall be followed [NOAA/NWS: Heat Wave Alert issued for _____]:

_____
_____
_____
_____
_____

All site personnel have read the above plan and are familiar with its provisions.

	Name	I.D. #	Signature	Company
Project Team Leader				
Site Safety Officer				
Other Site Personnel				

The following Appendices are attached to this plan:
Appendix A - Instrument Calibration Log
_____ Results of On-Site Monitoring
_____ On-Site Sample Collection Log
_____
_____
_____
_____

Source: Occupational Safety and Health Guidance Manual for Hazardous Waste Site Activities, U.S. Department of Health and Human Services, Public Health Services, Centers for Disease Control, National Institute for Occupational Safety and Health, DHHS Publication 85-115, U.S. Government Printing Office, October, 1985.

FIGURE 6-1 (continued)

CONFINED SPACE ENTRY PERMIT

1. <u>Purpose</u>: (Check One)
   A. PRCS Entry: _____ C. CS Entry: _____
   B. PRCS Declassification and Entry: ____ D. Other: _____

3. <u>Project Information</u>:
   Location: _____ Duration of Permit: _____
   Date: _____ (not to exceed 24 hours, or the task listed below:)

4. <u>Task Description - Purpose</u>: _____

5. <u>Employee Assignments</u>:
   Supervisor: _____ Certified/Verified: _____
   Attendant(s):* _____ Certified/Verified: _____

   Entrant(s): _____ Certified/Verified: _____
   _____
   _____

6. <u>Rescue</u>
   <u>Services</u>:* _____ Certified/Verified: _____

Confirmed: (Date/Time)* _____
Emergency Contact # _____
Alarm Device Located At: _____
Lifting Device Located At: _____
Other Emergency Information: _____

* Required for PRCS Entry Only

7. <u>Air Monitoring</u>

Test	Device	Calibration	Results	Acceptable Range	By
O_2				>19.5% <23%	
Combust/ Vapors				<10% LEL	
LFL - Dusts				>5 ft/ vision	
Toxins				<PEL <PEL <PEL <PEL	
RAM				Background	
Other (Noise) (Light)				<85 dBA >25 ft. candles	

8. <u>Safety Procedures</u>:
   Energy lock-out:
   Double Blinds/Blanks: (isolation)
   Ventilation Requirements:
   Signs:
   Barricades:
   Trenching Requirements:
   Briefing (Emergency Action Plan):
   (Employee Alarm System [Communication]:
   Continuous Air Monitoring Requirements: (for/by)

FIGURE 6-2 Permit Required Confined Space Entry Permit.

9. <u>Protective Equipment</u>:
   Harnesses:*
   Lifelines:*
   Personal Protective Equipment: _____
   _____
   Other: _____

10. <u>Other Permits in Space (e.g. hot work)</u>: _____
    Other Contractors in space: _____

11. <u>Comments</u>:
    Permit Posted At: _____

12. <u>Certification</u>:
    All procedures required by this permit and (29CFR1910.146) will be enforced.

    _____
    Supervisor (sign)   Date   Time

    All procedures required by this permit and (29CFR1910.146) were followed; the permit removed and voided, and the confined space locked/closed.

    _____
    Supervisor (sign)   Date   Time
    Equipment Problems: _____
    Incidents/Accidents: _____
    Comments: _____
    _____

13. Annual Review:
    A copy of this permit will be on file at:
    _____
    _____

    for a one (1) year period.
    Annual Review:

    _____
    Date
    Comments: _____

FIGURE 6-2 (continued)

HOT WORK PERMIT

Seek an alternative/safer method if possible!

Before initiating hot work, ensure precautions are in place as required by NFPA 51B and ANSI Z49.1.
Make sure an appropriate fire extinguisher is readily available.

This Hot Work Permit is required for any operation involving open flame or producing heat and/or sparks. This work includes, but is not limited to, welding, brazing, cutting, grinding, soldering, thawing pipe, torch-applied roofing, or chemical welding.

Date	Hot work by ❏ employee ❏ contractor
Location/Building and floor _____	Name (print) and signature of person doing hot work
Work to be done _____	I verify that the above location has been examined, the precautions marked on the checklist below have been taken, and permission is granted for this work.
Time started _____ Time completed _____	Name (print) and signature of permit-authorizing individual (PAI)
THIS PERMIT IS GOOD FOR ONE DAY ONLY	

- ❏ Available sprinklers, hose streams, and extinguishers are in service and operable.
- ❏ Hot work equipment is in good working condition in accordance with manufacturer's specifications.
- ❏ Special permission obtained to conduct hot work on metal vessels or piping lined with rubber or plastic.

Requirements within 35 ft (11 m) of hot work
- ❏ Flammable liquid, dust, lint, and oily deposits removed.
- ❏ Explosive atmosphere in area eliminated.
- ❏ Floors swept clean and trash removed.
- ❏ Combustible floors wet down or covered with damp sand or fire-resistive/noncombustible materials or equivalent.
- ❏ Personnel protected from electrical shock when floors are wet.
- ❏ Other combustible storage material removed or covered with listed or approved materials (welding pads, blankets, or curtains; fire-resistive tarpaulins), metal shields, or noncombustible materials.
- ❏ All wall and floor openings covered.
- ❏ Ducts and conveyors that might carry sparks to distant combustible material covered, protected, or shut down.

Requirements for hot work on walls, ceilings, or roofs
- ❏ Construction is noncombustible and without combustible coverings or insulation.
- ❏ Combustible material on other side of walls, ceilings, or roofs is moved away.

Requirements for hot work on enclosed equipment
- ❏ Enclosed equipment is cleaned of all combustibles.
- ❏ Containers are purged of flammable liquid/vapor.
- ❏ Pressurized vessels, piping, and equipment removed from service, isolated, and vented.

Requirements for hot work fire watch and fire monitoring
- ❏ Fire watch is provided during and for a minimum of 30 min. after hot work, including any break activity.
- ❏ Fire watch is provided with suitable extinguishers and, where practical, a charged small hose.
- ❏ Fire watch is trained in use of equipment and in sounding alarm.
- ❏ Fire watch can be required in adjoining areas, above and below.
- ❏ Yes ❏ No   Per the PAI/fire watch, monitoring of hot work area has been extended beyond the 30 min.

© 2008 National Fire Protection Association   NFPA 51B

FIGURE 6-3 Ready for Hot Work Permit.

NOTIFICATION OF DEMOLITION AND RENOVATION

Operator Project #	Postmark	Date Received	Notification #
I. TYPE OF NOTIFICATION (O=Original R=Revised C=Cancelled):			
II. FACILITY INFORMATION (Identify owner, removal contractor, and other operator)			
OWNER NAME:			
Address:			
City:	State:	Zip:	
Contact:	Tel:		
REMOVAL CONTRACTOR:			
Address:			
City:	State:	Zip:	
Contact:	Tel:		
OTHER OPERATOR:			
Address:			
City:	State:	Zip:	
Contact:	Tel:		
III. TYPE OF OPERATION (D=Demo O=Ordered Demo R=Renovation E=Emer. Renovation):			
IV. IS ASBESTOS PRESENT? (Yes/No)			
V. FACILITY DESCRIPTION (Include building name, number, and floor or room number)			
Bldg Name:			
Address:			
City:	State:	County:	
Site Location:			
Building Size:	# of Floors:	Age in Years:	
Present Use:	Prior Use:		
VI. PROCEDURE, INCLUDING ANALYTICAL METHOD, IF APPROPRIATE, USED TO DETECT THE PRESENCE OF ASBESTOS MATERIAL:			

VII. APPROXIMATE AMOUNT OF ASBESTOS, INCLUDING: 1. Regulated ACM to be removed 2. Category I ACM Not Removed 3. Category II ACM Not Removed	RACM To Be Removed	Nonfriable Asbestos Material Not To Be Removed		Indicate Unit of Measurement Below	
		Cat I	Cat II	UNIT	
Pipes				LnFt:	Ln m:
Surface Area				SqFt:	Sq m:
Vol RACM Off Facility Component				CuFt:	Cu m:
VIII. SCHEDULED DATES ASBESTOS REMOVAL (MM/DD/YY) Start: Complete:					
IX. SCHEDULED DATES DEMO/RENOVATION (MM/DD/YY) Start: Complete:					

FIGURE 6-4 Notification of Demolition and Renovation.

NOTIFICATION OF DEMOLITION AND RENOVATION (continued)

X. DESCRIPTION OF PLANNED DEMOLITION OR RENOVATION WORK, AND METHOD(S) TO BE USED:			
XI. DESCRIPTION OF WORK PRACTICES AND ENGINEERING CONTROLS TO BE USED TO PREVENT EMISSIONS OF ASBESTOS AT THE DEMOLITION AND RENOVATION SITE:			
XII. WASTE TRANSPORTER #1			
Name:			
Address:			
City:	State:		Zip:
Contact Person:		Telephone:	
XII. WASTE TRANSPORTER #2			
Name:			
Address:			
City:	State:		Zip:
Contact Person:		Telephone:	
XIII. WASTE DISPOSAL SITE			
Name:			
Location:			
City:	State:		Zip:
Telephone:			
XIV. IF DEMOLITION ORDERED BY A GOVERNMENT AGENCY, PLEASE IDENTIFY THE AGENCY BELOW:			
Name:		Title:	
Authority:			
Date of Order (MM/DD/YY):		Date Ordered to Begin (MM/DD/YY):	
XV. FOR EMERGENCY RENOVATIONS			
Date and Hour of Emergency (MM/DD/YY):			
Description of the Sudden, Unexpected Event:			
Explanation of how the event caused unsafe conditions or would cause equipment damage or an unreasonable financial burden:			
XVI. DESCRIPTION OF PROCEDURES TO BE FOLLOWED IN THE EVENT THAT UNEXPECTED ASBESTOS IS FOUND OR PREVIOUSLY NONFRIABLE ASBESTOS MATERIAL BECOMES CRUMBLED, PULVERIZED, OR REDUCED TO POWDER.			
XVI. I CERTIFY THAT AN INDIVIDUAL TRAINED IN THE PROVISIONS OF THIS REGULATION (40 CFR PART 61, SUBPART M) WILL BE ON-SITE DURING THE DEMOLITION OR RENOVATION AND EVIDENCE THAT THE REQUIRED TRAINING HAS BEEN ACCOMPLISHED BY THIS PERSON WILL BE AVAILABLE FOR INSPECTION DURING NORMAL BUSINESS HOURS. (Required 1 year after promulgation) _____ _____ (Signature of Owner/Operator) (Date)			
XVII. I CERTIFY THAT THE ABOVE INFORMATION IS CORRECT. _____ _____ (Signature of Owner/Operator) (Date)			

FIGURE 6-4 (continued)

Project ID	_____
Date	_____
	CL8

LEAD ABATEMENT NOTIFICATION FORM

TYPE OF NOTIFICATION (Check One)
☐ Initial ☐ Revision (Fill in any sections that need revision and highlight) ☐ Cancellation

FACILITY DESCRIPTION
Building Name _____
Street Address _____
City _____ **PA** Zip Code _____
Building Size _____ (sq. ft.) No. of Floors _____ Building Age (in years) _____
Present Use _____ Prior Use(s) _____
Will the building be occupied while abatement occurs? ☐ Yes ☐ No

ABATEMENT CONTRACTOR
Company Name _____
PA Certification # _____
Street Address _____
City _____ State ____ Zip Code _____
Contact Person _____ Telephone _____

OTHER CONTRACTOR
Company Name _____
Street Address _____
City _____ State ____ Zip Code _____
Contact Person _____ Telephone _____

FACILITY OWNER
Company Name _____
Street Address _____
City _____ State ____ Zip Code _____
Owner Name _____ Telephone _____

FACILITY INSPECTION OR RISK ASSESSMENT
Inspector or Risk
Assessor Name _____ PA Certification # _____
Company Name _____ Telephone _____
Street Address _____
City _____ State ____ Zip Code _____
Date of Inspection/Risk Assessment _____
Was any type of lead-based paint present? ☐ Yes ☐ No
Specify procedures followed (below) and attach copy of results to this application.
_____
_____
_____
_____
_____

L&I USE ONLY Date Postmarked _____ Date Received _____ Project ID# _____

LIBI-600L REV 3-06 (Page 1)   COMMONWEALTH OF PENNSYLVANIA   DEPARTMENT OF LABOR & INDUSTRY

FIGURE 6-5 Lead Abatement Notification Form.

Project ID _____
Date _____
CL8

PROJECT DESCRIPTION

Description of Material	Location of Material (Room #, Floor # or Area)	Amount and Units of Lead-Based Paint	Abatement Type

OPERATION SCHEDULE
Start Date _____ Completion Date _____
Days of Week ☐ Mo ☐ Tu ☐ We ☐ Th ☐ Fr ☐ Sa ☐ Su
Daily Hours of Operation (Circle AM or PM): _____ AM PM to _____ AM PM

DESCRIPTION OF PLANNED WORK
_____
_____
_____
_____
_____
_____

DESCRIPTION OF WORK PRACTICES AND ENGINEERING CONTROLS TO BE USED DURING LBP REMOVAL
_____
_____
_____
_____
_____
_____
_____

WASTE TRANSPORTER(S)
Transporter # 1
Company Name _____
Street Address _____
City _____ State _____ Zip Code _____
Contact Person _____ Telephone _____

LIBI-600L REV 3-06 (Page 2)

FIGURE 6-5 (continued)

Project ID _____
Date _____ **CL8**

Transporter # 2
Company Name _____
Street Address _____
City _____ State _____ Zip Code _____
Contact Person _____ Telephone _____

WASTE DISPOSAL SITE
Landfill Name _____
Street Address _____
City _____ State _____ Zip Code _____
Contact Person _____ Telephone _____

PENNSYLVANIA CERTIFICATIONS
Inspector Name _____ Certification # _____
Risk Assessor Name _____ Certification # _____
Contractor Name _____ Certification # _____
Supervisor Name _____ Certification # _____

OWNER/OPERATOR CERTIFICATIONS

I hereby certify that an individual trained in the provisions of 40 CFR part 745 will be on-site during the lead-based paint abatement, as well as documentation that this person has received the training required by law. This documentation will be available for inspection during all normal working hours. I further certify that all work will be done in accordance with all applicable state and municipal rules and regulations.

Owner/Operator Name (Printed) _____ Title _____

Owner/Operator Name (Signed) _____ Date _____

I hereby certify that the foregoing statements and the information contained in this notification form are true. This certification is made subject to the penalties set forth in 18 Pa C.S. §4904 relating to unsworn falsification to authorities.

Owner/Operator Name (Printed) _____ Title _____

Owner/Operator Name (Signed) _____ Date _____

INSTRUCTIONS

This form must be mailed to the following address at least 5 working days (Monday-Friday) before the date that abatement will begin. The postmark on the envelope serves as proof of compliance with this requirement.

    PA DEPARTMENT OF LABOR & INDUSTRY
    CERTIFICATION, ACCREDITATION AND LICENSING DIVISION
    ROOM 1623, L&I BUILDING
    HARRISBURG, PA 17120

This notification requirement may be waived in emergency situations if approved by the Department. Call (717) 772-3396, between 8:00 AM and 5:00 PM, Monday-Friday, to obtain this waiver. If approved, a fully completed copy of this form must be faxed to the Department by 8:00 AM on the following business day.

LIBI-600L REV 3-06 (Page 3)

FIGURE 6-5 (continued)

Chapter Summary

Once hazardous environments have been identified, written plans are required to protect the public adjacent to the sites, public safety responders to the sites, on-site workers, and the environment. These site-specific plans include HASPs; permits (e.g., confined space permits, hot work permits, or permits for work with specified toxic materials such as asbestos or lead); and identification of safe procedures, protocols, and equipment for individuals who will operate in the hazardous environments.

Terms

Emergency Planning Zone (EPZ): The area adjacent to and including an identified target hazard where emergency protocols are developed for public protection.

Health and Safety Plan (HASP): A written document identifying protocol to prevent accidents and illnesses at hazardous sites, for groups who work at hazardous locations. HASPs may take the form of confined space entry permits, ready for hot work permits, or toxic material work permits (asbestos abatement and lead-based-paint removal).

References

Cocciardi, J.A. *Emergency Planning: Medical Services-1: A Manual for Medical Services Personnel Addressing Contaminated or Irradiated and Otherwise Physically Injured Persons and Equipment Following a Nuclear Power Plant Accident.* Mechanicsburg, PA: Cocciardi and Associates, 1996.

———. *A Health and Safety Manual.* Mechanicsburg, PA: Cocciardi and Associates, 2011.

Environmental Protection Agency, Office of Solid Waste. *RCRA Orientation Manual.* Washington, D.C.: Government Printing Office, 1990. U.S. EPA/530 SW-90-036, Washington D.C. 20460.

LIBI-600, Rev. 1/99. Asbestos Abatement and Demolition Renovation Notification Form. Harrisburg, PA: Commonwealth of Pennsylvania, Department of Labor and Industry, 2010.

LIBI-607L Lead Abatement Notification Form. Harrisburg, PA: Commonwealth of Pennsylvania, Department of Labor and Industry, 2010.

National Institute for Occupational Safety and Health, United States Coast Guard, Occupational Safety and Health Association, Environmental Protection Agency. *Occupational Safety and Health Guidance Manual for Hazardous Waste Site Activities,* DHHS-85-115. Washington, D.C.: Government Printing Office, 1985.

NFPA. Code #51B: Fire Prevention in Use of Cutting and Welding Processes. *National Fire Codes.* Quincy, MA: NFPA.

On-Site Calibration Log Form. Mechanicsburg, PA: Cocciardi and Associates, 2011.

Results of On-Site Monitoring Form. Mechanicsburg, PA: Cocciardi and Associates, 2011.

Transportation Office of the Federal Register, National Archives and Records Administration. *Title 49 CFR.* Washington D.C.: Government Printing Office, 1991.

U.S. Department of Housing and Urban Development. *Guidelines for the Evaluation and Control of Lead Based Paint Hazards in Homes.* Washington, D.C.: Government Printing Office, October 1, 1996.

U.S. Department of Labor, Occupational Safety and Health Administration. *General Industry Safety Regulations Permit Required Confined Spaces; 29 CFR 1910.146.* Washington, D.C.: Government Printing Office, 1998.

U.S. Federal Emergency Management Agency, National Fire Academy, National Emergency Training Center. *Hazardous Materials: Incident Analysis.* NFA-SMHMIA/TET, 1985.

Exercise 6

Protective plans safeguard employees and the public in and around hazardous areas, sites, and projects. Identify the appropriate plan developed to protect individuals in the following areas:

Population **Protective Plan**

1. Workers at Hazardous Waste Sites. _____

2. The Public, residing in the 10-mile area around Nuclear Power Plants. _____

3. Workers entering a Confined Space (PRCS). _____

4. Individuals adjacent to worksites where Extremely Hazardous Substances are used onsite. _____

5. Occupants of housing projects during lead paint removal. _____

6. Employees who will evacuate or provide first aid actions during an emergency. _____

7. Employees, trained to respond to hazardous areas and initiate emergency actions. _____

8. Individuals who may be exposed to asbestos during demolition or renovation of structures. _____

9. Firefighters who may respond to spills or fires in PCB electrical equipment. _____

10. Workers in factories where Hot Work is scheduled. _____

7 Personal Protective Equipment

Key Concepts

- Engineering Controls
- Administrative Controls
- Personal Protective Equipment

Introduction

This chapter discusses the selection, use, care, and maintenance of personal protective equipment (PPE). PPE must be chosen as a protective mechanism for employees subsequent to the use of **engineering and administrative controls** for employee protection or the elimination of the work activity. This prioritized format is clearly identified in the Occupational Safety and Health Administration (OSHA) regulations and the National Institute of Occupational Safety and Health (NIOSH) recommendations.

While job elimination is self explanatory, engineering and administrative controls are exemplified by the following:

- Manufactured equipment guards that arrive with products or engineered systems designed to eliminate or reduce personal risk, despite hazard.
- Administrative controls (such as maintaining distance from hazardous areas, or reduction of the amount of time in a hazardous areas), designed to reduce **or** eliminate employee exposure to hazards.

When job elimination, engineering or administrative controls are infeasible, or in the time necessary to implement these controls, PPE may be used. PPE is chosen after job and site analyses by a competent person[1] to ensure proper selection as well as the authority to implement and enforce the employer's PPE Program. PPE reduces employee exposure to illness-causing agents (below hazardous levels) and reduces or eliminates the possibilities of physical hazards that may cause employee accidents.

The Requirements for Selection of Personal Protective Equipment

Job Site Review

OSHA and various national consensus standard groups (e.g., the **American National Standards Institute [ANSI]**, the **American Society for Testing Materials [ASTM]**, and NIOSH) require that employers have competent person(s) review job sites before work is conducted. The purpose of this review is to determine potential occupational exposures (accident- and illness-causing agents), and subsequently identify appropriate PPE for employee-protection purposes (again, subject to the engineering and administrative control requirements previously discussed). This requirement is found in the current OSHA General Industry regulations at 29 CFR 1910.132(d) and the OSHA Construction Industry regulations at 29 CFR 1926.28.

Under the General Industry Safety regulations, employers must also select and provide PPE and appropriate training to employees in the use, care, and maintenance of the equipment. Under the OSHA general industry standard, employers are required to certify that an initial site survey has been made to identify the need for PPE and the type of PPE necessary. This certification must be in writing, identifying the workplace evaluated, the person certifying that the evaluation has been performed, and the date the hazard assessment was completed.

In the construction industries, engineering and administrative controls must also be used. However, it is generally accepted that certain hazards cannot be completely eliminated from work sites by these two mechanisms. Consequently, requirements for PPE, such as hard hats or safety shoes, normally exist at construction work sites.

Responsibilities for Provision of Equipment

The provision of PPE is generally the employer's financial responsibility. However, the U.S. Department of Labor has determined that employers may require employees to purchase PPE that may be used off the job site, such as safety shoes. Other types of PPE (specialized equipment used only at job sites), such as safety glasses and hard hats, which are traditionally not used off the job sites, are required to be provided by employers.

PPE selection and enforcement of use by employees is the employer's responsibility. In general, employees cannot indemnify employers against this requirement. The maintenance and care of PPE may be administratively assigned to employees.

Agencies Certifying Personal Protective Equipment

OSHA and various consensus national standards groups (e.g., the American National Standards Institute [ANSI]; the American Society for Testing and Materials; the International Safety Equipment Association [ISEA]) require certification of PPE in a variety of ways, and only certified equipment should be used. In general, regulatory agencies such as OSHA do not certify PPE but allow other governmental organizations such as NIOSH, or the consensus standards groups mentioned, to develop certification procedures. Subsequently, PPE is certified and labeled according to the listed national standards by the manufacturers.

---

[1] OSHA defines a competent person as one who has the knowledge through experience and/or education to select/identify safety related equipment or issues and the authority to implement remedies.

TABLE 7-1 Personal Protective Equipment Requirements

Equipment	Current Concensus National Standard	Current OSHA Requirement
Head Protection	ANSI/ISEA Z89.1 (2009) American National Standard for Industrial Head Protection	29 CFR 1910.135 (ANSI Z89.1, 2003; ANSI Z89.1, 2009)
Eye and Face Protection	ANSI/ISEA Z87.1 (2010) American National Standard for Occupational and Educational Personal Eye and Face Protection Devices	29 CFR 1910.133 (ANSI Z87.1, 2003; ANSI Z87.1, 2010)
Foot Protection	ASTM F2412-05 Standard Test Methods for Foot Protection, and ASTM F2413-05 Standard Specification for Performance Requirements for Foot Protection	29 CFR 1910.136 (ANSI Z41; 1999; ASTM F2413-05 and ASTM F2412-05)
Hand Protection	ANSI/ISEA 105 (2011) American National Standard for Hand Protection Selection Criteria	29 CFR 1910.138
Firefighters' PPE	NFPA 1971: Standard on Protective Ensembles for Structural Fire Fighting, and Proximity Fire Fighting (2013) and NFPA 1851: Standard on the Selection, Care and Maintenance of Protective Ensembles for Structural Fire Fighting and Proximity Fire Fighting (2008)	29 CFR 1910.156 • Protective Clothing for Structural Firefighting NFPA 1971 (1975) • Criteria for Firefighters Gloves (NIOSH) (1976) • Model Performance Criteria for Structural Firefighting Helmets: U.S. Fire Administration (1977)

Regulatory requirements, as illustrated in **TABLE 7-1**, may not be as current as those of the consensus standards groups addressing PPE. Employers are well advised to meet the standards set in the current edition of national standards for the referenced PPE, whether or not the current edition is cited in the regulatory text (e.g., OSHA standard). In rare instances, current consensus standards may require less stringent PPE; however, the body of evidence considered by consensus standards organizations generally justifies any reduction in the level of protection provided by the PPE.

Table 7-1 identifies common styles of PPE found in the workplace, current regulatory requirements, and the most recent consensus standard, which should be considered in the selection of PPE.

Personal Protective Equipment Applicable to Most Hazardous Environments

Certain types of PPE may be applicable to most hazardous environments. An example is the selection of head, foot, hand, and eye protection for work sites. These pieces of equipment are generally included in all ensembles of PPE for most hazardous work sites.

This basic PPE may be selected and integrated into the levels of protection chosen for hazardous waste workers (Level A, B, C, and D protection) or the levels of protection recommended for individuals working in laboratories performing research with biologically hazardous materials (Biosafety Levels BL1, BL2, BL3, and BL4).

Employees requiring higher levels of protection in hazardous environments may require a specialized selection mechanism. This is necessary to ensure equipment compatibility and the ability of the employee to continue to work within the ensemble of PPE, both physically and psychologically. All PPE must be selected to ensure that additional hazards—such as slips, trips, obscured vision due to the use of PPE, and excessive physiological strains—are not introduced to the work regimen.

This is important, as over-selection of PPE may lead to the following complications:
- More hazardous conditions at the worksite
- A subsequent lack of use of the PPE on the part of the employees.

General Personal Protective Equipment Selection

Each employer must evaluate and select appropriate PPE in the areas of head, eye, face, hand, and foot protection. The areas of hearing and respiratory protection require specialized evaluation and selection criteria and are covered later in this chapter. The selection of chemical, biological, and radiological protective ensembles, including body coverings, also requires special selection criteria.

Head Protection

A Head protection selection criterion begins with the determination that falling objects or "bumps" may affect employees. If such is the case, employers should provide and require head protection to be worn, as shown in **TABLE 7-2**. Of note, hard hats and head protection manufactured after 1993 are labeled under the current ANSI standard in a manner different from older head protection (manufactured after 1969 but before 1993).[2]

Bump caps meeting the requirements of ANSI Z89.1 may be used when bumps are the only hazards anticipated in the workplace.

The most recent American National Standard (ANSI Z89.1-2009) should be used for the selection of head protection. Head protection is now described by impact protection and electrical class. Type I helmets are designed to reduce the force of impact resulting only from a blow to the top of the head, while Type II helmets provide protection from force applied to the top or sides.

Helmets are also rated for electrical protection. Class G (General) helmets reduce the danger of contact with low voltage conductors (2200 volts—phase to ground). These are labeled as Class A under the 1986 ANSI standard still accepted by OSHA at this time, if the hard hat was placed in service prior to 2012. Since 2012, OSHA has accepted the two most recent versions of the consensus national PPE standards. Class E (Electrical) helmets reduce the danger of contact with higher voltage conductors (20,000 volts, phase to ground). These are labeled as Class B under the 1986 ANSI standard. Class C (Conductive) helmets do not provide protection from electrical hazards. These are labeled as Class C under the 1986 ANSI standard. Reverse wearing helmets have passed all tests in both a forward and backward position and are so marked with a reverse donning pictogram (**PHOTO 7-2**). In addition to the type, class and reverse wearing markings, low temperature (marked "LT" and tested to −22°F for four [4] hours, while typical head protection is tested to 0°F for two [2] hours) and high visibility (marked "HV" indicating high luminance) are indicated on the product label.

As with all PPE, head protection (hard hats) must be assembled, used, cleaned, maintained, and worn in compliance with the manufacturer's recommendations. This includes installation of all components (such as all components of suspension ensembles or chinstraps and cups) in an appropriate manner. To be protective, they must be used as directed (worn with the appropriate sections of the suspension contacting the appropriate part of the head [e.g., the hard hat may not be worn backwards, unless so designed, tested and labeled with the reverse donning pictogram]).

In addition, maintenance procedures listed by manufacturers must be followed, such as cleaning and storage. Hard hats meeting these general industry requirements are not specified for certain hazardous operations such as firefighting[2], rock climbing, or other recreational uses. Other testing specifications and certifications apply to head protection used in these situations.

Eye and Face Protection

Eye and face protection is required to be selected in compliance with recognized standards, such as the most recent ANSI standards, found in **TABLE 7-3**. Eye protection offers protection for the eye and mucus membranes from both physical particles and gases, vapors, or mists. Face

PHOTO 7-1 ANSI-compliant head protection can be full or front brimmed. Firefighters' helmets meet both NFPA and ANSI standards. Bump caps protect against bumps in the workplace.

[2] Head protection used for interior structural firefighting must meet the requirements of NFPA 1971; Standard on Protective Ensembles for Structural Fire Fighting and Proximity Fire Fighting (2013).

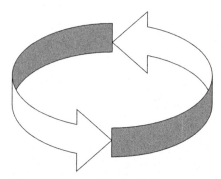

PHOTO 7-2 Reverse Donning Pictogram.

TABLE 7-2 Head Protection Labeling and Selection Criteria

	ANSI Z89.1, 1986	ANSI Z89.1, 2003, 2009[1]
Electrical		
Low Voltage Protection	Class A	Class G (General)
High Voltage Protection	Class B	Class E (Electrical)
No Electrical Protection	Class C	Class C (Conductive)
Impact		
Top of Head		Type I
Top and Side of Head		Type II
Falling Objects	Class A, B, or C	
Brim		
Full	Type I	
None	Type II	

[1]Additionally, since 2009, hardhats that have passed applicable tests are labeled for reverse wearing, (Photo 7-2), for low temperatures (LT); or for high visibility certifications.
Adapted from: American National Standards Institute. *American National Standard Z89.1: Head Protection.* New York, NY, 1986
American National Standards Institute. *American National Standard Z89.1: Head Protection.* New York, NY, 1997, 2003, 2009.

protection is designed to protect against flying objects, dust, fumes, and mists and is generally not airtight. Full-face piece or hood respirators may have face pieces meeting the eye/face protection requirements.

Since 2003, "low impact" eye protection has been certified under the ANSI Z87.1 Standard. Low impact certified eye and face protection now carries a written warning attached to the equipment, while "high impact" equipment is labeled "Z87+". High impact resistance equipment is always recommended. All eye and face protection manufactured prior to 2003 is high impact resistant.

In addition to the "+" sign that indicates impact resistant testing, all lenses and frames, which have passed testing, are marked "Z87", and those fitting small head sizes are marked "H". When tested, lens types W (Welding) and a shade number; U (UV Filter) and a scale number; L (Visible Light) and a scale number; R (Infra Red) and a scale number; V (Variable Tint); S (Special Purposes); and use (D3 [splash/droplets]; D4 [Dust]; D5 [Fine Dust]) may also be marked.

Foot Protection

Industrial and occupational foot protection should be selected in compliance with ASTM F2412-05 and ASTM F2413-05, which has replaced, since 2005, the ANSI Z41 standards. All ASTM labeled footwear must meet impact and compression testing and additional tested/certified protection may also be listed. Labeling occurs in one shoe.

Foot protection labeling is similar under current ASTM and older ANSI standards and is described below (see **TABLE 7-4**):

- Standard and year of the test to which the footwear complies (Line 1)
- Male or female footwear (Line 2)
- The compression tests performed listed as **Class 30, Class 50, or Class 75** (identifying 30 lb, 50 lb, and 75 lb compression tests from toe to heel) (Line 2)
- Impact resistance by drop test on the toe listed as **Class 30, Class 50, or Class 75** (identifying a 1000 lb, 1500 lb, or 2000 lb drop test from a 4 foot height) (Line 2)

PHOTO 7-3 Eye and face protection is selected based on the anticipated hazard.

TABLE 7-3 American National Standards Institute: Z87.1 (2010) Eye and Face Protector Selection Chart

ANSI/ISEA Z87.1-2010
Eye & Face Protector Selection Chart

Protective devices do not provide unlimited protection. This information is intended to aid in identifying and selecting the types of eye and face protectors that are available, their capabilities and limitation for the hazards listed. This guide is not intended to be the sole reference in selecting the proper eye and face protector.

Hazard	Protectors	Limitations	Marking[1]
IMPACT - Chipping, grinding, machining, masonry work, riveting, and sanding			
Flying fragments, objects, large chips, particles, sand, dirt, etc.	• Spectacles with side protection • Goggles with direct or indirect ventilation • Faceshield worn over spectacles or goggles • Welding helmet	Caution should be exercised in the use of metal frame protective devices in electrical hazard areas. Metal frame protective devices could potentially cause electrical shock and electrical burn through contact with, or thermal burns from exposure to the hazards of electrical energy, which include radiation from accidental arcs. Atmospheric conditions and the restricted ventilation of a protector can cause lenses to fog. Frequent cleaning may be required.	Impact rated: + (spectacle lens) Z87+ (all other lens) Z87+ (plano frame) Z87.2+ (Rx frame)
HEAT - Furnace operations, pouring, casting, hot dipping, gas cutting, and welding			
Hot sparks	• Spectacles with side protection • Goggles with direct or indirect ventilation • Faceshields worn over spectacles or goggles • Full-facepiece respirator • Loose-fitting respirator worn over spectacles	Spectacles, cup and cover type goggles do not provide unlimited facial protection. Operations involving heat may also involve optical radiation. Protection from both hazards shall be provided.	
Splash from molten metal	• Faceshields worn over goggles • Full-facepiece respirator • Loose-fitting respirator worn over spectacles		
High temperature exposure	• Screen faceshield over spectacles or goggles • Reective faceshield over spectacles or goggles		
CHEMICAL - Acid and chemical handling, degreasing, plating			
Splash and irritating mists	• Goggles with indirect ventilation (eyecup or cover type) • Faceshield worn over spectacles or goggles • Full-facepiece respirator	Atmospheric conditions and the restricted ventilation of a protector can cause lenses to fog. Frequent cleaning may be required.	Splash / droplet: D3

Hazard	Protectors	Limitations	Marking[1]
DUST - Woodworking, buffing, general dusty conditions			
Nuisance dust	• Goggles with direct or indirect ventilation (eyecup or cover type) • Full-facepiece respirator	Atmospheric conditions and the restricted ventilation of a protector can cause lenses to fog. Frequent cleaning may be required.	Dust: D4 Fine dust: D5
OPTICAL RADIATION			
Welding: Electric Arc	• Welding helmet over spectacles or goggles	Protection from optical radiation is directly related to filter lens density. Select the darkest shade that allows adequate task performance. Note: Filter lenses shall meet the requirements for shade designations in table 6 of ANSI/ISEA Z87.1-2010.	Welding: W shade number UV: U scale number Glare: L scale number IR: R scale number Variable tint: V Special purpose: S
Viewing electric arc furnaces and boilers	• Handshield over spectacles or goggles TYPICAL FILTER LENS SHADE: 10-14		
Welding: Gas	• Welding helmet over spectacles or goggles		
Viewing gas-fired furnaces and boilers	• Welding goggles • Welding faceshield over spectacles or goggles TYPICAL FILTER LENS SHADE: 4-8		
Cutting	• Welding goggles • Welding helmet over spectacles or goggles • Welding faceshield over spectacles or goggles		
Torch brazing	• Welding respirator TYPICAL FILTER LENS SHADE: 3-6		
Torch brazing	• Welding goggles • Welding helmet over spectacles or goggles • Welding faceshield over spectacles or goggles TYPICAL FILTER LENS SHADE: 3-4		
Torch soldering	• Spectacles • Welding faceshield over spectacles • Welding respirator TYPICAL FILTER LENS SHADE: 1.5-3		
Glare	• Spectacles with or without side protection • Faceshield over spectacles or goggles	Shade or special purpose lenses, as suitable. Note: Refer to definition of special purpose lenses in ANSI/ISEA Z87.1-2010.	

1. Refer to ANSI/ISEA Z87.1-2010 table 4a for complete marking requirements.

Source: American National Standard for Occupational and Educational Personal Eye and Face Protection Devices, ANSI Z87.1-2010. Copyright 2010. International Safety Equipment Association.

TABLE 7-4 Typical Labeling Components for Foot Protection	
Line One	Standard and Year of Test to Which the Footwear Complies.
Line Two	Gender, Impact, and Compression Ratings
	M—Male
	F—Female
	I-75—2000 lb drop test
	I-50—1500 lb drop test
	I-30—1000 lb drop test
	C-75—75 lb compressor test
	C-50—50 lb compressor test
	C-30—30 lb compressor test
	Metatarsal Protection: Mt75/50/30
Lines Three and Four	Cd: Electronically dissipative
	CS: Chainsaw cut resistant
	DI: Dielectric insulation
	EH: Electrical Hazard: Reduce hazards due to contact with electrical parts
	PR: Puncture resistant sole
	SD: Reduce the accumulation of excess static electricity by conducting a charge to ground, while maintaining a high level of resistance to protect the worker from live electrical circuits

- Metatarsal protection (**Class 30, Class 50, or Class 75**) (Line 2 under the ASTM standard and Line 3 or 4 under the ANSI standards, if provided)
- Puncture resistance of soles (identified by **PR**) (Line 3 or 4, if provided)
- Electrically protected, a designation useful for individuals who may come in contact with and need to be insulated from electric sources (designated **EH**) (Line 3 or 4, if provided)
- Static dissipative protection, which may be necessary for individuals operating in areas where the accumulation of static charges may cause a problem (identified by **SD**) (Line 3 or 4, if provided)
- Electrically dissipative footwear, useful in situations in which it is necessary to dissipate a charge from the individual (identified by the **Cd** insignia) (Line 3 or 4, if provided)

Additionally, the ASTM standard identifies footwear (Line 3 or 4) that has been tested and provides chainsaw cut resistance (**CS**), and dielectric insulation (**DI**).

Safety shoes are now produced in all colors and configurations, including sneakers. It is important to note that neither the ANSI or ASTM standards allow for the use of "add-on" devices, strap on foot, toe or metatarsal guards, as a substitute for protective footwear, however, if an employer can provide documentation that these are equally effective, they are acceptable under the OSHA standards.

Once a foot hazard is identified, the above criteria provide information to assist the employer's competent person with the PPE selection process.

PHOTO 7-4 Safety shoes are produced in a variety of configurations.

Hand Protection

Hand protection is selected based on physical hazards observed. In addition to gloves, various manufacturers currently produce creams and salves that may be applicable for hand protection in certain industries.

Hand protection (gloves) is classified/certified under American National Standard Institute/International Safety Equipment Association (ANSI/ISEA) 105-2011 (a revision of the 2005 Standard). Hand protection includes gloves, mittens, partial gloves, and other items covering the hand.

Gloves utilized for special applications (e.g., ionizing or non-ionizing radiation, welding, or firefighting [emergency response]) are certified under separate standards. Gloves used as a part of a glove box mechanism, biological/safety cabinet or fume hood configuration are not covered.

Gloves may be tested and identified under the 2011 ANSI/ISEA standard for the following properties: cut resistance (0–5); puncture resistance (0–5); abrasion resistance (0–6); chemical permeation (0–6); chemical degradation (0–4); detection of holes (pass/fail); heat and flame protection (0–4); heat degradation resistance (0–4); conductive heat (0–5); anti-vibration or vibration reduction (pass/fail); dexterity (1–5). Under this system, low numbers indicate low protection and high numbers indicate high protection, in compliance with the tests listed in the standard.

Glove manufacturers are permitted to report this information, and additionally provide name/glove designation, size, and expiration dates (if applicable) on the package containing the smallest number of glove elements sold. Manufacturers are also required to describe whether the product contains natural rubber/latex.[3]

Additional labels providing information referenced above may also be provided.

Again, the selection process is a requirement placed on the employer and generally includes the following six-step approach prior to the employer's certification of appropriate PPE selection in this area:

1. Review the task and conduct a hazard assessment;
2. Match performance standards referenced above to the hazards identified;
3. Consider hand protection features necessary (e.g., length, surface);
4. Choose the glove offering optimal performance/features;
5. Select gloves of the appropriate size (to allow for function, fit and comfort);
6. Re-evaluate periodically or after incident/accidents related to hand protection.

Specialized testing (e.g., National Fire Protection Association [NFPA] 1971: Protective Clothing for Structural Firefighting) and certification is required for hand protection used in firefighting operations.

In certain cases, double gloving may be recommended for either the prevention of cross-contamination or protection from dissimilar hazards. Of note, many gloves have common inconsistencies, such as a lack of thermal protection while providing chemical protection, or the lack of abrasion and cut resistance while providing heat or cold protection.

The terms "penetration" and "permeation" rates are used in the identification of glove characteristics, as well as other types of body protection. These terms identify rates that materials or physical agents may be absorbed into or through gloving materials. Job safety analysis will identify appropriate gloving or body protection time frames when these characteristics are considered. Gloves used as part of glove-box mechanisms, biological safety cabinets, or fume hood configurations are not covered under this evaluation process and should be selected in consultation with manufacturers.

Any PPE evaluation should take into consideration the health requirements as discussed in the previous chapters. These include a review of employees' ability to work safely relative to themselves and surrounding employees while the protective equipment is worn. The options for other equally protective mechanisms that would alleviate the need for PPE should also be considered.

---

[3] Gloves containing natural rubber/latex are labeled under the ASTM D5712-16 Standard: Standard Test Method for Analysis of Aqueous Extractable Protien in Natural Rubber and its Products Using the Modified Lowery Method; 2010.

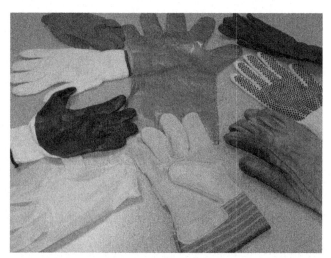

PHOTO 7-5 Double or triple gloving may be required for multiple hazards or to prevent cross-contamination in a hazardous environment.

Specialized Equipment Used in Hazardous Environments

Hazardous environments generally require an ensemble of PPE or procedures. Elaborate configurations of equipment, restrictive use and maintenance requirements, and specific testing requirements before use may apply. Respiratory protection and hearing protection equipment fall into this category. Ensembles of PPE used for protection against chemical, biological, radiological, nuclear or explosive (CBRNE) agents fall into this category as well and some of these ensembles are now certified and tested for CBRNE use by NIOSH. When tested, this equipment is so certified and labeled for CBRNE use.

Hearing Protection Equipment

Any hearing protection equipment recommendations should begin with the identification of a "noisy area" and a noise survey. Engineering and administrative controls are applied first to reduce sound pressure levels, and hearing protection used in the timeframe necessary to take these actions or if actions are insufficient (such as at construction sites). Selection of hearing protection by employees should be based on anticipated noise levels, as monitored by an industrial hygienist or safety professional. Hearing protection is generally recommended at 85 decibels measured on the "A" weighting scale (dBA) averaged over an eight (8) hour period. Lower averages will apply to longer timeframes. NIOSH recommends mandatory use of hearing protection at the 85-dBA average.

The "A" weighting scale most accurately represents sound pressure level actions in the human hearing system, and it is used for many environmental and industrial noise measurements. "B", "C", "D" and "Z" weightings have also been developed. "D" weightings are no longer in use (they were used to describe specialized noise, such as high impact jet engine sound pressures). The "C" weighting scale mimics ear response to high frequencies. "Z" or zero weighting was introduced in 2003 to allow for individual cut points on noise curves.

Again, before requiring hearing protection, administrative and engineering controls must be attempted.

Sound pressure levels are measured in decibels. Decibels behave logarithmically, requiring difficult calculations for time averaging. Therefore, employers are cautioned to use the previously identified professionals for noise surveys. From an administrative standpoint, a Hearing Conservation Program is required by OSHA at sound pressure levels of 85 dBA averaged over an eight-hour day (50% dose). Mandatory sound pressure level reduction is required at 90 dBA (100% dose). However, certain groups, such as the U.S. Air Force, have more stringent requirements than those of OSHA.

Once sound pressure levels and employee exposures are determined, employers may select noise attenuation equipment to reduce employee exposure to below the 85 dBA required Hearing Conservation Program level.

Manufacturers of hearing protection devices specify **noise reduction ratings (NRR)** to characterize equipment subsequent to testing, and they label these devices according to U.S. Environmental Protection Agency requirements. NRR identify the noise reduction in decibels each hearing protector may afford. Multiple hearing protectors may be used. OSHA currently specifies that a safety factor of 7 dBA be applied to (i.e., subtracted from) all selected hearing protection device ratings, and identifies that half of remaining noise will pass through the protective devices in field use.

Of note, NRR greater than 30 dBA may not be accurate. Although some manufacturers may specify NRR of greater than 30 dBA, it is the maximum NRR per device that should be applied at the present time.

Recently (since 2009), the U.S. EPA has proffered new testing and labeling of hearing protection devices. The new protocol calls for more stringent testing and labeling a range of NRR protection (with the low number protective of 80% of the use population and the high number protective of the most skilled and motivated user, or 20% of the use population). This label number would then be applied directly to field use situations, without requiring a rating reduction.

While a variety of scales are used to characterize sound pressure levels, the "A" weighted scale, which mimics the response of the human ear to sound pressure levels, is accepted (and is specified by OSHA) for sound pressure level readings. The OSHA requirement for a 7 dB reduction is applied to the "A" weighted scale readings.

Noise surveys should be provided initially, when procedures change or are modified, and at routine intervals (such as annually) to guarantee employee protection. The requirements of a Hearing Conservation Program as promulgated by OSHA and NIOSH are listed in **TABLE 7-5**.

Respiratory Protection

Mechanics of Respiration

All selection of respiratory protection equipment is based on air quality: The constituency of sufficient oxygen in air (generally recognized as being at least 19.5%), and

TABLE 7-5 Hearing Conservation Program Requirements		
	OSHA	NIOSH
Optional Protection	85 dBA	N/A
Mandatory Program	90 dBA (Protection) 85 dBA (Hearing Conservation Program)	85 dBA
Baseline Audiograms	85 dBA	85 dBA
Annual Audiograms	Yes	Yes
Area Noise Surveys or Personal Dosimetry	Noisy areas (i.e. 85 dBA)	Noisy areas
Employee Hearing Protection Use	85 dBA (optional) 90 dBA (mandatory) or if a threshold shift has occurred	85 dBA (mandatory) or if a threshold shift has occurred
Training	Annually	Annually

the absence of toxic materials. The toxicology of such materials has been reviewed in previous sections. Once air (oxygen) has entered the respiratory system, it is efficiently distributed through the body by the respiratory and cardiovascular systems in 3- to 30-second cycles. Components of the Human Respiratory System are depicted in **FIGURE 7-1**.

Both internal and external mechanisms may affect respiration. Interferences to human respiration include blockages in air passageways, particularly in the small sections of the lower respiratory system; introduction of toxins that are accepted by the respiratory system and circulated prior to the circulation of oxygen; and introduction of toxins that are distributed by the respiratory system and may subsequently damage other body parts or systems.

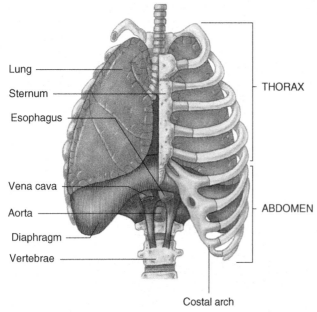

FIGURE 7-1 The human respiratory system.

Interferences to Respiration

The primary interference to respiration is lack of oxygen or oxygen pressures within air. Typically, events such as fire or oxidation/reduction reactions may reduce the amount of oxygen available for human respiration (below 19.5%). Refer to Table 3-7 for an identification of human physiological responses to various oxygen levels and pressures.

In addition, at high altitudes (such as altitudes in excess of 8000 feet), the partial pressure of oxygen (versus nitrogen) found in air is less (less than 100 mm Hg) and consequently makes breathing more difficult. Therefore, at higher altitudes higher oxygen levels are required for human respiration. Another blockage to the uptake of oxygen by the blood is excessive pressures. This allows the passage of nitrogen (79% of air) through the respiratory system to the blood.

In underwater diving, two conditions—nitrogen narcosis (diver's euphoria) or decompression sickness (the bends)—may occur as a result of the body's acceptance of nitrogen and failure to release nitrogen "bubbles" from the bloodstream. These situations (low or high pressures) can be characterized as "ordinary air inspired under abnormal conditions."

A second category of respiratory interferences is those that cause blockages to or malfunctions in the human respiratory system. Bronchitis, asthma (both physiologically and chemically induced), and pneumonia are examples of systemic malfunctions.

Also, temperature extremes, biological/fungal agents, radioactive materials, and hazardous chemicals may cause irritation to the respiratory system (e.g., pneumonia).

Chemicals, such as cyanide, that interfere with the enzymes allowing oxygen to release from the lung to the bloodstream are severe interferences to the respiratory system. Certain sensitizers (such as hydrogen sulfide) are included in this category. Hydrogen sulfide deadens the sense of smell and eliminates warning properties. This allows higher doses to enter the body undetected.

A final category of interference involves materials that are transported by the cardiovascular system and cause toxic effects in other parts of the body. Carcinogens, mutagens, and teratogens fall into these categories (e.g., carcinogens: benzene; mutagens: DDT and 2, 4, 5-trichlorophenoxicacetic acid [2, 4, 5, T][4]; teratogens: polychlorinated biphenyls [PCBs]).

Carcinogens, mutagens, and teratogens are special cases because current research does not identify safe levels of exposure to many of these agents (i.e., there is no level below which adverse effects may not occur). Current governmental policies, including selecting high levels of respiratory protective equipment, reflect the philosophy that exposure should be kept as low as reasonably achievable through all mechanisms. Although body systems are constantly combating carcinogens through internal mechanisms, causes and levels of causal agents are undefined.

Once potential interference to the mechanisms of respiration has been identified, appropriate respiratory protection may be selected.

Air Purifying Respirators

Air purifying respirators (APRs) provide filtering protection to users. APRs should be selected only if three conditions are met:

1. Sufficient oxygen is present in air (generally considered to be at least 19.5%).
2. Known levels of toxic materials have been identified through monitoring or modeling, and appropriate filtering media is available for the level of toxin.

    In addition, a cartridge change schedule must be developed and implemented for cartridge respirators used in gas/vapor atmospheres. This is particularly important and required for filters without end-of-service-life indicators or for filters used against substances with poor warning properties.
3. The APR **protection factor (PF)** (assigned to the specific class of respirator) identifies the level of contaminants that are assumed to be inside respirators as below the level chosen as acceptable.

---

[4] The herbicide "Agent Orange", most notorious of the "Rainbow Herbicides" used during the Vietnam era by the U.S. Military, was a 50-50 mixture of 2, 4, 5, T and 2, 4, D.

Supplied Air Respirators

Conversely, **supplied air respirators (SARs)** should be selected and used in the following situations:

- Less than 19.5% oxygen is present in the atmosphere to which the employee may be exposed.
- High or unknown levels of toxic materials or interferences to respiration are present, or atmospheres are present for which filters are not approved.
- Unknown hazards are present (such as in a fire or emergency rescue situation).

Advantages of APRs include smaller weight and size, reduced cost, and ease of maintenance and training. APRs may be of the half- or full-face piece style, offering different protection factors by the mask.

Advantages of SARs are based on their ability to be used in most atmospheres, including unknown concentrations of hazardous materials and low oxygen levels. SARs may be of the self-contained variety, such as a self-contained breathing apparatus (SCBA). A SCBA may be open-circuit style (i.e., it discharges exhaled air through the unit to the atmosphere) or the closed-circuit style (exhaled breath is scrubbed and oxygen added so that it may be rebreathed). In addition, SARs may be of the airline style, providing air from remote locations or remote air compressors to a distance of 300 feet. A variety of requirements are applicable to the use of SAR equipment such as use of Grade D (minimum) breathing air for any units, carbon monoxide safety devices on oil-containing air compressors, and auxiliary (escape) cylinders carried with airline units. SARs have high protection factors. Characteristics of Grade D and better breathing air are found in **TABLE 7-6**.

PHOTO 7-6 SCBA is typically used in emergency response situations; airline-connected SARs allow individuals longer times in hazardous environments. Many include emergency escape SCBA of 5 to 10 minutes' duration; emergency escape masks are certified for "escape only."

TABLE 7-6 Characteristics of Grade D and Better Breathing Air						
Characteristics	D	E	F	G	H	I
% O_2 (v/v) Balance predominately N:	19.5-23.5	20-22	20-22	20-22	20-22	20-22
Acetylene[1]						0.05
CO_2	10	10	5	5	5	1
CO_2[1]	1000	500	500	500	0.5	
Gaseous Hydrocarbons (as methane)[1]			25	15	10	0.5
Halogenated Solvents				10	1	0.1
Hydrocarbons (condensed)[2] at NTP	5	5				
Nitrogen Dioxide[1]				2.5	0.5	0.1
Nitrous Oxide[1]						0.1
Odor	*	*	*	*	*	*
Sulfur Dioxide[1]				2.5	1	0.1
Water (dew point)[3]	−65 or a dew point corresponding to 10°F below the lowest expected temperature					
Footnote[1] = Parts per million (volume) Footnote[2] = mg/m³ Footnote[3] = Degrees Fahrenheit * = No pronounced order						

A variety of safety requirements, in addition to training, apply to the use of SARs in immediately dangerous to life or health (IDLH) situations. These include the use of standby personnel for lifesaving and emergency rescue purposes (generally two rescuers). Of note, this requirement for a rescue team may be reduced to one person in certain situations, such as fully characterized hazardous atmospheres and confined spaces.

Written Respirator Programs

Any work using respiratory protection must be accomplished in compliance with a written respirator program. OSHA requirements for written respirator programs are summarized in **TABLE 7-7**.

Respirator Protection Factors

The protection factor (PF) of a respirator identifies leakage into the respirator of contaminants other than through filtering cartridges. PFs are calculated using the following formula:

$$PF = \frac{\text{Concentration of contaminants outside face piece}}{\text{Concentration of contaminants inside face piece}}$$

When a PF for a respirator is known, the concentration of the toxin inside the respirator can be compared with a predetermined exposure limit.

PFs assigned by OSHA are listed in **TABLE 7-8**. Both NIOSH and ANSI have issued protection factors for respiratory protection; however, the OSHA PF table published in 2006, subsequent to the OSHA respirator standard publication, is now in general use.

Respirator Filters, Adsorbents, and Absorbents

Each filter, adsorbent canister, or absorbent cartridge used in conjunction with a respirator will have an expected service life. Some filters have a signal or color change indicator that identifies when the filter has outlived its filtering capability.

If an indicator is not present, a cartridge change schedule must be calculated and enforced for each filter in use. Humidity, temperature, and other contaminants in an air stream may affect filter efficiency.

In addition, each cartridge used in gas/vapor environments will have maximum concentrations listed on it. These are concentrations above which the filter will not be effective.

Particulate filters used for protection against dusts, fumes, mists, and aerosols are tested and rated by NIOSH according to test criteria published in June

TABLE 7-7 OSHA: Respiratory Protection Standards Information

OSHA: RESPIRATORY PROTECTION STANDARD INFORMATION

- Applies to Construction and General Industries
- Covers both voluntary and required respirator use (except for voluntary dust masks use).
- Establishes recordkeeping practices for documents relative to the Respirator standard.
- Requires a written Respirator Program, specific to job sites and practices, in compliance with the standard (9 mandatory components).
- Written programs must include specific selection criteria for respirators, schedules for cartridge changes, if cartridge respirators are used, and quality (GRADE D) breathing air, if supplied air respirators are used.
- <u>Eliminates</u> annual Physical requirements, but requires an initial medical checklist, the availability of health consultation and health care professional clearance in certain situations.
- Requires use of the OSHA protection factors for respirator selection criteria.
- Requires annual <u>quantitative</u> and/or <u>qualitative</u> fit testing <u>for</u> all tight fitting respirator users.
- Requires initial and annual training.
- Requires designation of a qualified program administrator, and an annual program evaluation and audit, including discussions with respirator users.
- Identifies tagging and check/maintenance protocol for respirators and compressors.
- Gives specific emergency and IDLH requirements for respirator use, and maintenance of respirators used during IDLH or emergency response work.

TABLE 7-8 OSHA: Assigned Protection Factors[5]

Type of Respirator[1,2]	Quarter Mask	Half Mask	Full Facepiece	Helmet/ Hood	Loose-fitting Facepiece
1. Air-Purifying Respirator	5	[3]10	50	.....	.....
2. Powered Air-Purifying Respirator (PAPR)	.....	50	1,000	[4]25/1,000	25
3. Supplied-Air Respirator (SAR) or Airline Respirator					
• Demand mode	.....	10	50	.....	.....
• Continuous flow mode	.....	50	1,000	[4]25/1,000	25
• Pressure-demand or other positive-pressure mode	.....	50	1,000	.....	.....
4. Self-Contained Breathing Apparatus (SCBA)					
• Demand mode	.....	10	50	50	.....
• Pressure-demand or other positive-pressure mode (e.g., open/closed circuit)	.....	.....	10,000	10,000	.....

Notes:

[1]Employers may select respirators assigned for use in higher workplace concentrations of a hazardous substance for use at lower concentrations of that substance, or when required respirator use is independent of concentration.

[2]The assigned protection factors in Table 1 are only effective when the employer implements a continuing, effective respirator program as required by this section (29 CFR 1910.134), including training, fit testing, maintenance, and use requirements.

[3]This APF category includes filtering facepieces, and half masks with elastomeric facepieces.

[4]The employer must have evidence provided by the respirator manufacturer that testing of these respirators demonstrates performance at a level of protection of 1,000 or greater to receive an APF of 1,000. This level of performance can best be demonstrated by performing a WPF or SWPF study or equivalent testing. Absent such testing, all other PAPRs and SARs with helmets/hoods are to be treated as loose-fitting facepiece respirators, and receive an APF of 25.

[5]These APFs do not apply to respirators used solely for escape. For escape respirators used in association with specific substances covered by 29 CFR 1910 subpart Z, employers must refer to the appropriate substance-specific standards in that subpart. Escape respirators for other IDLH atmospheres are specified by 29 CFR 1910.134 (d)(2)(ii).

TABLE 7-9 NIOSH: Particulate Filter Approval (by Test)	
N-Series (not resistant to oil)	• Solid and nonoil-based particulates • Filters only extended past eight hours of use after work place evaluation
R-Series (resistant to oil)	• For use against any particulate contaminant • Filters only extended past eight hours of use after work-place evaluation
P-Series (oil proof)	• For use against any particulate contaminant • No service time limitations
Each series will have three levels of efficiency, 95%, 99%, and 99.97% (100%) efficiency denoted on the filter. All nine classes are appropriate for protection against tuberculosis.	

1995 (42 CFR 84). The three certifications for filtering respirators carry their own requirements for end of service life. "P" series filters (**TABLE 7-9**) have no service time limitations; however, users must be aware of rationale for end of service life for these respirators (e.g., breathing resistance or the detection of "breakthrough properties"). Previous approvals (30 CFR 11) had been issued based on hazards tested (e.g., dusts, fumes, mists, pesticides, paint sprays). The newest testing methods rate filters based on efficiency and the type of particle present. Three series of filters have been classified and are listed in Table 7-9. Additional NIOSH Filtering Selection Guides are found in **TABLE 7-10**.

Personal Protective Equipment Ensembles

For hazardous atmosphere work, a variety of equipment ensembles, such as the levels A through D series of PPE ensembles recommended by the U.S. EPA, NIOSH, OSHA, and the U.S. Coast Guard, can be selected. These ensembles combine the need for respiratory and personal protective equipment, as previously discussed. A full explanation of each level with its recommendation for use is found in **TABLES 7-11** and **7-12**.

Selection guidelines, including PPE, as published by CDC and the National Institute of Health for work with biohazardous materials in laboratories, are found in **TABLE 7-13**.

The U.S. Department of Defense has published Mission Oriented Protective Posture (MOPP) levels, ensembles of PPE designed to protect warfighters from chemical, biological, radiological or nuclear events. **TABLE 7-14** identifies MOPP levels 0–4 and the associated ensemble components.

Personal Protective Equipment Medical Requirements for Use of Ensembles

Prior to the use of any PPE, medical approval for the use of the devices should be obtained. This medical

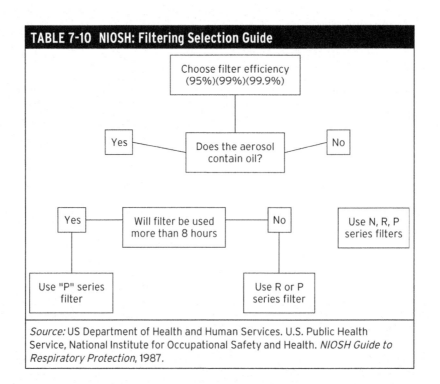

TABLE 7-10 NIOSH: Filtering Selection Guide

Source: US Department of Health and Human Services. U.S. Public Health Service, National Institute for Occupational Safety and Health. *NIOSH Guide to Respiratory Protection*, 1987.

TABLE 7-11 Levels of Protection: Chemical Protective Ensembles

	Respiratory Protection	Skin Protection	Criteria
A	Positive pressure SAR	Positive pressure full body covering	Hazardous materials present are IDLH by skin contact or are capable of harm by absorption through the skin
B	Positive pressure SAR	Yes	Hazardous materials present, less than 19.5% oxygen in event, IDLH situations present, or unknown levels of hazardous materials in air
C	Filter respirator	Yes	Hazardous materials present; 19.5% oxygen present
D	No	Yes	Hazardous materials present but below permissible exposure limits in air

Note: Table 7-13 contains selection guidelines for individuals working with etiologically hazardous substances.

TABLE 7-12 Levels of Protection: NIOSH Guidelines: Hazardous Substances

Levels of Protection

I. *Level A:* Level A protection should be used when:
   1. The hazardous substance has been identified and requires the highest level of protection for skin, eyes, and the respiratory system based on either the measured (or potential for) high concentrations of atmospheric vapors, gases, or particulates; or the site operations and work functions involve a high potential for splash, immersion, or exposure to unexpected vapors, gases, or particulates of materials that are harmful to skin or capable of being absorbed through the skin;
   2. Substances with a high degree of hazard to the skin are known or suspected to be present, and skin contact is possible; or
   3. Operations are being conducted in confined, poorly ventilated areas, and the absence of conditions requiring Level A have not yet been determined.

II. *Level B:* Level B protection should be used when:
   1. The type and atmospheric concentration of substances have been identified and require a high level of respiratory protection, but less skin protection;
   2. The atmosphere contains less than 19.5% oxygen; or
   3. The presence of incompletely identified vapors or gases is indicated by a direct-reading organic vapor detection instrument, but vapors and gases are not suspected of containing high levels of chemicals harmful to skin or capable of being absorbed through the skin.

NOTE: This involves atmospheres with IDLH concentrations of specific substances that present severe inhalation hazards and that do not represent a severe skin hazard; or that do not meet the criteria for use of air-purifying respirators.

III. *Level C:* Level C protection should be used when:
   1. The atmospheric contaminants, liquid splashes, or other direct contact will not adversely affect or be absorbed through any exposed skin;
   2. The types of air contaminants have been identified, concentrations measured, and an air-purifying respirator is available that can remove contaminants; and
   3. All criteria for the use of air-purifying respirators are met.

IV. *Level D:* Level D protection should be used when:
   1. The atmosphere contains no known hazard; and
   2. Work functions preclude splashes, immersion, or the potential for unexpected inhalation of or contact with hazardous levels of any chemicals.

Source: DDHS NIOSH/EPA/OSHA/USCG Health and Saftey Guidelines for Hazardous Waste Site Activities.

approval is required by OSHA standards before using items such as respirators and hearing protectors, as well as for individuals who work with specific toxic materials and with hazardous wastes. Triggers for these medical surveillance requirements are codified by OSHA, and they are generally related to time working above action or permissible exposure limits.

Medical surveillance requirements include preplacement physical examinations to ensure that employees are capable of using PPE safely and without risk to themselves and other employees, routine (such as annual) physical examinations to ensure that personal health has not changed and to review employee exposure characteristics, physical examination after occurrences

TABLE 7-13 Selection Guidelines for Personal Protective Equipment (Biohazards) from U.S. Department of Health and Human Services, Centers for Disease Control and Prevention, and National Institutes of Health

Level	Recommendations	Required PPE or Other Equipment
Universal Precaution	Use in all situations in which uncontrolled blood or body fluids are present. Equipment should be changed if soiled, or torn, or between patient handling.	Protective Equipment Recommended Under "Universal Precautions": Gloves: • Disposable; protects against blood and fluids. Masks (such as dust masks), Eye, and Mucus Membrane Protection: • Masks and eyewear are worn together if you anticipate splashes. Gowns: • Gowns or apron are worn to avoid soaking of clothes with potentially infectious fluids.
BL1–(P1)	Minimal hazard (e.g., culturable microorganisms)	• Designed for universal precautions and decontamination, such as sinks.
BL2–(P2)	Moderate risk, such as those associated with human disease	• Use of warning signs. • Prohibit entry to "at risk" individuals. • Protective body covering and gloves.
BL3–(P3)	Severe risk hazard, with serious or lethal consequences (e.g., mycobacterium tuberculosis)	• Work performed in biosafety cabinets • Respirators and PPE required • Specialized work area construction (filtered HVAC system and air flow)
BL4–(P4)	Dangerous and exotic agents that pose a high risk of life-threatening disease (e.g., lassa virus)	• Class III biosafety cabinets, or Class I and II biosafety cabinet work in conjunction with one piece positive pressure suits and life support systems

TABLE 7-14 Military Oriented Protective Posture (MOPP)

MOPP Level	Equipment	Use
0	PPE available for donning	During periods of heightened alert
1	Chemically protective over garment worn (overboots, mask, gloves carried)	During periods of heightened alert, when CBRN attack could occur with little/no warning or when CBRN contamination is present but higher levels of protection are not warranted
2	Chemically protective over garment, overboots, and field gear worn (protective mask and gloves carried)	Pre/post attack, when CBRN attack could occur with little or no warning or when CBRN contamination is present but higher levels of protection are not warranted
3	Chemically protective over garment, overboots, field gear, and protective mask are worn (gloves are carried)	Pre/post attack, when CBRN attack could occur with little or no warning or when CBRN contamination is present but higher levels of protection are not warranted
4	Chemically protective over garment, overboots, field gear, mask, and gloves	Post attack or when CBRN attack is in progress, this is the highest level of CBRN MOPP

Note: M8/Mp indicator paper, M291/M295 decontamination kits and nerve agent antidotes are carried at MOPP 1-4.
Source: United States Air Force. *Airman's Manual AFPAM 10-100.* 1 March 2009; revised 24 June 2011.

that produce symptomatic exposures to ensure that appropriate medical procedures have been provided and to ensure that employees remain physically capable of using protective equipment, and physical examinations at the completion of any hazardous activity to ensure the PPE has worked appropriately.

Of note, several federal standards allow medical surveillance provided to employees 6 to 12 months prior to termination dates as clearance physical examinations (i.e., at the completion of hazardous work activities). Examination requirements for work in hazardous areas generally require a physical examination, a medical history check, and certain medical procedures for both diagnostic and baseline purposes.

The most current U.S. Department of Labor Respirator Standard (29 CFR 1910.134) allows the use of a medical questionnaire, when reviewed by a licensed healthcare professional, in lieu of an actual physical examination for medical approval to use respirators in some situations. Further medical monitoring information is found in Chapter 11.

Physical examinations for employees working with toxic materials may include a review of both physical capabilities (e.g., lung capacity as measured by spirometery tests or cardiovascular system rates) and psychological capabilities of using equipment (e.g., the ability to work inside a closed-face space, such as a respirator or protective suit—contraindicated in employees suffering from claustrophobia).

Chapter Summary

PPE may be chosen as a protective mechanism for employees subsequent to the use of engineering and administrative controls to protect employees. PPE may consist of single devices, such as hard hats or eye protection, or ensembles that are chosen and designed to protect against thermal, radiological, asphyxiant, chemical, etiological, or mechanical hazards. Levels A through D typically identify ensembles of chemical protective equipment. Biosafety Levels 1 through 4 include PPE requirements and typically identify ensembles of biologically protective equipment.

Although certain types of PPE are selected with relative ease, the selection of respiratory protection equipment or hearing protection equipment generally requires pretesting of the environment to be entered in order to determine toxin level, medical preapproval before using devices, and written safety protocols for equipment use.

Terms

ANSI: The American National Standards Institute—a consensus national standards making organization.

Air Purifying Respirator (APRs): A respirator that functions either under negative or positive pressure by purifying the atmosphere in which the user is present.

ASTM (ASTM International): The American Society for Testing and Materials: a consensus national standards making organization.

Noise Reduction Rating (NRR): The reduction in sound pressure levels (noise attenuation) provided by a hearing protector, tested and labeled in compliance with U.S. EPA requirements.

Protection Factor (PF): The rating assigned to a respirator that identifies the ability of the respirator to prevent leakage into the face piece. The PF is determined by using the following equation:

$$PF = \frac{\text{Concentration of contaminants outside face piece}}{\text{Concentration of contaminants inside face piece}}$$

Supplied Air Respirator (SAR): A respirator that functions by supplying an atmosphere to the user (generally Grade "D" breathing air) either under a positive or negative pressure. SARs may be a self-contained breathing apparatus (**SCBA**) or may be attached to a compressor or remote air supply.

References

American National Standards Institute/International Safety Equipment Association. *American National Standard for Occupational and Educational Personal Eye and Face Protection Devices*. Z87.1, ISEA. Arlington, VA, 2010.

———. *American National Standard Z89.1: American National Standard for Industrial Head Protection*. ISEA. Arlington, VA, 2009.

———. *American National Standard for Hand Protection Selection Criteria*. 105-2011. ISEA. Arlington, VA, 2011.

———. *American Society for Testing and Materials: ASTM F2412-05: Standard Test Methods for Foot Protection*. Conshohocken, PA, 2005.

———. *ASTM F2413-05: Standard Specifications for Performance Requirements for Foot Protection*. Conshohocken, PA, 2005.

Compressed Gas Association. *Air Specification G-7.1*. Washington, D.C.: Government Printing Office, 1992.

National Fire Protection Association. Code #1971: Standard on Protective Ensembles for Structural Fire Fighting and Proximity Fire Fighting. In *National Fire Codes*. Quincy, MA: NFPA, 2013.

———. Code #18511: Standard or Selection, Care and Maintenance of Protective Ensembles for Structural Fire Fighting and Proximity Fire Fighting. In *National Fire Codes*. Quincy, MA: NFPA, 2008.

National Institute for Occupational Safety and Health, U.S. Department of Health, Education and Welfare, U.S. Public Health Service. *Criteria for Firefighters Gloves*. Washington, D.C.: Government Printing Office, 1976.

U.S. Department of Health and Human Services, U.S. Public Health Service. *NIOSH/EPA/OSHA/USCG Health and Safety Guidelines for Hazardous Waste Site Activities*. Washington, D.C.: Government Printing Office, 1989.

U.S. Department of Health and Human Services, U.S. Public Health Service, Centers for Disease Control and Prevention, National Institute for Occupational Safety and Health. *Recommended Criteria for a Hearing Conservation Program*. Morgantown, WV: DDHS, 1998.

———. *NIOSH Guide to Respiratory Protection*. Washington, D.C.: Government Printing Office, 1987.

U.S. Department of Labor, Occupational Safety and Health Administration. 29 CFR 1910.95 Noise. In *General Industry Safety Standards*. Washington, D.C.: Government Printing Office, Current.

———. 29 CFR 1910.130-137: Sub Part I: Personal Protective Equipment. In *General Industry Safety Standards*. Washington, D.C.: Government Printing Office, Current.

———. 29 CFR 1910.134: Respiratory Protection. In *General Industry Safety Standards*. Washington, D.C.: Government Printing Office, Current.

———. 29 CFR 1910.1001/29 CFR 1926.1101: Sub Part Z: Toxic Materials: Asbestos. In *General and Construction Industry Safety Standards*. Washington, D.C.: Government Printing Office, Current.

———. 29 CFR 1910.1028/29 CFR 1926.62: Lead. In *General and Construction Industry Safety Standards*. Washington, D.C.: Government Printing Office, Current.

U.S. Air Force, Secretary of the Air Force. *Air Force Pamphlet 10-100* (revised 24 June 2011); *Airman's Manual AFPAM-10-100* (revised 24 June 2011).

U.S. Fire Administration. *Model Performance Criteria for Structural Firefighters Helmets*. Emmitsburg, MD: 1977.

Exercise 7

Complete the exercise by describing the level and criteria for each style of chemical protective equipment and the skin/body and respiratory protection equipment required.

1. **Level** _____.

   Criteria:
   _____
   _____
   _____
   _____
   _____

   Skin/Body Protection:
   _____
   _____
   _____
   _____

   Respiratory Protection:
   _____
   _____
   _____
   _____
   _____

2. **Level** _____.

   Criteria:
   _____
   _____
   _____
   _____
   _____

   Skin/Body Protection:
   _____
   _____
   _____
   _____

   Respiratory Protection:
   _____
   _____
   _____
   _____
   _____

3. **Level** _____ .

Criteria:
_____
_____
_____
_____
_____

Skin/Body Protection:
_____
_____
_____
_____
_____

Respiratory Protection:
_____
_____
_____
_____
_____
_____
_____

4. **Level** _____ .

Criteria:
_____
_____
_____
_____
_____

Skin/Body Protection:
_____
_____
_____
_____
_____

Respiratory Protection:
_____
_____
_____
_____
_____
_____

5. Using Table 7-2 and the head protection criteria discussed in this chapter (ANSI-Z89.1, 2009), identify the protection provided by a hardhat labeled Class G; Type II, HV

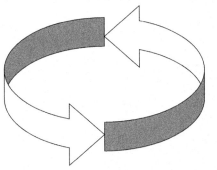 :

_____
_____

6. Using Table 7-3, identify the appropriate eye protection for a worker mixing liquid chemicals:

_____

7. Referencing Table 7-4, identify the protection provided by "safety shoes" labeled under the ASTM F2412-05/F2413-05 and ANSI Z.41.1 standards:

> ASTM F2412-05/13-05
> M, 1-75, C-75, Mt
> PR; CS

_____
_____
_____
_____
_____

8. Hand protection (gloves), optimally certified under ANSI/ISEA 105-2011 for cut resistance and puncture resistance should have a rating of _____.

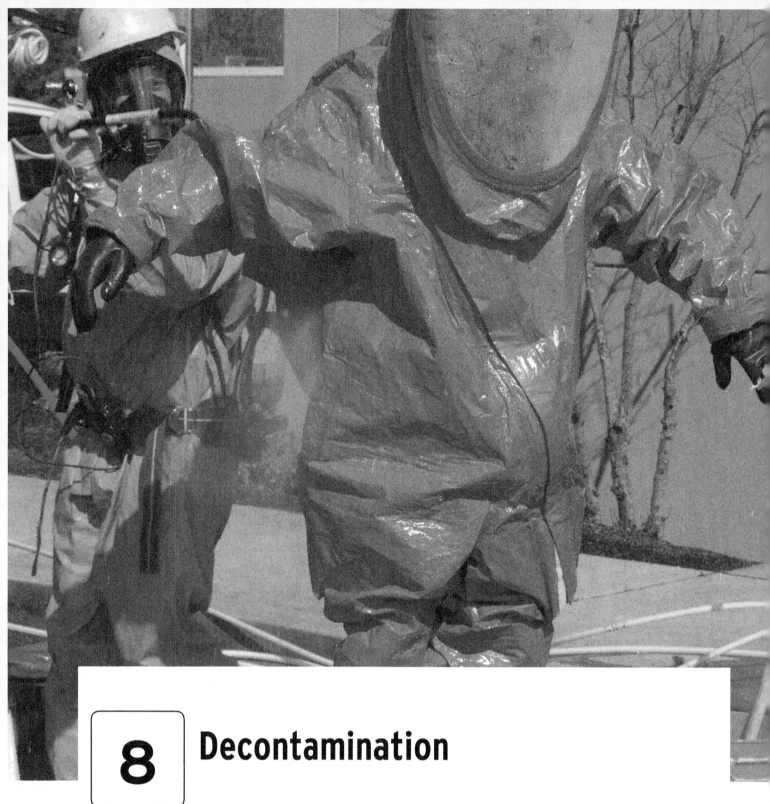

8 Decontamination

Key Concepts

- Mass Decontamination
- Emergency Decontamination
- Technical Decontamination

Introduction

Decontamination (decon) is the process of removing or neutralizing contaminants or other unwanted items that have accumulated on personnel and/or equipment. The purpose of orderly decontamination is to ensure that materials or personnel can be returned to use safely, to ensure that contaminants have not damaged materials or harmed personnel, and to ensure contaminants are appropriately packaged and disposed of without cross-contaminating personnel or "clean" items.

Decontamination Plans

Any decontamination procedure initiates with a written plan, developed prior to entry into the hazardous atmosphere or site. As with all written Health and Safety Plans, this plan must be understandable to users. Often, information concerning the written plan is posted at key locations to alert and remind personnel of decontamination procedures. All site contaminants must be identified and quantified to ensure that appropriate decontamination can be accomplished. Decontamination plans and procedures for both fixed sites and emergency responses should be included in Standard Operating Procedures, written Safety and Health Plans and programs, site-specific Safety and Health Plans and programs, and emergency response plans.

The establishment of written, posted standard operating procedures, many times abbreviated for site postings, is essential for any decontamination plan.

Theoretically, decontamination procedures should remove articles or items that are most contaminated first and work through to least-contaminated items or items that contain an acceptable level of contaminants. Decontamination plans should include the following information:

- The number and location of decontamination stations or procedures (order is critical here!)
- Any equipment needed to perform the decontamination process
- A written methodology for performing decontamination
- Written instructions that preclude or prevent **cross-contamination** (i.e., the movement of contaminants from a primary to a secondary area or the spreading of contaminants through inappropriate decontamination procedures)

Decontamination procedures should minimize personal contact with contaminants and be performed using engineering or administrative controls when possible. Decontamination plans must include procedures for the disposal of contaminants as well as materials, equipment, and protective devices that cannot be decontaminated, including those used in the decontamination process.

The decontamination plan must be a workable document and should be revised as equipment changes occur, levels of contaminants are decreased or increased, or site conditions are reassessed and new information acquired. In addition, changes in levels of personal protective equipment and styles of respirators in use at contaminated locations may require amendments to the decontamination plan.

PHOTO 8-2 Emergency decon equipment generally includes quick rinse or drench equipment and tools for lifesaving stabilization of patients.

PHOTO 8-1 Decon plans are often posted to illustrate procedures.

Decontamination Design

Location

Decontamination areas should be identified to minimize the spread of contamination. They should be located in safe areas adjacent to hazardous zones (Danger Zones) in concert with the chemical, physical, and toxicologic properties of the contaminants at the location. Whether the contaminant is heavier or lighter than air, heavier or lighter than water, its volatility, and its physical hazards (e.g., is it slippery?) should also be considered in the design phase of decontamination procedures. The procedures should also consider the amount of material to be decontaminated and the amount of the contaminants. The exposure potential to individuals performing decontamination and other individuals within the hazardous area, as well as the compatibility of the material with other wastes to be generated, stored, or disposed of, must be assessed for any decontamination process.

Movement into areas where decontamination is occurring must be controlled as with any hazardous area (i.e., the decontamination zone itself is a hazardous area that may pose equal or greater hazards than the hazardous environments). This stops the spread of contaminants from the controlled location, limits access to hazardous areas, and allows for control of the decontamination process. Potential emergencies within the decontamination zone and emergencies requiring **emergency decontamination** (within "Danger or Hazardous Zones") must be considered in the process of decontamination planning.

The decontamination area is typically located adjacent to any designated "hot zone."

Prevention of Contamination

FIGURES 8-1 and **8-2** identify both a basic and more elaborate decontamination protocol. **FIGURE 8-3** illustrates a three-stage decontamination area required for Class 1 asbestos abatement activities. **FIGURE 8-4** illustrates a typical decontamination layout for lead-based paint removal activities.

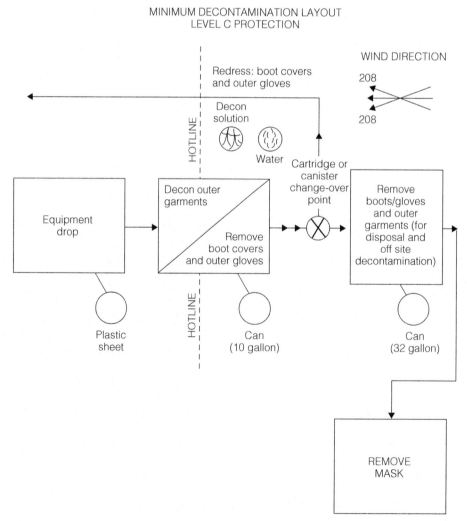

FIGURE 8-1 USCG: FSOP: decontamination (minimum): Level C.
Adapted from: U.S. Coast Guard Standard Field Operating Procedures.

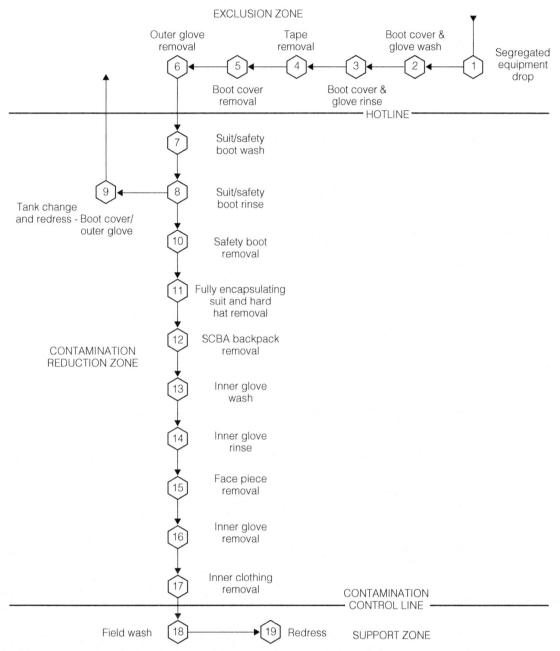

FIGURE 8-2 USCG: FSOP: decontamination (maximum): Level A.
Adapted from: U.S. Coast Guard Standard Field Operating Procedures.

Minimizing Contact with Contaminants

Minimizing contact with contaminants aids in safe decontamination and may occur in several ways. Work practices using engineering and administrative controls, such as remote equipment handling, may be used. Equipment itself may be protected from contaminants by bagging, shielding, or creating inert atmospheres. Where possible, disposable decontamination equipment is beneficial.

In some cases, equipment decontamination is not possible, and encasement or enclosure of the materials and subsequent disposal after appropriate testing is recommended. In other cases, such as with radioactive materials, allowing materials to remain in a stable environment (and go through a decay process) may bring contaminated items into a "safe" state.

Worker Protection During Decontamination Activities

To maximize worker protection during the decontamination phase of any operation, workers should dress in the appropriate levels of protective equipment before

CHAPTER 8: Decontamination

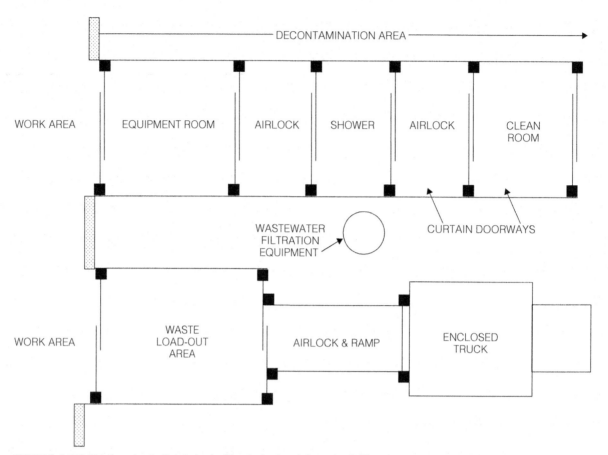

FIGURE 8-3 U.S. EPA: decontamination design for Class I asbestos abatement activities.
Adapted from: U.S. EPA, Model Curriculum for Asbestos Abatement Contractors and Supervisors.

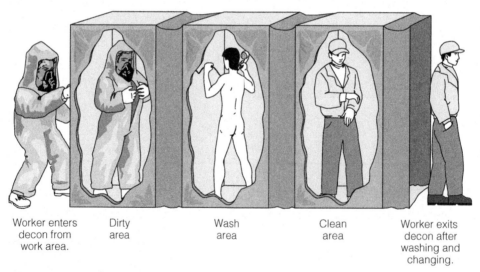

FIGURE 8-4 U.S. EPA/U.S. HUD decontamination design for lead-based paint removal.
Adapted from: U.S. EPA/U.S. Hud. "Lead Based Paint in Public and Indian Housing." Washington D.C., 1995.

entering decontamination areas or locations. All skin surfaces should be protected if the substance within the hazardous environment may affect or be absorbed through the skin (i.e., contact with the skin may greatly add to body load of the toxin). Other health and safety hazards to be addressed include the compatibility of decontamination solutions or materials with other materials in the decontamination process. Also, direct hazards to employees from decontamination procedures, practices, and substances should be addressed. This is of particular concern if ionizing radiation or chemical sterilants are used for decontamination purposes.

The plan should select personal protective equipment (PPE) for employees performing decontamination in conjunction with the anticipated hazards. PPE for decontamination workers may be equal to that required for employees in hazardous areas. In all cases, decontamination workers must decontaminate or undergo an approved decontamination procedure before leaving the decontamination site. The process of dismantling decontamination sites becomes a remedial activity in itself; however, as with site remediation, appropriate planning allows dismantling of decontamination equipment and areas with minimal personal hazards.

Mass (Gross) Decontamination

Mass (gross) decontamination is a procedure used to remove large quantities of hazardous substances from large populations in as little time as possible. Frequently utilized as a lifesaving measure, mass decontamination removes gross contamination quickly, thus reducing patient/victim exposures. Typically, two steps occur: (1) Removal of clothing (where contaminants may lodge) and (2) Rapid showering with large quantities of water. In fixed locations (such as large public arenas or critical government sites), pre-piped decontamination locations are in place.

In other locations, mass decontamination may be affected by first responders. A typical mass decontamination layout is pictured in **FIGURE 8-5**.

Subsequent to mass decontamination, medical procedures may be affected or continued.

Emergency Decontamination

Emergency decontamination must be established before entry into hazardous areas to address the need for decontamination prior to lifesaving activities. Before performing any complete personal decontamination, provide lifesaving stabilization aid to persons requiring emergency assistance. In general, the requirements of the *Emergency Response Guidebook (ERG-2012)* can be used for emergency decontamination. In most cases, copious amounts of water are acceptable for emergency decontamination. It is essential to protect emergency workers during the emergency decontamination phase of any operation.

Decontamination Equipment

Decontamination equipment must be appropriate for the procedures to be performed. Items such as poly drop cloth, collection containers and retaining area building materials, wash and rinse containers, brushes, paper towels and cloths, or facilities for personal hygiene are appropriate. Emergency first aid supplies should be accessible, including eyewash stations or drench showers.

For heavy equipment decontamination, pressurized sprays, curtain enclosures, and applicable pumps may be necessary. As with minor equipment, appropriate waste disposal may prove more economical than equipment decontamination and testing. Decontamination materials and procedures may also include dissolving materials, surfactants, solidification processes, rinsing of materials, or disinfection and sterilization as appropriate.

PHOTO 8-3 Chemical decontamination removes contaminants in a progressive manner.

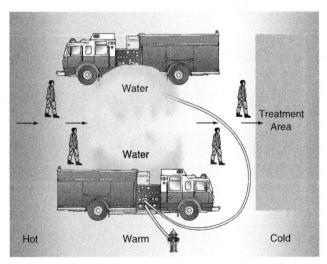

FIGURE 8-5 An example of a simple mass decontamination corridor using two fire engines.

Testing Decontaminated Materials and Equipment

Cleanliness and sanitation levels for equipment and personnel to be decontaminated, cleaned, or washed should be established before the decontamination process initiates. This may include a visual observation in natural or artificial light to confirm that contaminants have been removed. In some cases, ultraviolet light examinations may be appropriate. For certain styles of equipment, wipe sampling may be performed to document removal of hazardous substances. Again, it should be ensured that appropriate levels identifying cleanliness are established before samples are taken.

Testing materials for permeation (the introduction of the toxin into the materials) and penetration (the allowance of the toxin through the material) may also occur. In some cases, testing the decontamination solution, such as the waters used to decontaminate personnel or equipment, may be appropriate to determine whether decontamination is being performed effectively.

Chapter Summary

Planning and preparation for both routine and emergency decontamination must occur before entering any hazardous environment. It is not acceptable to prepare decontamination protocol after entry. The decontamination procedure identifies a process for people or equipment to move through a series of steps (most to least contamination) in an organized and safe manner. When performed effectively the physical and chemical properties of the contaminants are considered in the decontamination process. Finally, decontamination areas are dismantled in an organized manner, leaving a clean environment.

Terms

Cross-contamination: The movement of contaminants from a primary to a secondary area or the spreading of contaminants through inappropriate decontamination procedures.

Decontamination: The organized removal of contaminants from personnel and/or equipment, according to a plan, to achieve predetermined levels of contaminants (such as levels, which may pose no harm).

Emergency Decontamination: The rapid removal of contaminants from a victim requiring lifesaving stabilization or procedures. Emergency decontamination permits lifesaving procedures to be performed without harm to resources and may be followed by patient decontamination once rescue or stabilization has occurred.

Mass (Gross) Decontamination: The rapid removal of contaminants from large populations with equipment or procedures designed to handle multiple victims simultaneously.

References

Cocciardi, J.A. *Asbestos Abatement Manual for Supervisors and Contractors*. Mechanicsburg, PA: Cocciardi and Associates, 2011.

———. *Emergency Planning: Medical Services-1: A Manual for Medical Personnel Addressing Contaminated or Irradiated and Otherwise Physically Injured Persons and Equipment Following a Nuclear Power Plant Accident*. Prepared for Commonwealth of Pennsylvania: Emergency Management Agency. New Cumberland, PA: Cocciardi and Associates, 2004.

USCG, National Institute for Occupational Safety and Health, Environmental Protection Agency, Occupational Safety and Health Administration. *Guidance Manual for Hazardous Waste Site Activities*. Washington, D.C.: Department of Health and Human Services, 1989.

U.S. Environmental Protection Agency. *EPA 560: Controlling Asbestos in Buildings*. Washington, D.C.: Government Printing Office.

U.S. Environmental Protection Agency, U.S. Housing and Urban Development. *Lead Based Paint in Public and Indian Housing*. Washington, D.C.: Government Printing Office, 1995.

Exercise 8

Identify the types of decontamination and appropriate uses for the pictured decontamination processes below. Describe the decontamination plan/process and equipment necessary for each. Identify any concerns.

1. Asbestos Decontamination

2. Hazmat Decontamination

3. Emergency Decontamination

4. Mass Casualty Decontamination

5. Mass Decontamination

9 Responding to Emergencies

Key Concepts

- Emergency Action Plan (EAP)
- Emergency Coordinator
- Emergency Response Plan (ERP)
- Incident Commander/Incident Command System
- Reportable Quantity (RQ)
- Terrorism

Introduction

Activities that may produce hazardous environments require emergency preplanning. Areas or facilities where hazardous environments may occur are required to prepare for likely emergencies within their boundaries. Although some emergencies may be handled by removing persons from the problem (evacuations, such as in the case of a fire) and notifying outside emergency services, others require immediate on-site actions (e.g., medical emergencies). Barriers to hazardous environment entry may exist (such as those keeping the public away from these areas), which further lengthen the timeframe for an offsite emergency response and identify a need for on-site emergency response groups.

Emergency Planning

As a minimum, operators within hazardous environments or facilities where hazardous environments occur need to develop and administer a written **emergency action plan (EAP)** that identifies the basic actions individuals should take in the event of unusual or emergency incidents and that identifies the individuals responsible for implementing the plan(s). In general, the emergency action is the timely removal of occupants from the hazardous area and actions that may aid this timely removal. Providing additional information or training for all individuals with responsibilities in these EAPs is also necessary. In addition to multiple OSHA standards, many state and local building and/or occupancy codes require EAP development and posting of emergency contact numbers.

PHOTO 9-1 Basic emergency action plans require facility evacuations. Required elements include notification of individuals, providing immediate assistance to those who need help with the evacuation process, and providing information to responding emergency services.

TABLE 9-1 The Four-Step Emergency Planning Process

Step 1	Identify potential emergencies and unwanted events, in prioritized order.
Step 2	Identify expected employee/public actions.
Step 3	Identify equipment necessary to implement actions.
Step 4	Provide training necessary to implement actions and exercise the actions routinely.

Conversely, **emergency response plans (ERP)** identify actions individuals may take in response to an emergency for the purpose of mitigating the hazards or effects of the emergency. They generally require that responders remain within or enter hazardous areas.

The Four-Step Emergency Planning Process

Any emergency action or response planning and preparedness process initiates with a vulnerability study or risk assessment that identifies expected or likely emergencies (preferably in a prioritized manner) and the expected actions once these unwanted events occur. Subsequently, information necessary to complete the planning process includes the expected actions employees will take, and the equipment and training necessary for them to implement the expected actions in response to these emergent events. This planning process is presented in **TABLE 9-1**, the Four-Step Emergency Planning Process. An example of this process is found in **TABLE 9-2**.

The primary responsibility for developing emergency actions lies with the owner, employer, or the political jurisdiction (city, county, state) within which the hazardous atmosphere or site resides. Many jurisdictions have emergency planning procedural requirements for facilities. Examples include the County of Los Angeles written Business Plan requirements (written identification of on-site hazards and emergency action, required of all businesses within the county), or the OSHA EAP Planning requirements for hazardous activities, such as those in facilities or sites covered by the Hazardous Waste Operations and Emergency Response or Process Safety Management standards.

For facilities with certain hazardous substances on-site, emergency planning and community right-to-know notifications are required by the Superfund Amendments and Reauthorization Act (SARA) of 1986.

Response Options

In general, three options are available to employers and planners in the emergency preparedness and planning process to protect employees and property:

| TABLE 9-2 Sample Four-Step Emergency Planning Document XYZ Site ||||
Unwanted Event	Expected Action	Equipment	Training
Medical Emergency	1. Provide first aid 2. Notify EMS 3. Contain infectious wastes	1. First aid kid with CPR valves 2. Blood-borne pathogen kit 3. Posted EMS number on work phones	1. Basic first aid 2. Adult CPR 3. Annual BBP training
Fire	1. Notify occupants 2. Notify fire department 3. Evacuate the building	1. Alarm system and phone with emergency numbers posted	1. Annual fire drill and fire warden training
Severe Weather Event	1. Shelter occupants in place 2. Maintain capabilities to call for information and emergency assistance	1. NOAA radio 2. Flashlights 3. Cell phones and cell phone lists	1. Annual severe weather drill

1. Evacuation during times of emergencies and contact with outside resources (private or public) to respond in order to control the emergency. This is an EAP.
2. Evacuation during times of emergencies and response to and handling of the emergency with resources totally within the confines of a facility or area. This is an ERP.
3. A combination of the two, such as a mutual aid agreement.

Each option has benefits. The first option minimizes site training and equipment responsibilities. This option, however, may not provide sufficient or timely emergency response.

The second option may be necessary where facilities are remote from public or private emergency response services. The second option generally allows for quick response and anticipates that the effects of many emergencies can be minimized due to this quick response

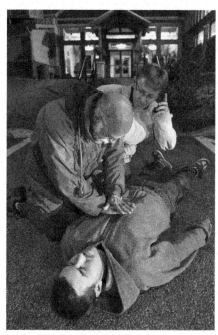

PHOTO 9-3 Medical responses typically use internal (first aid teams) and external (paramedic or hospital) responses.

effort. Training, equipment, and procedural requirements are increased for this option. Typically, facilities remote from public or private response agencies (not necessary geographically, but remote due to barriers or restricted access, such as correctional facilities), or those intending to handle problems with a quick response effort will choose this option.

Finally, the third option, a combination of internal and external response resources, may be chosen. This option allows for facilities to handle small incidents before they become major and to continue with outside response efforts.

Once facilities have identified expected employee actions, these must be placed in written form.

PHOTO 9-2 Some industrial plants provide emergency response teams capable of internal emergency management.

Emergency Action and Emergency Response Plans

Key Elements of Emergency Action and Response Plans

While EAPs focus attention on immediate excavation and the actions performed by employees present when an emergent event occurs, ERPs add the response of personnel and equipment from outside the immediate area where the emergency has occurred. Individuals present when an incident occurs are acutely aware of specific hazards and events, while individuals responding from outside the area must make a safe response and site evaluation. Key EAP requirements are found in **TABLE 9-3**. Components of ERPs are described in subsequent response options to specific events in latter sections of this chapter.

Warning: Alarming and Alerting

Each plan must identify a primary and secondary means of warning employees or the public. The alarm mechanism, commonly referred to as the fire alarm system or emergency evacuation signal, must be different for different types of emergencies (chemical alarm, fire alarm, evacuation alarm). Each alarm must be distinct from other communications mechanisms. The alarm system must be tested routinely (e.g., monthly), and fire alarm systems installed to meet code requirements require additional testing, both upon installation and at routine intervals (e.g., annually). Installation of these systems is required according to building codes, national fire codes, and insurance requirements.

The primary purpose of the evacuation alarming mechanism is to alert employees and the public to the fact that an emergency exists that requires them to take appropriate actions (e.g., evacuate).

Second, alerting mechanisms should be in place. Alerting may be tied to the alarm system, but the purpose of the alerting systems is to notify emergency responders (either on-site or off-site) that an emergency exists and that they should take appropriate actions. Alerting protocols are included in Emergency Response Plans. Typically, on-site emergency responders are notified through communication systems, whereas off-site emergency responders are notified through direct links to communication centers such as 9-1-1 information centers.

The option for alerting emergency responders, in most cases, lies with the employer and should be determined in the facility analysis previously discussed.

The Facility Emergency Coordinator

All facilities should identify an emergency coordinator. An alternate emergency coordinator should also be identified, so that this position is continually available. This individual must be knowledgeable of facility emergency procedures and, as a minimum, act as a liaison during emergencies or potential emergencies, both within the site and for emergencies that may affect off-site areas. The emergency coordinator should also be responsible for notification of agencies of public or environmental safety in the event a **Reportable Quantity (RQ)** of a hazardous substance is released. Each employer should use segments of the **incident command system (ICS)** for handling emergency operations. The ICS, developed by the federal government in response to the need to coordinate large emergencies, is a generic term that identifies a business approach to emergency operations. Simply stated, the ICS identifies one individual or organization "in charge" of emergency operations once an emergency is declared. This individual may initially be a site or facility employee (in particular for emergencies handled at their incipient stages). Typically, incident command is transferred across jurisdictional bounds as additional resources are notified. However, individuals with legal responsibility for hazardous environments or materials will maintain some command and control responsibilities, such as financial responsibilities (payment for response costs).

The ICS was developed as a result of fires that consumed large portions of wildlands, including structures, in Southern California in the 1970s. A need was identified to develop a system that allowed responders from intrajurisdictional agencies to work together toward a common goal in an efficient manner.

The National Incident Management System (NIMS)

On February 29, 2003, Homeland Security Presidential Directive #5 (HSPD-5) was issued, which called for the Secretary of the newly formed Department of Homeland

---

TABLE 9-3 Emergency Action Plan Requirements

1. Procedures for reporting a fire or other emergency.
2. Procedures for evacuation, including exit route assignments and safe areas of refuge.
3. Procedures for employees who will operate or shut down critical plant systems prior to evacuation.
4. Accountability procedures for staff, residents, patients, and visitors.
5. Procedures to be followed by employees performing rescue or medical duties.
6. Names of individual(s) to contact to acquire additional plan information.

Source: U.S. Department of Labor. *OSHA General Industry Safety Regulations: 29 CFR 1910.38: Emergency Action Plans.* OSHA, December 2002.

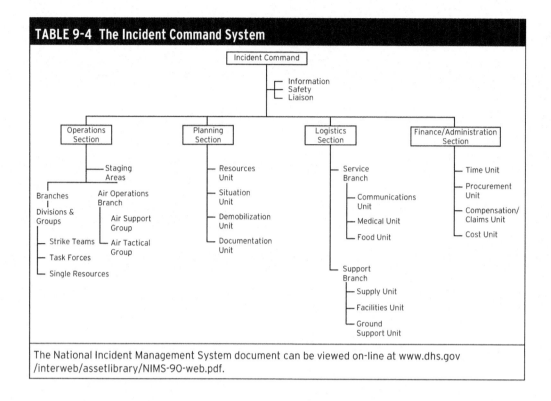

TABLE 9-4 The Incident Command System

The National Incident Management System document can be viewed on-line at www.dhs.gov/interweb/assetlibrary/NIMS-90-web.pdf.

Security to develop a system for a consistent, nationwide approach for entities to work together to prepare for, prevent, respond to, and recover from domestic terrorist incidents. **NIMS** was subsequently issued on March 1, 2004, consolidating in-place approaches into a comprehensive national approach to domestic incident management. Flexible and comprehensive, NIMS provides a framework for interoperability and compatibility among resources with response functions. The need for NIMS was additionally defined in the recommendations of the National Commission on Terrorist Attacks upon the United States (the 9/11 Commission). NIMS is now utilized to command and control all hazard events.

Although large scale Incident Management/Command Systems had been in-place since the 1970s addressing the needs of multi-agency response efforts, the system morphed into a national fire service model by the 1990s, and subsequently the NIMS.

Basic to NIMS is an Incident Command System under the control of a single individual, with responsibilities for the core command functions of safety, public information and liaison with coordinating and assisting agencies, and the staff functions of Operations, Logistics, Planning, and Administration/Finance.

Since its inception, advanced ICS coursework has been promulgated by the DHS, and NIMS compliance is required for all entities receiving federal funding and/or support. The key structural elements of NIMS ICS, designed to be utilized for all scales of emergency response, are identified in **TABLE 9-4**.

Terrorism

In the 21st century, the potential for **terrorism** and terrorist acts should be considered by all individuals who may enter hazardous environments and respond to actual or potential emergencies. Although terrorist acts may cause emergency actions (many times a hazardous substance release incident), terrorist acts are also crime scenes. Hazardous environments and responders to these events may be the target of terrorist acts. Therefore, individuals who operate around these environments should be trained to identify the potential for hazardous substance releases to be terrorist acts. Once identified, these areas must be secured for investigative (i.e., police) review. In addition, the potential for secondary devices or continuing releases of hazardous substances should be anticipated in the case of a suspected terrorist act. Terrorist acts may be masked as fires, hazardous material incidents, or medical emergencies.

As in any emergency action, personal safety should be considered first. The potential hazards of a terrorist act add an additional dimension to employee safety concerns due to the severity of events and the potential for rapid escalation and multi-agency responsibilities (i.e., the need for NIMS/ICS). Worldwide, nearly 70%

PHOTO 9-4 The incident command system allows integration of all emergency resources while identifying individuals in command and control.

Medical Emergencies

According to the U.S. Department of Labor, more than 5,000 people die annually from workplace incidents, and more than 3.7 million suffer on-the-job injuries or illnesses. Seven workers are injured every minute. The U.S. Department of Labor, as well as many National Consensus standards, requires all employers and public sites to prepare for medical emergencies. These requirements identify the need for first aid supplies, first aid training, emergency communications, and transport arrangements to hospitals or other medical facilities. These items are essential for preplanning for medical emergencies (all facilities are required to have minimal lifesaving preplans).

Merely notifying the emergency medical services (EMS) system is usually insufficient for individuals operating in hazardous atmospheres or environments, because remote locations, barriers, control systems for

PHOTO 9-5 Once an incident has been identified as a potential act of terrorism, all responders must be aware of the potential for secondary hazardous devices. Often a terrorist incident is an intentional hazardous substance release, and therefore a crime scene.

PHOTO 9-6 Hazardous environments, including large public gatherings, generally require onsite EMS and transport capabilities. (Note: Several state EMS agencies require this level at a gathering of 5,000 or more people at one time [PAOT]).

of all terrorist incidents involve explosives, and most explosive incidents (80%) involve bombs. Bomb threats may be hoaxes but are nonetheless criminal events, and facilities and emergency coordinators should plan for these events. Evacuation appears to be an appropriate response to any bomb threat situation, and subsequently law enforcement officials should be notified. Facility search procedures should be performed only by trained personnel.

CHAPTER 9: Responding to Emergencies

TABLE 9-5 ANSI/ISEA Z308.1-2009 First Aid Kit Label

ANSI/ISEA Z308.1-2009 Type I, II, III or IV

Caution! This kit meets ANSI/ISEA Z308.1-2009 only when required minimum fill is maintained with first aid products marked "ANSI/ISEA Z308.1-2009."

Required Minimum Fill	Recommended Supplies
1 First Aid Guide	Analgesic (Oral)
1 Absorbent Compress 4 × 8 in. min.	Bandage Compress 2 × 36 in. min.
16 Adhesive Bandages 1 × 3 in.	Breathing Barrier, single use
1 Adhesive Tape 2.5 yd.	Burn Dressing 12 sq. in. min.
10 Antiseptic Treatment Applications 0.5 gm. each	Cold Pack 4 × 5 in. min.
6 Burn Treatment Applications 0.9 gm. each	Eye Covering 1/4 in. thick min.
4 Sterile Pads 3 × 3 in. min.	Eye/Face Wash, sterile 4 fl. oz. min
2 Pair Medical Exam Gloves	Roller Bandage 2 in. × 4 yd. min.
1 Triangular Bandage 40 × 40 × 56 in. min.	Hand Sanitizer, 0.9 gm. min
6 Antibiotic Treatment Applications 0.5 gm. each	

The described kit may be suitable for some businesses. However, the adequacy of the contents for hazards of each work environment should always be evaluated by competent personnel. For a variety of operations, employers may find that additional first aid supplies and kits are needed.

hazardous atmospheres, and security procedures may preclude quick emergency medical services assistance.

The provision of EMS at large gatherings, assemblies, and events is required by some jurisdictions, again due to the anticipated delay in calling for help and responding efficiently at large gatherings. In many cases, such as large assemblies, the maximum people at one time (PAOT) is the driving factor identifying the need for on-site EMS or other forms of emergency action assistance (e.g., trained evacuation personnel). Typically, special events, such as sporting events, may require these services on site (in the case of a sporting event, the hazardous atmosphere is the large assembled crowd).

First Aid and CPR Information and Training

First aid, cardiopulmonary resuscitation (CPR), and management of obstructed airways are situations typically handled by employees prior to the arrival of EMS systems. Management of severe bleeding incidents falls into this category as well. OSHA requirements specify a 3- to 4-minute response time for life-threatening emergencies, and a 15-minute response time for non-emergent situations. While OSHA allows employers to select site first aid supplies, these supplies must be adequate, should be responsive to the type of injuries that may occur, and must be securely stored in an area where they are readily available for emergency access. An AED should be considered when selecting first aid supplies and equipment and is discussed below. The American National Standards Institute (ANSI) lists four types of first aid kits:

- Type I (Indoor Use); mountable (e.g., cabinet);
- Type II (Indoor Use); portable (e.g., soft pack);
- Type III (Indoor/Outdoor Use); portable, mountable and water resistant (e.g., plastic kit);
- Type IV (Indoor/Outdoor Use); portable, mountable and waterproof (e.g., metal kit).

Contents of kits (minimum required supplies) are listed in **TABLE 9-5**.

Some national providers certify individuals in disciplines such as first aid and CPR. However, the responsible party is free to choose from these or internally developed first aid training segments. This training, in addition to emergency rescue and decontamination capabilities, should be provided to individuals typically monitoring (though not necessarily entering) hazardous environments, to meet the referenced timeframes.

Automated External Defibrillators (AED)

Using an AED as soon as possible (i.e., within 3–5 minutes) after a sudden cardiac arrest caused by ventricular fibrillation can lead to a 60% survival rate. CPR supports the circulation and ventilation of a victim, while an AED delivers an electric shock that can restore the fibrillating heart to normal.

While all worksite and public venues are candidates for AED programs, OSHA has identified some worksites of particular concern for sudden cardiac arrest. These include sites that may aggravate or contribute to cardiovascular disease, such as sites with certain chemicals (carbon monoxide, carbon disulfide, halogenated hydrocarbons), extreme heat or cold, and shift work. Additionally, electrical shocks may produce cardiac arrest. Situations that may produce high blood pressures, such as excessive noise, lead or arsenic, are sites where an AED should be considered.

AED programs require physician oversight, although many communities have trained first responders and public access AED programs. The American College of Occupational and Environmental Medicine supports the placement of AED in workplaces, and it has published a guidance of 12 points for use of AED in workplace settings (see References).

Fire Response and Fire Suppression

All work places have the potential for unintentional fires. Most fires are handled in their incipient stage (i.e., when they are small) prior to the need for organized fire protection services. The National Fire Protection Association (NFPA) reports that seven of eight fires are handled in this manner, with makeshift or minor fire extinguishing and suppression devices. This is a testimony to quick actions and the usefulness of employee training in fire suppression measures. Employers have five options for site employees in the event of a fire emergency. These are identified in **TABLE 9-6**.

TABLE 9-6 OSHA Employer Fire Options

1. **Total evacuation** [Emergency Action Plan]
2. Evacuation of some employees and training of other site employees in the use of fire extinguishers or minor fire suppression devices (annual fire extinguisher discharge for employees assigned this responsibility is required) **[Emergency Action Plan]**
3. Use of a fire extinguisher or other minor fire suppression devices by all employees at their work locations (annual fire extinguisher discharge for employees is required) **[Emergency Action Plan]**
4. Incipient Fire Brigade: Training of select employees to respond to fires within their facilities prior to the need for PPE for the purpose of fire suppression (initial and annual live fire training required) **[Fire Brigade-Emergency Response Plan]**
5. Structural Fire Brigades: Training of employees in the use of PPE to perform fire suppression and rescue operations in hazardous environments (initial and quarterly fire education required; live fire training required annually) **[Fire Brigade-Emergency Response Plan]**

Source: 29 CFR 1910.38:29 CFR 1920.156-457.

Fire Extinguisher Use

Most jurisdictions require installation of fire extinguishers to comply with building codes. Employers should identify fire hazards on their sites and ensure that fire extinguishers are placed appropriately for Class A and Class B hazards. Work places designed and certified to current building codes will meet these requirements. However, the introduction of hazardous environments into general buildings, with associated equipment, decontamination areas, and security restrictions may increase travel distances or block previously placed extinguishers. In general, a travel distance of no more than 75 feet to a fire extinguisher from any site in a work place (placed at or adjacent to exits and exit passageways to ensure individuals who may use fire extinguishers can evacuate from an emergency area) is required. In hazardous areas, additional extinguishing capabilities (e.g., an extinguisher with a 4A versus a 2A rating) must be available. Class B, C, D, and K extinguishing or suppression agents should be placed adjacent to the potential hazards they are designed to protect against (i.e., no more than 30 feet travel distance).

For hazardous atmospheres, multiple (at least two) extinguishers should be present, and some employees should be annually trained in their use. **TABLES 9-7** and **9-8** identify fire extinguisher placement and travel requirements.

PHOTO 9-7 Typical fire extinguishers include dry chemical, carbon dioxide, and pressurized water extinguishers; Employees assigned to use extinguishers must know the classification and rating prior to use.

TABLE 9-7 Fire Extinguisher Size and Placement for Class A Hazards			
	Light (Low) Hazard Occupancy	Ordinary (Moderate) Hazard Occupancy	Extra (High) Hazard Occupancy
Minimum rated single extinguisher	2A	2A	4A
Maximum floor area per unit of A	3000 ft^2 (278.7m^2)	1500 ft^2 (139.35m^2)	1000 ft^2 (92.9m^2)
Maximum floor area for extinguisher	11,250 ft^2 (1095 m^2)	11,250 ft^2 (1095m^2)	11,250 ft^2 (1045m^2)
Maximum travel distance to extinguisher	75 ft. (22.9m^2)	75 ft. (229m^2)	75 ft. (22.9m)

For SI units: 1 ft. = 0.505 m; 1 ft^2 = 0.0929 m^2
*Two 2 1/2 gal (9.46 l.) water-type extinguishers can be used to fulfill the requirements of one 4A rated extinguisher.
[Up to two water-type extinguishers, each with 1-A rating, can be used to fulfill the requirements of one 2-A rated extinguisher.
Source: Portable Fire Extinguishers, NFPA #10. National Fire Protection Association, Batterymarch Park, Quincy, MA 2010.

The Fire Triangle: A Review

Any discussion concerning fire suppression should initiate with an understanding of the fire triangle (the fire tetrahedron), presented in Chapter 4. This is necessary for both the purposes of understanding fire prevention practices as well as fire suppression procedures. In all cases, sides of the fire triangle, or their ability to combine, are removed to eliminate initial ignition or to extinguish a fire. Figures 4-1 and 4-2 in Chapter 4 identify the components of the fire triangle and the concept of flammable ranges of oxygen and fuel. Classes and characteristics of fire are presented in **TABLE 9-9**.

Fire Extinguisher Ratings

Fire extinguishers are rated by class and the quantity of fire that they may suppress. Numerical quantifiers are associated with Class A and Class B fires. A Class 1A fire is a 10-foot by 10-foot by 10-foot, wood crib fire constructed of 1-inch lumber. A 1A rated fire extinguisher will suppress this fire. Testing has determined that 1¼ gallons of water have a 1A rating (extinguished due to the reduction of the heat of the burning material below its flash point [elimination of fuel]). A Class 1B fire is a 1 square foot area of a flammable liquid. Quantification of Class B fire suppression takes into account that untrained operators of extinguishers may use these devices (i.e., this rating is 40% of the actual amount of fire that is suppressed in tests). To extinguish a Class B fire, the oxygen is eliminated from the fuel and the area above the flammable liquid pool. Class C (electrical) fires do not have a quantifiable rating associated with them but are identified due to the nonconductive nature of the agent used to suppress the fire (in general, Class C fire suppression initiates with shutdown of power [source of ignition] thus creating a Class A or B fire). Class D fire suppression agents, used for the suppression of combustible metals, are specific to the combustible materials they are designed to "blanket." In most cases, these agents separate the materials from other combustibles in the environment as they continue to burn.

Class K fire extinguishing agents are identified for their applicability in kitchen (i.e., grease) fires to blanket grease-covered appliances in order to prevent reignition of the Class K hazard. Class K extinguishers (wet chemical extinguishers) are used subsequent to fixed Class B extinguishing systems, and the Class B systems must be activated prior to application of the Class K agent.

Fire Extinguisher Inspections

Most fire prevention codes, as well as NFPA #10 (Fire Extinguishers), require the inspection of fire extinguishers on a monthly basis. It is highly recommended that employees within work areas (those who may use fire extinguishers) perform monthly fire extinguisher inspections according to the manufacturer's inspection requirements. These requirements are identified in **TABLE 9-10**.

TABLE 9-8 Fire Extinguisher Size and Placement for Class B Hazards		
Type of Hazard	Basic Minimum Extinguisher Rating	Maximum Travel Distance to Extinguishers
Light (low)	5-B	30 ft. (9.15 m)
	10-B	50 ft. (15.25 m)
Ordinary (moderate)	10-B	30 ft. (9.15 m)
	20-B	50 ft. (15.25 m)
Extra (high)	40-B	30 ft. (9.15 m)
	80-B	50 ft. (15.25 m)

Source: Portable Fire Extinguishers, NFPA #10, National Fire Protection Association, Batterymarch Park, Quincy, MA, 2010.

TABLE 9-9 Classes and Characteristics of Fires

Class	Flashpoint(s)	Burning Temperature	Typical Extinguishing Mechanisms
A: Ordinary Combustibles	> 200°F (>93°c)	600–1,000°F	Water, ABC Dry Chemicals, Foams
B: Flammable and Combustible Liquids	< 200°F (<93°c)	1,200°F (649°c)	ABC Dry Chemicals, Class B Foams, CO_2
C: Energized Electrical Equipment	N/A	N/A	ABC Dry Chemicals, CO_2, Removal of Energized Source
D: Combustible Metals	Varied	>1,500°F (> 816°c) (Note: Some Also Provide Their Own O_2)	Specialized Procedures or Agents (CLASS D)
K: Cooking Oils & FATS	> 500°F	Varied	Specialized fire extingushing agents (CLASS K) required in conjunction with fixed Class B extinguishing systems. These agents react with oils to create a soapy foam over super heated materials (saponification) to prevent re-ignition.

Note: Class E fires in radioactive materials are referenced in some literature, however are not identified in the typical classification system proffered by NFPA upon which Table 9-9 is based (NFPA 10: Portable Fire Extinguishers).

TABLE 9-10 Fire Extinguisher Requirements: Inspection (Monthly) and Maintenance

Frequency: Fire extinguishers shall be inspected when initially placed in service and thereafter at approximately 30-day intervals by personnel. Fire extinguishers shall be inspected at more frequent intervals when circumstances require.

Procedures: Periodic inspection of fire extinguishers shall include a check of at least the following items:

a. Location in designated place
b. No obstruction to access or visibility
c. Operating instructions on nameplate legible and facing outward*
d. Safety seals and tamper indicators not broken or missing*
e. Fullness determined by weighing or "hefting"
f. Examination for obvious physical damage, corrosion, leakage, or clogged nozzle
g. Pressure gauge reading or indicator in the operable range or position
h. Condition of tires, wheels, carriage, hose, and nozzle checked (for wheeled units)*
i. Recording/documentation maintained

Maintenance (Annual)**

a. Conductivity test CO_2 hose assemblies and document results
b. Disassemble loaded stream extinguishers
c. Test wheeled extinguisher pressure regulators per manufacturer's instructions and uncoil/inspect hose
d. Remove (pull) tamper seal (Note: It is not necessary to internally inspect nonrechargeable, CO_2, or stored pressure extinguishers, except as noted)
e. Disassemble boots, foot rings, and attachments to facilitate a thorough inspection, then perform a thorough extinguisher physical inspection
f. Verify operating instructions and HMIS information is present and facing forward
g. Attach tag, identifying month/year of maintenance and identity of individual performing the work (and organization, as applicable)

*Required if the location has experienced a high frequency of fires; exhibits severe hazards; is located where it is susceptible to mechanical injury or physical damage or experiences exposure to abnormal temperatures or corrosive atmospheres

**Maintenance is required to be performed by certified individuals who have passed a certification test acceptable to the authority having jurisdiction

Source: Adapted from *Portable Fire Extinguishers, NPFA #10.* National Fire Protection Association. Batterymarch Park, Quincy, MA, 2010.

It is recommended that semiannual and annual fire extinguisher inspections/maintenance be provided by trained employees who may use the devices. These employees are required to be certified (by test, acceptable to the Authority Having Jurisdiction [AHJ]) to perform maintenance activities. These inspections may require weighing of extinguishers or inspection of extinguisher shells for hydrostatic (pressure) test date. All inspections should be performed in compliance with manufacturer's instructions, which are provided with each fire extinguisher. Finally, maintenance, which addresses the internal extinguisher components, must be provided only by individuals certified or insured to perform these operations (e.g., six-year maintenance of extinguishers). Typically, these operations include internal extinguisher inspections and hydrostatic testing of extinguisher shells.

Fire Extinguisher Use and Training

Only trained employees should be assigned the use of fire extinguishers. OSHA requires the discharge of fire extinguishers on an annual basis to ensure employees are trained in their appropriate use. Annual live fire training in addition to fire extinguisher discharge is required for both incipient and structural fire brigades.

Typical fire extinguisher use instructions include the following:

1. Surveying the scene and accessing the closest extinguisher (after ensuring that it is appropriately classified), ensuring a way out of the immediate area before attacking the fire, calling for help, then pulling the pin or emergency activation device to allow use of the extinguisher;
2. Testing the extinguisher a safe distance from the fire to ensure it is operable;
3. Applying the agent in the extinguisher to the fire (directly to the base of the fire for cooling purposes and in a sweeping motion for oxygen elimination purposes);
4. Backing away from the fire (observing the fire at all times) and placing used extinguishers on their sides so as not to mistake them for full extinguishers; best practices dictate the discharge of the entire extinguisher contents.

If compatible extinguishers and more than one trained individual are present, it is more beneficial to use two extinguishers simultaneously than one after another.

Outdated and Illegal Extinguishers

Certain types of fire extinguishers are currently outdated or illegal. These include the following:

- Soda acid extinguishers (those that are turned upside down to activate), brass shell ribbed extinguishers, and extinguishers that do not have a device for stopping or starting the flow of the extinguishing agent;
- Extinguishers containing the chemical carbon tetrachloride;
- Certain ozone depleting (HALON) extinguishers.

Standpipe and Hose Systems

Standpipe and hose systems for use in fire suppression (hazardous areas) are classified by the NFPA into three types:

1. **Type I systems (Class I Standpipes):** for fire department or fire brigade use;
2. **Type II systems (Class II Standpipes):** for occupant use;
3. **Type III systems (Class III Standpipes):** for both fire service and occupant use.

Basic maintenance requirements for standpipe systems include annual flow tests and reracking of hose.

PHOTO 9-8 Presently illegal extinguishers include those containing soda acid, carbon tetrachloride, or those without a mechanism for the starting or stopping of the extinguishing agent.

Fire hose may be classified for occupant use, and fire nozzles are classified and listed for occupant standpipe system. Although a typical fire hose is double jacketed (i.e., two layers of material woven in opposite directions), occupant standpipe hose may be single layered or constructed with a sprayed interior. As a minimum, nozzles and hose lines should flow 25 gallons of water per minute to meet certain regulatory standards, although many flow in excess of this quantity. Training in the use of standpipe hose applications is required on an annual basis if these systems are to be used in the workplace. All nozzles and applicators associated with standpipe hose systems, as with fire extinguishers, need to be able to be turned off or on by the operator.

Fire Brigades

Employers may choose to assign employees from areas outside of their immediate work locations to respond to fire emergencies. This may be done for the following purposes:

- Lack of available off-site (municipal) fire protection (public or private);
- Increased hazards on site;
- Specialized equipment necessary for fire suppression (e.g., Class D fire extinguishers).

Incipient Fire Brigades

Incipient fire brigades are designed to suppress fires in their beginning stages. Typically, incipient fire brigades use fire extinguishers or minor fire suppression devices and fire hoses of less than or equal to 1½ inches in diameter (i.e., capable of being handled by one person). Training is required on an annual basis, including live fire training. Fire brigade leaders receive additional training. An organizational statement addressing standard operating procedures (SOPs), in compliance with the OSHA Fire Brigades standards, found at 29 CFR 1910.156, is required for incipient fire brigades, as well as a review by the employer of the medical ability of employees to participate in fire emergency response activities. In addition, incipient fire brigades may operate systems such as sprinkler systems, water control devices, or other extinguishing agent application systems. Structural fire brigades are required to be more qualified (i.e., receive levels of higher training) than incipient fire brigades because they will enter and operate in immediately dangerous to life and health (IDLH) atmospheres.

Standard Operating Procedures

The written fire brigade SOPs (ERP) must identify chains of command such as the appropriate ICS to be used during fire emergencies. In addition, special procedures or hazards on site are listed in this written operating procedure. The OSHA Fire Brigade Standard requires employers to ensure that employees are medically fit to perform firefighting functions prior to assignment to fire brigade duties.[1]

In many situations, incipient fire brigades are combined with emergency action personnel to provide response to other work or community site emergencies (e.g., the Emergency Action Team).

Fire teams entering IDLH environments are classified as structural fire brigades by OSHA.[2]

Advanced Exterior Fire Brigades

The NFPA identifies a third level of fire brigade response personnel, the advanced exterior fire brigade, in which individuals use thermal protective clothing and may be required to use a self-contained breathing apparatus (SCBA) to extinguish fires that have advanced past the incipient stages from outside structures.

Structural Stage Fire Brigades (Additional Requirements)

Structural stage fire brigades respond to facility fire and rescue emergencies and operate in hazardous atmospheres beyond the incipient stage. These brigades require personal protective equipment (PPE), including respiratory protection due to the IDLH atmospheres that structural stage fire personnel enter. These individuals use tools and devices beyond those typically available to the incipient fire brigade. Qualifications for structural stage industrial fire brigades and firefighters are found in NFPA Codes 600 and NFPA 1081.

In addition to the incipient fire brigade requirements, the following apply to structural stage fire brigades:

- Training such as that recognized nationally for industrial firefighters must be provided four times per year, with at least one live fire session provided annually.
- Full PPE must be provided. This includes fire fighters' helmets with earflaps, face shields, and chinstraps; SCBA of at least 30-minute duration, and firefighters' "turnout gear" (heat and flame resistant coat, gloves, and trousers with calf-length boots or 3/4 length boots).

---

[1] The current U.S. Department of Labor standard requires employers to ensure that employees performing interior structural fire suppression are medically fit to perform this duty. It requires employees to identify whether situations such as known heart disease, emphysema, or epilepsy are present and subsequently disallow these employees from participating in firefighting duties.

[2] Note: Airport Fire Service and Wildland Firefighting are not covered by the OSHA Fire Brigade Standard.

Employers placing this equipment into service should ensure that equipment meets the specifications of the latest NFPA code for the applicable equipment specified (See Chapter 7).

In addition to the OSHA required PPE, NFPA and NIOSH recommend a personal alert safety system (PASS) for municipal and industrial firefighting, both within and outside hazardous areas. A PASS device is a motion detector that communicates a shrill sound (typically 90 decibels, directed) in the event the user remains motionless for 30 seconds or more. In addition, PASS devices have activation points that allow an individual to manually activate the alarm and call for assistance.

PASS devices may additionally be used in emergency rescue, IDLH, and a variety of other hazardous situations to call for rescue or assistance when the individual in the hazardous atmosphere may be incapable of calling on his or her own. PASS devices do not meet the requirements for direct communication within the **buddy system** as required by OSHA for entry into IDLH environments.

The use of the buddy system and placement of a standby crew (two additional qualified emergency responders) for emergency rescue purposes at interior structural firefighting or rescue situations is required by OSHA under the Respiratory Protection Standard and recommended by most national standards. This OSHA requirement is applicable for entry into all IDLH atmospheres.

Although rapid intervention and standby teams are now common in many situations (such as public fire protection), they have been required by OSHA and national consensus standards for individuals who enter hazardous atmospheres for nearly 40 years.

Medical approval to use respiratory protection (in addition to the employer's survey to ensure the assigned industrial firefighter is physically capable of performing duties) is required by the OSHA Respiratory Protection Standard.

Industrial firefighters who respond as part of mutual aid agreements to situations for which they are not typically familiar (e.g., to assist municipal fire departments) should meet the requirements of NFPA 1500/NFPA 1082, Health and Safety Programs for Fire Departments and Professional Qualifications for Fire Fighters, outside of industrial areas.

Maintenance of Fire Safety and Emergency Response Equipment

Fire safety equipment for facilities or hazardous sites, including both life safety and suppression equipment, should be maintained in compliance with local building codes, local fire prevention codes, and the requirements of the NFPA. It is typical for a facility to assign a fire marshal to maintain this equipment, particularly if they are placed, retrieved, or inspected at the initiation of each workday. Typical preventive maintenance schedules are found in **TABLE 9-11**.

Intrinsically Safe and Explosion-Proof Equipment

Individuals entering flammable or potentially flammable atmospheres should be cognizant of the use of intrinsically safe or explosion-proof equipment. Explosion-proof equipment is, by design, not capable of producing the energy necessary to provide a source of ignition

TABLE 9-11 Typical Preventative Maintenance Procedures: Emergency Response Equipment	
Item	Requirement
Communications, alarming and alerting systems, and communication devices	Test when installed and check daily for accessibility and operability. Test according to manufacturer's requirements (29 CFR 1910.165).
Fire extinguishers	Ensure accessibility; Inspect monthly; Maintain annually (NFPA 10). [See TABLE 9-10]
Standpipe/hose systems	Ensure accessibility daily. Ensure operability monthly. Flow test annually (29 CFR 1910.158 and NFPA 25).
PPE for Emergency Response	Inspect and maintain per manufacturer's instructions and inspect monthly.
First Aid Kits and Supplies	Check prior to each job and weekly there after (29 CFR 1926.50 for CONSTRUCTION SITES)
PASS Devices	Test weekly (NFPA 1500), and prior to each use.
SCBA (30 minute minimum duration)	Inspect and tag monthly (29 CFR 1910.134).

TABLE 9-12 Classification of Intrinsically Safe Equipment: Equipment May Be Classified by Class, Division and Group
Class I: Flammable vapors
Class II: Combustible vapors
Class III: Flying brands
Division I: Continuous hazardous atmosphere
Division II: Intermittent hazardous atmosphere
Group A: Acetylene
Group B: Hydrogen and equivalents (e.g., ethylene oxide)
Group C: Ethyl ether and equivalent
Group D: Ammonia, gasoline, natural gas, and propane (or equivalent)
Group E: Combustible metal dusts (Class II)
Group F: Carbonaceous dusts (Class II)
Group G: Wood dusts (Class II)
Source: NFPA #70, The National Electrical Code. National Fire Protection Association, Batterymarch Park, Quincy, MA.

and subsequently a fire or explosion. Intrinsically safe equipment is designed to meet National Electrical Code Standards[3] and rated by a class, division, and group criteria. These criteria are found in **TABLE 9-12**. In addition, all individuals in these areas should be cognizant of grounding and bonding procedures to minimize the chance of sources of ignition during the movement of fluids or personnel.

Emergency Rescue

Some sites may require emergency rescue teams due to hazardous atmospheres (the complex equipment that they possess, the style of operations they perform, or the products they process). Certain emergency rescue situations are described in U.S. Department of Labor standards, and specific rescue training should be provided for these situations. Standards such as the OSHA Tunneling Standard, the OSHA Standard for Power Propelled Scaffolds, the OSHA Permit Required Confined Spaces Standard, and the Mine Safety and Health Administration requirements for mine rescue teams typify these situations.

---
[3] NEC: Article 500, The National Electrical Code, NFPA.

Emergency Rescue from Immediately Dangerous to Life or Health Situations

In the event an IDHL situation has been identified (see the discussion in Chapter 5 regarding IDLH situations), various requirements apply, including the requirements for use of the buddy system, communications from within to the exterior of hazardous areas, and the use of standby rescue personnel who are trained and equipped to provide emergency rescue.

The purpose of an emergency rescue team is to remove any victims as quickly and as safely as possible from a hazardous area to an area where lifesaving procedures can be performed. The patients are subsequently transferred to EMS- and hospital-based treatment systems. Consequently, although the requirement to have first aid trained personnel on site appears in some standards, it is essential for the protection of those who work in hazardous atmospheres. This requirement also protects rescuers in addition to victims. In all cases in which a victim is identified in an IDLH situation, the emergency medical system should be activated in compliance with the facility EAP. Various standards, such as the Hazardous Waste Operations and Emergency Response standard, require that **standby personnel** be available while individuals are operating in hazardous atmospheres. Transport capabilities to hospitals or healthcare facilities are also required.

Ropes, Knots, and Harnesses

Harnesses used for emergency rescue purposes (and lifting or fall protection) are required to be full-body harnesses. An attachment point in the middle of the upper torso is required, although some harnesses have wristlets that place a victim in a vertical position in the event narrow access passageways are necessary for emergency rescue and retrieval. Harnesses are required to be inspected routinely (e.g., monthly) and in compliance with manufacturer's specifications. Harnesses are required to carry a nameplate certifying fitness for use and should be removed from service as obvious degradation of materials occurs. Inspections evidencing a 1/4-inch tear in harness material show obvious degradation.

All lifesaving ropes must also be inspected as life-safety equipment. Some manufacturers have published a four-year replacement schedule for ropes. The NFPA requires lifesaving rope that is shock loaded to be removed from service. OSHA requires that personal fall arrest systems that have sustained an impact be removed from service unless inspected and certified fit for service by a competent person. Lifesaving and life-safety ropes require a minimum tensile breaking strength of

5000 pounds when placed in service. Synthetic materials are now required for all lifelines and harnesses. This material should be inspected routinely (e.g., monthly) and according to manufacturer's specifications. As with all life-safety equipment, records and maintenance should be maintained with the equipment. Typically, static kernmantle rope is used for these applications. Static kernmantle rope is double braided synthetic material, with little stretch.

Knots

Individuals involved with emergency rescue should be capable of connecting ropes to each other and to objects for the purpose of attaching or lifting by placing a minimum amount of stress on the rope. Properly tied knots are easily installed and removed. Five knots are recommended for use by emergency response personnel:

PHOTO 9-9 A full-body harness is required for rescue and fall protection. A properly maintained life-safety rope is kept out of both sunlight (UV exposure) and the weather.

PHOTO 9-10 Body belts may be used as positioning devices.

PHOTO 9-11 Rescue teams that provide services at permit required confined spaces must practice rescue skills annually.

1. **Square knot:** Used for connecting ropes of the same size (**FIGURE 9-1**).
2. **Sheet bend:** Used for connecting ropes of different sizes (**FIGURE 9-2**).
3. **Clove hitch:** Used for connecting a rope to an object when a pull on the object will be in one direction (**FIGURE 9-3**).
4. **Bowline:** Used for connecting a rope to an object when the pull may be in various directions (**FIGURE 9-4**).
5. **Figure Eight Knot:** Used as a substitute for the bowline knot (**FIGURE 9-5**).

Confined Spaces Rescue

Facilities that allow individuals to enter permit-required confined spaces are required to have standby personnel immediately available. The entry requirements concerning confined spaces are discussed in Chapter 11, and a typical permit required for entry into a confined space is presented in Figure 6-2. OSHA has determined that municipal emergency services off site in general do not meet this requirement and that a standby rescue person or team is required on site (at or adjacent to the point of entry) during these entries. Typically, a lifting device and attachment to individuals within permit-required confined spaces is required for vertical entries greater than five feet to allow for an external rescue. Permit-required confined space rescue teams are required to practice annually on the spaces from which they may need to remove victims.

Lifts, Drags, and Carries

Emergency standby personnel should be capable of using a variety of lifts, drags, and carries in a manner that is protective of the victim and ergonomically proficient. They must also be capable of removing the victim in a

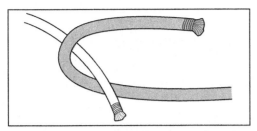
Step 1: Form a bight in one of the ends to be tied (if two ropes of equal diameter are being tied).
Step 2: Pass the end of the second rope through the bight.

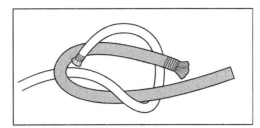
Step 3: Bring the loose end around both parts of the bight.

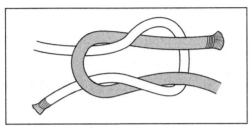
Step 4: Tuck this end through the bight.
Step 5: Pull the knot snug.

FIGURE 9-1 The square knot.

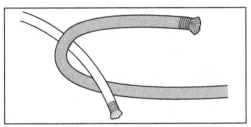
Step 1: Form a bight in one of the ends to be tied (if two ropes of unequal diameter are being tied, the bight always goes in the larger of the two).
Step 2: Pass the end of the second rope through the bight.

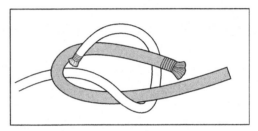
Step 3: Bring the loose end around both parts of the bight.

Step 4: Tuck this end under its own standing part and over the bight.

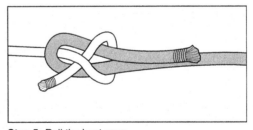
Step 5: Pull the knot snug.

FIGURE 9-2 The sheet bend knot.

quick manner. One- and two-person shoulder lifts and drags should be practiced by all standby rescue personnel if the potential to move victims exists in the rescue scenario. In addition, rolling victims and placing them onto devices such as Stokes baskets or backboards may be required. All personnel, including emergency rescue personnel, should be familiar with the stop, drop, and roll procedure, a means of extinguishing fires on one's person or on an adjacent individual.

Hazardous Materials Emergencies

Facilities that use hazardous materials and substances may be required to respond to hazardous material emergencies. Under the OSHA requirements (specified at

CHAPTER 9: Responding to Emergencies 171

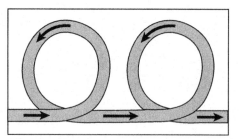

Step 1: Form a loop in your left hand with the working end to the right crossing under the standing part.
Step 2: Form another loop in your right hand with the working end crossing under the standing part.

Step 3: Slide the right-hand loop on top of the left-hand loop.
NOTE: This is the important step in forming the clove hitch knot.

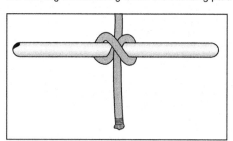

Step 4: Hold the two loops together at the rope forming the clove hitch.
Step 5: Slide the knot over the object.
Step 6: Pull the ends in opposite directions to tighten.

FIGURE 9-3 The clove hitch knot.

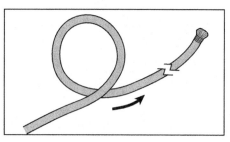

Step 1: Select enough rope to form the size of the knot desired.
Step 2: Form an overhand loop in the standing part.

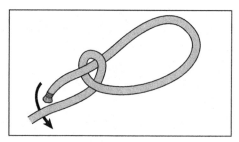

Step 3: Pass the working end upward through the loop.

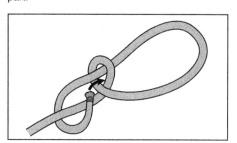

Step 4: Pass the working end over the top of the loop under the standing part.

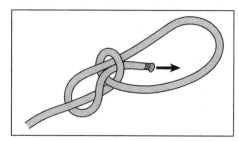

Step 5: Bring the working end completely around the standing part and down through the loop.
Step 6: Pull the knot snugly into place, forming an *inside* bowline with the working end on the inside of the loop.

FIGURE 9-4 The bowline knot.

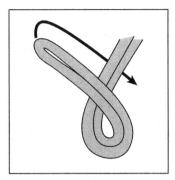  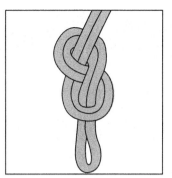

Step 1: Form a bight in the working end of the rope.
Step 2: Pass it over the standing part to form a loop.

Step 3: Pass the bight under the standing part and then over the loop and down through it; this forms the figure eight.

Step 4: Extend the bight through the knot to whatever size working loop is needed.
Step 5: Dress the knot.

FIGURE 9-5 The figure eight knot.

PHOTO 9-12 Rescue personnel must be capable of preparation for transport and safe movement of victims.

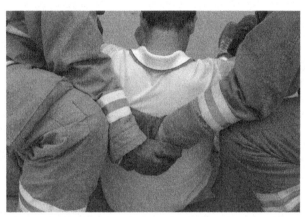

PHOTO 9-14 Similar to the one-person drag, the two-person drag allows rescuers to protect a victim's head and neck.

PHOTO 9-13 This one-person drag quickly moves a victim while protecting the head and neck.

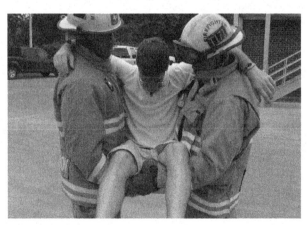

PHOTO 9-15 The two-person lift and carry quickly allows movement of non-ambulatory patients.

29 CFR 1910.120 and 29 CFR 1926.65), facilities where hazardous substances may be released from processes or containers in quantities capable of harming human health or the environment beyond the stage that can be handled under hazard communication procedures by individuals in the areas of the release when it occurs must make preparations for these releases. All individuals at a site must be trained to recognize releases and communicate emergencies under the facility's EAP or ERP. **Facilities must designate an emergency**

CHAPTER 9: Responding to Emergencies

TABLE 9-13 Hazardous Materials Emergency Response Plan Requirements

1. Pre-emergency planning and coordination with outside parties
2. Personnel roles, lines of authority, training, and communication
3. Emergency recognition and prevention
4. Safe distances and places of refuge
5. Site security and control
6. Evacuation routes and procedures
7. Decontamination
8. Emergency medical treatment and first aid
9. Emergency alerting and response procedures
10. Critiques of responses and follow-ups
11. PPE and emergency equipment programs
12. Nonrepetition of other response plans to avoid duplication

PHOTO 9-16 The requirement for onsite rescue personnel at a PRCS can be identified by confined space signs and confirmed with atmospheric testing. PRCSs may be declassified, thereby eliminating the need for on-site rescue personnel, if there is no potential for a hazard in the space.

response coordinator. Subsequently, responsible parties or employers are free to decide whether to have employees respond to spill emergencies or to evacuate employees and notify emergency services and off-site personnel. In the event facilities do respond to emergencies, a written ERP is required that addresses the 12 areas found in **TABLE 9-13**.

An employer with employees who may respond to hazardous material incidents must provide training and certify that employees are trained and competent in the procedures they are expected to perform. The required procedures are presented in Table 2-4 of Chapter 2. Annual practices and retraining are required for these individuals. The ICS, as previously discussed, is required to control incidents in which releases of hazardous materials and substances have occurred.

The National Response Team (NRT), chaired by the Environmental Protection Agency (EPA), has made available an emergency response planning tool, the

TABLE 9-14 Federal Regulations Requiring Emergency Planning, Applicable to "One-Plan" Compliance

Agency	Regulation	Comment
EPA	40 CFR 112.7(d) 40 CFR 112.20.21	EPA's Oil Pollution Prevention Regulation Requiring Spill Prevention Control and Countermeasure and Facility Response Plans
Mineral Management Service (Dept. of Interior)	30 CFR 254	Mining Facility Response Plan Requirements
Research and Special Programs Administration (RSPA) (U.S. DOT)	49 CFR 194	Pipeline Response Plan Regulations
U.S. Coast Guard (U.S. DOT)	33 CFR 154	Facility Response Plan Regulations
EPA	40 CFR 68	Risk Management Programs
EPA	40 CFR 264, "D" 40 CFR 265, "D" 40 CFR 279.52	Resource Conservation and Recovery Act: Hazardous Waste Contingency Planning Requirements
OSHA	29 CFR 1910.38(a)	Employee EAP Requirements
OSHA	29 CFR 1910.119	Process Safety Management Standard (EAP/ERP)
OSHA	29 CFR 1910.120	"HAZWOPER"
Adapted from Inspection Procedures for Hazardous Waste Operations and Emergency Response Standard 29 CFR 1910.120 "q" in *Emergency Response to Hazardous Substance Releases*, Instruction 2-2.59, Office of Health Compliance, Washington D.C., 1993.		

"One-Plan," to provide a mechanism for consolidating multiple plans that facilities have prepared to comply with various regulations. The "One-Plan" combines response plans required by a variety of federal agencies into one functional ERP or integrated contingency plan (ICP). The NRT (EPA, Mineral Management Service [Interior Department], Research and Special Programs Administration [U.S. Department of Transportation], and OSHA) is responsible for response and plan review requirements in preparation for federally controlled disasters. In addition to state or local emergency planning regulations, facilities may be subject to one or more of the federal emergency planning requirements found in **TABLE 9-14**. These sites can use the "One-Plan" format to demonstrate compliance.

Chapter Summary

Individuals who operate within hazardous environments require the protection of emergency action or response plans. The planning process initiates with a survey of potential risks and expected outcomes and culminates with a business-like procedure or SOP for emergency actions and responses.

Typically medical, fire, rescue, and hazardous materials emergencies are preplanned.

Responsible parties have the option of evacuation or response actions to safeguard individuals in hazardous environments during emergencies, although typically a mix of alarming (removing people from the additional hazard [evacuation]) and alerting trained individuals to respond to the emergency for the purpose of mitigating or reducing consequences is chosen. Various regulations address required elements of EAPs and ERPs. Equipment used for emergencies must be continuously available to be usable during emergency response timeframes.

Emergency responders in today's environment should be aware of the activities of terrorists, whose actions may resemble emergencies to which they respond. In this case, the scene must be preserved, police agencies notified, and precaution taken against potential follow-up terrorism activities or secondary devices at the site.

Terms

Buddy System: A system of organizing employees into work groups so that each employee is designated to observe at least one other employee in the work group. The purpose of the buddy system is to provide rapid assistance in the event of an emergency, then remain in visual or voice contact at all times.

Emergency Action Plan (EAP): Identifies basic actions individuals should take in the event of unusual or emergency incidents.

Emergency Response Plan (ERP): Identifies actions individuals may take in response to an emergency.

Incident Command System (ICS): An organized set of procedures for controlling personnel, facilities, equipment, and communication designed to be used in response to emergencies.

National Incident Management System (NIMS): A nationally recognized approach to emergency incident management that applies to all jurisdictions and functions, regardless of size or complexity of an incident. The NIMS applies to public and private response elements.

Reportable Quantity (RQ): Quantities of hazardous substances above which notification to agencies of public or environmental safety is required. RQ notifications may be for the purpose of pre-emergency planning (e.g., the amount of a hazardous substance on site, reportable for planning purposes under SARA), or for emergency response or investigation (e.g., the amount of a hazardous substance released during an incident). Many RQ reports are processed to the National Response Center (1-800-424-8802) and are subsequently processed to applicable governmental agencies.

Standby Personnel: Individuals who are trained, capable, and equipped to provide emergency rescue, located outside IDLH areas when individuals are in these areas.

Terrorism: The deliberate creation of a hazardous environment or unlawful use of force against persons or property to intimidate or harm a government or civilian population, in the furtherance of political or social objectives. A terrorist act is a crime.

Type I Systems (Class I Standpipes): Systems usable by the fire department or by structural stage fire brigades. These standpipe systems are generally located in protected areas such as stairwells and are designed to accept hoses larger than 1½ inches in diameter. Fire flows, fire pumps, and fire department connections are many times associated with these systems.

Type II Systems (Class II Standpipes): Systems that are designed to be used by occupants and may be located in structures in lieu of or in addition to fire extinguishers. Typically, hose sizes of 1½ inches or less are found on these systems.

Type III Systems (Class III Standpipes): Systems that are a combination of Class I and II systems and generally have a reducing adapter at the point where the smaller fire hoses are connected for occupant standpipe use.

References

American College of Occupational and Environmental Medicine. *Evidenced Based Statements ACOEM Guideline Automated External Defibrillation*. ACOEM, December 27, 2003 (online at www.acoem.org/guidelines/articles.asp?ID=41).

American National Standards Institute. *ANSI Z308.1: First Aid Supplies*. New York: ANSI, 2009.

———. *ANSI Z358.1: Emergency Eye Wash and Shower Equipment*. New York: ANSI, 2009.

American Red Cross. *First Aid: Responding to Emergencies*. St. Louis: Mosby, 2009.

———. *Automated External Defibrillators Can Save Lives During Cardiac Emergencies*. OSHA-3174, 2001.

Cocciardi, Joseph A. PhD. *Emergency Response Team Handbook*. Quincy, MA: National Fire Protection Association, 2004.

Directorate of Technical Support. *OSHA Instruction CPL 2-2.53: Guidelines for First Aid Training Program*. Washington, D.C.: Government Printing Office, January 1991.

National Commission on Terrorist Attacks. *The 9/11 Commission Report: Final Report of the National Commission on Terrorist Attacks Upon the United States*. New York: W. W. Norton and Company, 2004.

National Fire Protection Association. *NFPA #10: Portable Fire Extinguishers*. Batterymarch Park, Quincy, MA: NFPA, 2010.

———. *NFPA #70: National Electric Code*. Batterymarch Park, Quincy, MA: NFPA, 2011.

———. *NFPA #600: Industrial Fire Brigade*. Batterymarch Park, Quincy, MA: NFPA, 2010.

———. *NFPA #1081: Professional Competencies for Industrial Fire Brigade Members*. Batterymarch Park, Quincy, MA: 2012.

The National Response Team's Integrated Contingency Plan Guidance: Notice. *Federal Register* 61, No. 109 (June 5, 1996): 28641-31104.

Office of Health Compliance. 29 CFR 1910.120"q": Inspection Procedures for Hazardous Waste Operations and Emergency Response Standard. In *Emergency Response to Hazardous Substance Releases, Instruction 2-2.59*. Washington, D.C.: Government Printing Office, Current.

U.S. Department of Justice, Office of Justice Program, Bureau of Justice Assistance and Federal Emergency Management Agency, U.S. Fire Administration—National Fire Academy. *Emergency Response to Terrorism: Basic Concepts Instructor Guide*. NFA-ERT: BC-1G, June 1997.

U.S. Department of Labor, Occupational Safety and Health Administration. *Cardiac Arrest and Automated External Defibrillators (AED): Technical Information Bulleting*. January 12, 2007.

U.S. Department of Labor, Occupational Safety and Health Administration. 29 CFR 1910.38: Emergency Action Plans. In *OSHA General Industry Safety Regulations*. Washington, D.C.: Government Printing Office, 2010.

———. 29 CFR 1910.120: Fire Brigades. In *OSHA General Industry Safety Regulations*. Washington, D.C.: Government Printing Office, Current.

———. 29 CFR 1910.134: Respiratory Protection. In *OSHA General Industry Safety Regulations*. Washington, D.C.: Government Printing Office, Current.

———. 29 CFR 1910.146: Permit Required Confined Spaces. In *OSHA General Industry Safety Regulations*. Washington, D.C.: Government Printing Office, Current.

———. 29 CFR 1910.151: Medical Services and First Aid. In *OSHA General Industry Safety Regulations*. Washington, D.C.: Government Printing Office, Current.

———. 29 CFR 1910.156: Fire Brigades. In *OSHA General Industry Safety Regulations*. Washington, D.C.: Government Printing Office, Current.

———. 29 CFR 1910.157: Fire Extinguisher. In *OSHA General Industry Safety Regulations*. Washington, D.C.: Government Printing Office, Current.

U.S. EPA, Office of Solid Waste and Emergency Response. *U.S. EPA 8700-12: Notification of Hazardous Waste Activities. (Rev. 11/85)* Washington, D.C.: Government Printing Office, 1985.

The White House. *Homeland Security Presidential Directive-5 (HSPD-5)*. Washington, DC, February 28, 2003.

Exercise 9

1–4. Using Table 9-1 and Table 9-2 as guides, complete the four-step Emergency Planning Process for your current worksite.

Unwanted Event	Expected Action	Equipment	Training
1.			
2.			
3.			
4.			

5–9. Identify the 6 components of an Emergency Action Plan (EAP):

5. _____

6. _____

7. _____

8. _____

9. _____

10. Under NIMS, the individual in charge of an emergency response or event is designated as:

_____

11. To be effective for life-threatening emergencies, basic life support should be initiated in what timeframe?

_____.

12-14. Identify the location of first aid supplies in your workplace _____ . Do these supplies meet the minimum requirements identified in Table 9-5? _____ Can they be accessed from and brought to any location in your worksite within 15 minutes? _____

CHAPTER 9: Responding to Emergencies

15. An appropriate fire extinguisher for a Class A fire must be no more than _____ (distance) from any location in a worksite. Table 9-6 can be used to identify this distance (NFPA #10).

16. Class K (Wet Chemical) fire extinguishers are found _____ (location).

17. All fire extinguishers should be inspected _____ (timeframe).

18. Teams stationed for permit required confined space rescue must practice site specific rescue procedures _____ (timeframe).

19. Using this device within 3 to 5 minutes of a sudden cardiac arrest can lead to a 60% survival rate: _____.

20. Using the fire extinguishers pictured, identify which extinguisher is most appropriate to extinguish each small fire pictured:

a)

b)

c)

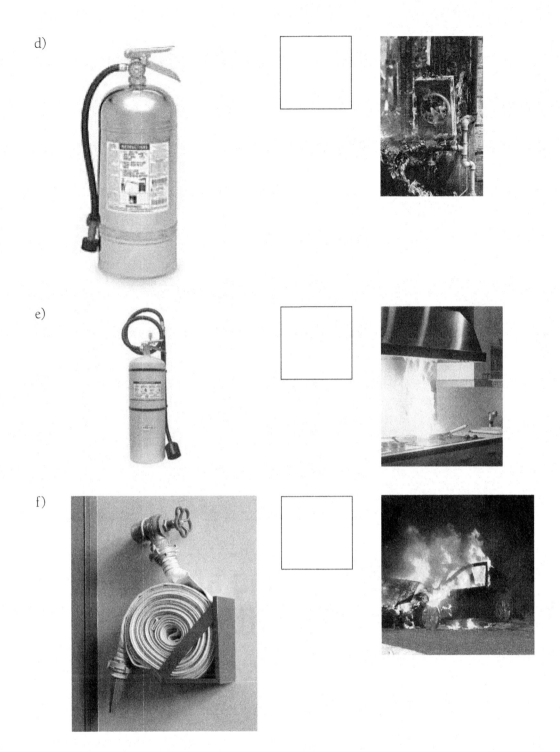

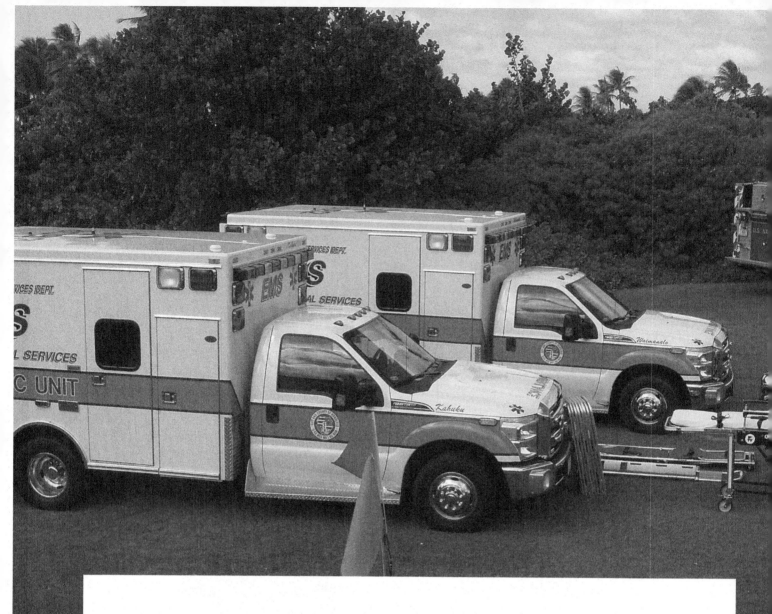

10 Medical Monitoring and Occupational Health

Key Concepts

- Initial Medical Monitoring
- Symptomatic Exposures
- Asymptomatic Exposures
- Physical Performance Testing
- Psychological Evaluations
- Routine (Annual) Monitoring
- Close-Out Medical Monitoring
- Medical Recordkeeping

Medical Monitoring Procedures

Medical monitoring is required for individuals entering hazardous areas. Medical monitoring identifies employees capable of performing hazardous tasks, identifies individuals capable of using personal protective equipment (PPE) safely (or required modifications to PPE regimens), and identifies body system functional levels for comparison purposes (pre and post exposure) for individuals who work in hazardous environments. In general, the time frames for required medical monitoring are referenced in standards, but they may include initial medical monitoring, symptomatic monitoring, routine (annual or repetitive) monitoring, and job close-out or final monitoring.

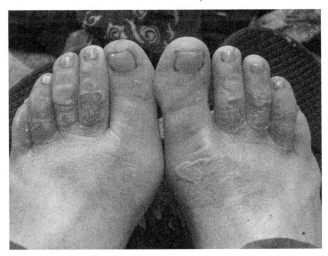

PHOTO 10-1 When symptoms of exposure are present, such as chloracne or skin rash, medical monitoring must be provided to employees.

Medical monitoring is performed by a health care professional who is licensed in the state or jurisdiction where the procedures are performed. Information concerning medical monitoring results and exposure characterizations must be kept by employers in separate confidential employee medical files. This requirement is found in the Occupational Safety and Health Administration (OSHA) standard 29 CFR 1910.1020. Employees must be notified annually of the existence of this medical data. In addition, employers who keep this data current can identify potential exposures to a licensed health care provider (LHCP) and subsequently the need for certain types of body system characterizations or investigations. The information is provided to the LHCP at annual medical reviews.

Initial Medical Monitoring

Initial medical monitoring is performed prior to employees entering hazardous environments. The purpose of this medical monitoring is to determine that work performed by the employees in hazardous environments or with protective equipment is not contraindicated, that employees can perform safely using the appropriate administrative controls or PPE, and to determine a baseline of body functions, prior to any exposure or use of restrictive equipment or procedures. To be able to make an initial determination, the following information should be provided to the physician or LHCP:

- An accurate description of the employees' job functions
- Information concerning hazardous environments to which the employee may be exposed
- Information concerning the types of PPE or administrative controls in place for employee protection

Medical monitoring generally will initiate with an employee occupational health history, as well as a physical examination performed by the LHCP. In the case of initial medical approval to use a respirator, current OSHA regulations require only a review of a brief initial medical evaluation (**TABLE 10-1**). Specific testing may additionally be required by regulation or may be recommended by the LHCP (e.g., chest x-rays, spirometry testing, or body system function testing). Baseline levels are established at this point. At the conclusion of initial medical monitoring, employers are notified of the acceptability for the employee to work under the conditions specified. The LHCP should notify employees, but not employers, of health-related conditions that do not affect on-the-job functions.

Under the requirements of the Medical Recordkeeping Standard, employee medical information must be kept for 30 years, but it is strongly advisable that employee medical records be maintained indefinitely.

Medical Monitoring for Individuals Who Work with Asbestos-Containing Materials

For individuals who will work with asbestos-containing materials above the permissible exposure limit (PEL) (0.1 f/cc TWA[8]) for 30 days or more per year, full medical monitoring includes a mandatory health history, chest x-ray, and spirometry testing. Asbestos-containing materials are lung toxins. X-rays reviewed under the Asbestos Standard must be reviewed by a qualified National Institute of Occupational Safety and Health (NIOSH) "B" licensed reader. This monitoring must be provided within 10 days of the 30th day of exposure.

TABLE 10-1 OSHA Respirator Medical Evaluation Questionnaire

OSHA Regulations (Standards–29 CFR)
Appendix C to Sec. 1910.134: OSHA Respirator Medical Evaluation Questionnaire (Mandatory)

To the employer: Answers to questions in Section 1, and to question 9 in Section 2 of Part A, do not require a medical examination.

To the employee:

Can you read? (circle one):   Yes/No

Your employer must allow you to answer this questionnaire during normal working hours, or at a time and place that is convenient to you. To maintain your confidentiality, your employer or supervisor must not look at or review your answers, and your employer must tell you how to deliver this questionnaire to the health care professional who will review it.

Part A. Section 1. (Mandatory) The following information must be provided by every employee who has been selected to use any type of respirator (please print).
1. Today's date: _____
2. Your name: _____
3. Your age (to nearest year): _____
4. Sex (circle one): Male/Female
5. Your height: _____ ft. _____ in.
6. Your weight: _____ lbs.
7. Your job title: _____
8. A phone number where you can be reached by the health care professional who reviews this questionnaire (include the Area Code): _____
9. The best time to phone you at this number: _____
10. Has your employer told you how to contact the health care professional who will review this questionnaire? (circle one):   Yes/No
11. Check the type of respirator you will use (you can check more than one category):
    a. _____ N, R, or P disposable respirator (filter-mask, non-cartridge type only)
    b. _____ Other type (for example, half- or full-facepiece type, powered air purifying, supplied-air, self-contained breathing apparatus).
12. Have you worn a respirator? (circle one):   Yes/No

If "yes," what type(s): _____

Part A. Section 2. (Mandatory) Questions 1 through 9 below must be answered by every employee who has been selected to use any type of respirator (please circle "yes" or "no").
1. Do you currently smoke tobacco, or have you smoked tobacco in the last month: Yes/No
2. Have you ever had any of the following conditions?
    a. Seizures (fits):   Yes/No
    b. Diabetes (sugar disease):   Yes/No
    c. Allergic reactions that interfere with your breathing: Yes/No
    d. Claustrophobia (fear of closed-in places):   Yes/No
    e. Trouble smelling odors: Yes/No
3. Have you ever had any of the following pulmonary or lung problems?
    a. Asbestosis: Yes/No
    b. Asthma: Yes/No
    c. Chronic bronchitis: Yes/No
    d. Emphysema: Yes/No
    e. Pneumonia: Yes/No
    f. Tuberculosis: Yes/No
    g. Silicosis: Yes/No
    h. Pneumothorax (collapsed lung):   Yes/No
    i. Lung cancer: Yes/No
    j. Broken ribs: Yes/No
    k. Any chest injuries or surgeries: Yes/No
    l. Any other lung problems that you've been told about: Yes/No

TABLE 10-1 (continued)

4. Do you currently have any of the following symptoms of pulmonary or lung illness?
   a. Shortness of breath: Yes/No
   b. Shortness of breath when walking fast on level ground or walking on a slight hill or incline: Yes/No
   c. Shortness of breath when walking with other people at an ordinary pace on level ground: Yes/No
   d. Have to stop for breath when walking at your own pace on level ground: Yes/No
   e. Shortness of breath when washing or dressing yourself: Yes/No
   f. Shortness of breath that interferes with your job: Yes/No
   g. Coughing that produces phlegm (thick sputum):   Yes/No
   h. Coughing that wakes you early in the morning: Yes/No
   i. Coughing that occurs mostly when you are lying down: Yes/No
   j. Coughing up blood in the last month: Yes/No
   k. Wheezing: Yes/No
   l. Wheezing that interferes with your job: Yes/No
   m. Chest pain when you breathe deeply: Yes/No
   n. Any other symptoms that you think may be related to lung problems: Yes/No
5. Have you ever had any of the following cardiovascular or heart problems?
   a. Heart attack: Yes/No
   b. Stroke: Yes/No
   c. Angina: Yes/No
   d. Heart failure: Yes/No
   e. Swelling in your legs or feet (not caused by walking):   Yes/No
   f. Heart arrhythmia (heart beating irregularly):   Yes/No
   g. High blood pressure: Yes/No
   h. Any other heart problem that you've been told about: Yes/No
6. Have you ever had any of the following cardiovascular or heart symptoms:
   a. Frequent pain or tightness in your chest: Yes/No
   b. Pain or tightness in your chest during physical activity: Yes/No
   c. Pain or tightness in your chest that interferes with your job: Yes/No
   d. In the past two years, have you noticed your heart skipping or missing a beat: Yes/No
   e. Heartburn or indigestion that is not related to eating: Yes/No
   f. Any other symptoms that you think may be related to heart or circulation problems: Yes No.
7. Do you currently take medication for any of the following problems:
   a. Breathing or lung problems: Yes/No
   b. Heart trouble: Yes/No
   c. Blood pressure: Yes/No
   d. Seizures (fits):   Yes/No
8. If you've used a respirator, have you ever had any of the following problems? (If you've never used a respirator, check the following space and go to question 9:)
   a. Eye irritation: Yes/No
   b. Skin allergies or rashes: Yes/No
   c. Anxiety: Yes/No
   d. General weakness or fatigue: Yes/No
   e. Any other problem that interferes with your use of a respirator: Yes/No
9. Would you like to talk to the health professional who will review this questionnaire about your answers to this questionnaire: Yes/No

   Questions 10 to 15 below must be answered by every employee who has been selected to use either a full-facepiece respirator or a self-contained breathing apparatus (SCBA). For employees who have been selected to use other types of respirators, answering these questions is voluntary.
10. Have you ever lost vision in either eye (temporarily or permanently):   Yes/No
11. Do you currently have any of the following vision problems?
    a. Wear contact lenses: Yes/No
    b. Wear glasses: Yes/No
    c. Color blind: Yes/No
    d. Any other eye or vision problems: Yes/No
12. Have you ever had an injury to your ears, including a broken ear drum: Yes/No
13. Do you currently have any of the following hearing problems?
    a. Difficulty hearing: Yes/No
    b. Wear a hearing aid: Yes/No
    c. Any other hearing or ear problem: Yes/No

(continues)

TABLE 10-1 (continued)

14. Have you ever had a back injury: Yes/No
15. Do you currently have any of the following musculoskeletal problems?
    a. Weakness in any of your arms, hands, legs, or feet: Yes/No
    b. Back pain: Yes/No
    c. Difficulty fully moving your arms and legs: Yes/No
    d. Pain or stiffness when you lean forward or backward at the waist: Yes/No
    e. Difficulty fully moving your head up or down: Yes/No
    f. Difficulty fully moving your head side to side: Yes/No
    g. Difficulty bending at your knees: Yes/No
    h. Difficulty squatting to the ground: Yes/No
    i. Climbing a flight of stairs or a ladder carrying more than 25 lbs.: Yes/No
    j. Any other muscle or skeletal problem that interferes with using a respirator: Yes/No

Part B. Any of the following questions, and other questions not listed, may be added to the questionnaire at the discretion of the health care professional who will review the questionnaire.

1. In your present job, are you working at high altitudes (over 5000 feet) or in a place that has lower than _____ normal amounts of oxygen? Yes/No

If "yes," do you have feelings of dizziness, shortness of breath, pounding in your chest, or other symptoms when you're working under these conditions? Yes/No

2. At work or at home, have you ever been exposed to hazardous solvents, hazardous airborne chemicals (e.g., gases, fumes, or dust), or have you come into skin contact with hazardous chemicals? Yes/No

If "yes," name the chemicals if you know them: _____
_____
_____

3. Have you ever worked with any of the materials, or under any of the conditions, listed below?
    a. Asbestos: Yes/No
    b. Silica (e.g., sandblasting): Yes/No
    c. Tungsten/cobalt (e.g., grinding or welding this material): Yes/No
    d. Beryllium: Yes/No
    e. Aluminum: Yes/No
    f. Coal (for example, mining): Yes/No
    g. Iron: Yes/No
    h. Tin: Yes/No
    i. Dusty environments: Yes/No
    j. Any other hazardous exposures: Yes/No

If "yes," describe these exposures: _____
4. List any second jobs or side businesses you have: _____
5. List your previous occupations: _____
6. List your current and previous hobbies: _____
7. Have you ever been in the military service? Yes/No

If "yes," were you exposed to biological or chemical agents (either in training or combat)? Yes/No

8. Have you ever worked on a HAZMAT team? Yes/No
9. Other than medications for breathing and lung problems, heart trouble, blood pressure, and seizures mentioned earlier in this questionnaire, are you taking any other medications for any reason (including over-the-counter medications)? Yes/No

If "yes," name the medications if you know them: _____

10. Will you be using any of the following items with your respirator(s)?
    a. HEPA Filters: Yes/No
    b. Canisters (for example, gas masks): Yes/No
    c. Cartridges: Yes/No
11. How often are you expected to use the respirator(s): (circle "yes" or "no" for all answers that apply to you):
    a. Escape only (no rescue): Yes/No
    b. Emergency rescue only: Yes/No
    c. Less than 5 hours per week: Yes/No
    d. Less than 2 hours per day: Yes/No
    e. 2 to 4 hours per day: Yes/No
    f. Over 4 hours per day: Yes/No

TABLE 10-1 (continued)

12. During the period you are using the respirator(s), is your work effort:
    a. Light (less than 200 kcal per hour):   Yes/No

If "yes," how long does this period last during the average shift: _____ hrs. _____ mins.
Examples of light work effort are sitting while writing, typing, drafting, or performing light assembly work; or standing while operating a drill press (1-3 lbs.) or controlling machines.

  b. Moderate (200 to 350 kcal per hour):   Yes/No

If "yes," how long does this period last during the average shift: _____ hrs. _____ mins.
Examples of moderate work effort are sitting while mailing or filing; driving a truck or bus in urban traffic; standing while drilling, nailing, performing assembly work, or transferring a moderate load (about 35 lbs.) at trunk level; walking on a level surface about 2 mph or down a 5-degree grade about 3 mph; or pushing a wheelbarrow with a heavy load (about 100 lbs.) on a level surface.

  c. Heavy (above 350 kcal per hour):   Yes/No

If "yes," how long does this period last during the average shift: _____ hrs. _____ mins.
Examples of heavy work are lifting a heavy load (about 50 lbs.) from the floor to your waist or shoulder; working on a loading dock; shoveling; standing while bricklaying or chipping castings; walking up an 8-degree grade about 2 mph; climbing stairs with a heavy load (about 50 lbs.).

13. Will you be wearing protective clothing and/or equipment (other than a respirator) when you're using your respirator? Yes/No

If "yes," describe this protective clothing and/or equipment: _____
_____

14. Will you be working under hot conditions (temperature exceeding 77 deg. F)?   Yes/No
15. Will you be working under humid conditions?   Yes/No
16. Describe the work you'll be doing while you're using your respirator(s):
_____
_____

17. Describe any special or hazardous conditions you might encounter when you're using your respirator(s) (for example, confined spaces, life-threatening gases):
_____
_____

18. Provide the following information, if you know it, for each toxic substance that you'll be exposed to when you're using your respirator(s):

Name of the first toxic substance: _____
Estimated maximum exposure level per shift: _____
Duration of exposure per shift: _____
Name of the second toxic substance: _____
Estimated maximum exposure level per shift: _____
Duration of exposure per shift: _____
Name of the third toxic substance: _____
Estimated maximum exposure level per shift: _____
Duration of exposure per shift: _____
The name of any other toxic substances that you'll be exposed to while using your respirator: _____
_____
_____

19. Describe any special responsibilities you'll have while using your respirator(s) that may affect the safety and well-being of others (for example, rescue, security):
_____
_____

Bring this completed questionnaire to your scheduled appointment!

OSHA Regulations (Standards – 29 CFR) – 1910. 134 App C

TABLE 10-2 OSHA: Blood Lead and ZPP Testing Requirements/Timeframes

Exposure Level	Time Frame
Initial determination	Prior to work
> 30 μg/m	Every six months
> 50 μg/m³	Every three months
> 200 μg/m³	Every two months

Source: 29 CFR 1926.62; 29 CFR 1910.1025.

Medical Monitoring for Individuals Who Work with Lead

For individuals who will work with leaded materials, initial blood lead and zinc protoporphyrin (ZPP) levels are required before work begins (i.e., before air monitoring can characterize actual exposures). Subsequently, full medical evaluations are required if exposure is above the PEL (50 mg/m³) for 30 days or more in a calendar year. In addition, repetitive blood lead and ZPP testing must occur on a specific schedule (TABLE 10-2).

Medical Monitoring for Individuals Who Perform Hazardous Waste Work, Emergency Responders to Releases of Hazardous Substances, and U.S. Department of Transportation Hazmat Drivers

Individuals at recognized hazardous waste sites are required to receive full medical monitoring if exposures greater than the PEL are anticipated for 30 days or more in any calendar year. In addition, Hazmat technicians and specialists, classified and certified under subsection "q" of the OSHA HAZWOPER Standard (29 CFR 1910.120/29 CFR 1926.65), must be medically preapproved prior to job assignment. Hazmat employees (U.S. Department of Transportation [DOT]) are required to receive medical preapproval (including drug testing) if they are classified as commercial drivers. Commercial driver Hazmat employees operate vehicles that are placardable (carry placarded quantities of hazardous materials) on public roadways (49 CFR 391.41). A DOT medical exam and commercial motor vehicle driver certification is valid for a two year period, however may be issued for less time based on physical findings.

Medical Monitoring (Other Toxins)

A variety of OSHA standards requires medical preplacement examinations. These are summarized in TABLE 10-3. In addition, the U.S. Public Health Service, Centers for Disease Control and Prevention, NIOSH recommends a medical preplacement examination for work in a variety of hazardous atmospheres, such as work in hot or cold environments.

Physical Performance Testing

Once medical preplacement has been accomplished, work in certain atmospheres may require physical performance testing to ensure that individuals are capable of the physical tasks required. Physical performance testing is generally accomplished before work and at routine intervals (such as semiannually). Public safety organizations (police, fire, and rescue services) routinely perform physical performance testing to document both initial and ongoing qualification. Physical performance tests are many times used to assess the correlation between an individual's physical capabilities and their ability to perform essential job tasks, the critical duties essential to job completion. Physical performance testing may use actual operations that are required as part of the job (e.g., climbing a ladder) or may use exercises that mimic necessary strengths (e.g., sit-ups, push-ups, aerobic runs).

Individuals should not participate in physical performance testing until medically preapproved for the activity (job) they are to perform.

Psychological Testing and Job Matching

Many occupations, in particular those that require entry into hazardous atmospheres, must ensure that individuals are psychologically capable of performing in those atmospheres in order to protect themselves and others.

There are many types of psychological testing. Initial testing may take the form of job or characteristic matching (e.g., the Minnesota Multi-Phase Personality Test), which is generally administered by a professional psychologist. Shorter versions of other tests may be administered and interpreted by employers (e.g., Wonderlic Testing Corporation).

Certain psychological testing regimens are also designed to identify potential problems with stressful environments (e.g., claustrophobia) or the inability of individuals to operate safely under certain stresses (e.g., hot or cold atmospheres).

Smoking Cessation

Tobacco use generally exacerbates exposures to toxic materials, either through its ability to anesthetize the cilia

TABLE 10-3 OSHA Medical Reference Chart

NAME OF SUBSTANCE	CAS NUMBER	SPECIFIC MEDICAL TESTS OR EXAMINATIONS (REFERENCES IN SUPERSCRIPT)
Acetaldehyde	00075-07-0	BL[3] ExA[3,4] U[3]
Acetone	00067-64-1	BL[3,5] BL(DE)[15] ExA[3-5] ExA(DE)[15] PFT[2] U[5] U(DE)[15] U(EOS)[3,18]
Acetonitrile	00075-05-8	BL(Lac pH)[7] BLC[2,3,16] BLP[2,3] BLP(Bicarb)[7]ExA[4] U[2,3]
Acetylaminofluorene, 2-	00053-96-3	InR(RIC Ster Preg Cig)[1,7] U[2]
Acetylene tetrabromide	00079-27-6	ExA[4] LFT[2,7]
Acrolein	00107-02-8	BGa[7] CXr[2,7] Ecg[7] ExA[4] PFT(FVC FEV1)[2,7] SpC[7] WBC[7]
Acrylamide	00079-06-1	NCS[7,15] Neur[7] RBC[5]
Acrylonitrile	00107-13-1	CXr[1] FOB[1] BL(Lac pH)[7] BLC[3] BLP[3] BLP(Bicarb)[7] BLS[15] ExA[4] PFT[2] RBC[5] U[3,5,15]
Aldrin	00309-00-2	BL[7] BLP[3] CBC[2] LFT[2] ReC[2] Ua[7]
Allyl chloride	00107-05-1	CBC[2] CXr[2,7] ExA[4] LFT[2] PFT(FVC FEV1)[2,7] Ua[2]
Aluminum metal/powder	07429-90-5	BLS[3,5,9,14] CXr[7] Rdg[7,14] SpC[7] TBS[7] U[3,5,9,17]
Aminodiphenyl, 4-	00092-67-1	InR(RIC Ster Preg Cig)[1,7] Cys[7] U[2] Ua[7]
Ammonia	07664-41-7	BGa[7] CXr[2,7] Ecg[7] ExA[4] PFT(FVC FEV1)[2,7] SpC[7] WBC[7]
Aniline	00062-53-3	BL[5] BL(COHb)[4] BL(Hgb)[7] BL(MHgb)[2-5,7,8,11,16] BL(MHgb EOS)[15] CBC[2] RBC[5] U[3-6,12] U(EOS)[2,11,15,18] U(EWW)[18] Ua(Cyt Hmt)[2]
Anisidine (o-,p-isomers)	Varies	BL(MHgb)[2,7,8,11] CBC[2,7] ReC[2] Ua[2]
Antimony and compounds	Varies	BL[9,17] BT[9] Cxr[7] Ecg[7] PFT(FVC FEV1)[2,7] U[3,5,9,13]
Arsenic, inorganic and compounds	Varies	CXr[1,7] Nas[1] Skn[1] BH[4,5,14] BL[5] CBC[7] NCS[14] PFT(FVC FEV1)[2,7] U[4,5,7,9,12-14] U(EWW)[11,15,16,18] U(PPS)[3]
Arsine	07784-42-1	BLP(Hgb)[7,16] CBC[7] LFT[7] PFT[2] ReC[16] U(Hgb)[7,16] Ua[7] WBC[16]
Asbestos	Varies	CXr[1,2,7] PFT(FVC FEV1)[1,2,7] Rdg[7] SpC[7] TBS[7]
Azinphos-methyl (Guthion®)	00086-50-0	BLS[9] CH(BLS RBC)[2] CH(RBC)[7,11]
Barium and compounds	Varies	BL[5] CXr[2,7] Ecg[2] PFT(FVC FEV1)[2,7] U[5,9,17]
Benomyl	17804-35-2	U[5]
Benzene	00071-43-2	CBC(Leuk Thrm Hct Hgb Ery)[1,2] PFT[1] BL[3,5,12] BL(PNS)[15] CBC[7] ExA[3-5,12] ExA(EWW)[18] ExA(PNS)[15] ExA(PTS)[18] U[3-5,9,10,12] U(DE)[5] U(EOS)[11,15,18] U(EWW)[18] U(PPS)[17] WBC[7]
Benzidine	00092-87-5	InR(RIC Ster Preg Cig)[1,7] Cys[7] PFT[2] U[2,3,17] Ua[7] Ua(Cyt[6M] Hmt[EM])[2] Ua(EM)[2]
Benzyl chloride	00100-44-7	CXr[2] ExA[4] PFT[2]
Beryllium and compounds	Varies	BGa[7] BT[9] CXr[2,7,16] Ecg[7] PFT(FVC FEV1)[2,7] SpC[7] U[3,5] WBC[7] WBC(LTT)[2]
Bis(chloromethyl)ether	00542-88-1	InR(RIC Ster Preg Cig)[1,7] PFT[2] SpC[15]
Boron trifluoride	07637-07-2	BGa[7] CXr[2,7] Ecg[7] PFT(FVC FEV1)[2,7] SpC[7] U(PPS)[2] WBC[7]
Bromine	07726-95-6	BGa[7] CXr[2,7] Ecg[7] ExA[4] PFT(FVC FEV1)[2,7] SpC[7] WBC[7]
Butadiene, 1,3-	00106-99-0	CBC[1] ExA[4] RBC[5] U[5]
Butoxyethanol, 2-	00111-76-2	BLS(Hgb)[7] CBC[2,7] RBC(Frg)[2] WBC[7]
Butyl alcohol, n-	00071-36-3	BL[3] ExA[4]
Butyl chromate, tert-	01189-85-1	CXr[2] U[2]
Butyl glycidyl ether, - (BGE)	02426-08-6	ExA[4] PFT[2]
Butyl mercaptan, n-	00109-79-5	BGa[7] Ecg[7] ExA[4] PFT[2] SpC[7] WBC[7]
Butyltoluene, p-tert-	00098-51-1	CBC[2] Ecg[2] ExA[4]
Cadmium dust and fume	Varies	BL[1,3-5,9-11,13,15,17,18] Ua(ß-2M)[1] BGa[7] BT[9] CBC[7] CXr[2,7] Ecg[7] LFT[2] PFT(FVC FEV1)[2,7] SpC[7] U[3-5,7,9-11,13-15,18] Ua[7] Ua(Cyt)[2]
Camphor, synthetic	00076-22-2	BL(BUN Ca CO$_2$ Glu)[7] BLP[3] U[3]
Carbaryl (Sevin®)	00063-25-2	BL(EOS)[15] BLS[5,9] CH(PPS)[2,3] CH(RBC PPS)[7] U[3-5,7] U(EOS)[15] Ua[2]
Carbon black	01333-86-4	PFT[2]
Carbon disulfide	00075-15-0	BUN[7] Ecg[2,7] ExA[4] LFT[2,7] Oph[2,7] U[5] U(Sed)[7] U(EOS)[11,15,18] U(ESW)[5] U(EWW)[18] U(PPS)[3] Ua[2,7]
Carbon monoxide	00630-08-0	BL[5] BL(COHb)[2-5,7,8,12] BL(COHb EOS)[11,18] BL(COHb EWW)[18] CBC[2] ExA[3-5,7] ExA(EOS)[11,18] ExA(EWW)[18]
Carbon tetrachloride	00056-23-5	BL[3,5] ExA[3-5] KFT[7] LFT[2,7] U[5] Ua[2,7]
Chlordane	00057-74-9	BL[15] BL(BUN Ca CO$_2$ Glu)[7] BLP[3] BLS[9] BT[3] Ua[2,7]
Chlorinated diphenyl oxide	55720-99-5	LFT[2]
Chlorine	07782-50-5	BGa[7] CXr[2,7] Ecg[7] PFT(FVC FEV1)[2,7,8] SpC[7] WBC[7]
Chlorine dioxide	10049-04-4	BGa[7] CXr[2,7] Ecg[7] PFT(FVC FEV1)[2,7] SpC[7] WBC[7]
Chlorine trifluoride	07790-91-2	BGa[7] CXr[2,7] Ecg[7] PFT(FVC FEV1)[2,7] SpC[7] WBC[7]
Chlorobenzene	00108-90-7	BL[3] U[5] U(EOS)[3,18] U(EWW)[18]
Chloroform (Trichloromethane)	00067-66-3	BL[3,5] ExA[3-5] LFT[2,7] Ua[2,7]
Chloromethyl methyl ether	00107-30-2	InR(RIC Ster Preg Cig)[1,7] CXr[7] PFT[2] SpC[15]
Chloropicrin	00076-06-2	BGa[7] CXr[2,7] Ecg[7] ExA[4] PFT(FVC FEV1)[2,7] SpC[7] WBC[7]
Chloroprene, beta-	00126-99-8	CXr[2,7] PFT[2]
Chromic acid and chromates	Varies	BGa[7] CBC[2] CXr[2,7] Ecg[7] LFT[2] PFT[2,7] SpC[7] U[7] Ua[2,7] WBC[7]
Chromium, metal and compounds	Varies	BL[3,9,12,17] BT[9] CXr[2,7] PFT[2] RBC[2] U[4,9,11-13,17] U(EOS)[3,15,18] U(ESW)[5,11] U(EWW)[15,18] U(PPS)[5]
Coal dust	68131-74-8	CXr[7] PFT(FVC FEV1)[7] Rdg[7] SpC[7] TBS[7]
Coal tar pitch volatiles	65996-93-2	CBC[2] CXr[2,7] PFT(FVC FEV1)[2,7] PpT[2,5] SpC[2,7] Ua[2] Ua(Cyt Hmt)[2]
Cobalt metal, dust, and fume	07440-48-4	BL[4,5,9,11,13,17] BL(EOS)[18] BL(ESW)[11] BL(EWW)[18] BLS[5] BT[9] CXr[2,7] PFT(FVC FEV1)[2,7] U[5,9,13] U(EOS)[18] U(ESW)[3,11] U(EWW)[18]
Coke oven emissions	—	CXr[1,7] PFT(FVC FEV1)[1,7] Skn[1,7] U(Cyt)[1,7] Ua(Glu Alb Hmt)[1] Cys[7] SpC[7] Ua[7]
Copper dust and mist	07440-50-8	BL[4,9,17] BLS[3,5] BT[9] U[3,5,7,9,17] U(24H)[3]
Cotton dust	—	PFT(FVC FEV1 FEV1/FVC)[1] CXr[2] PFT(FVC FEV1)[7] PFT(PPS)[2]
Cresol, p-	00106-44-5	LFT[2] PFT[2] U[9] U(PPS)[3,17]
Cumene	00098-82-8	BL[5] ExA[5] U(L2H)[3] U[5]
Cyanide	00057-12-5	BL[5] BL(Lac pH)[7,16] BLC[3,7,16] BLP[3,5] BLP(Bicarb)[7] CXr[2] PFT[2] U[3,5,15]
Cyclohexane	00110-82-7	BL[3] BL(DE)[15] ExA[3,4] ExA(DE)[5,15] U[3,5,10] U(EOS)[3] U(L4H)[15]

(continues)

TABLE 10-3 (continued)

NAME OF SUBSTANCE	CAS NUMBER	SPECIFIC MEDICAL TESTS OR EXAMINATIONS (REFERENCES IN SUPERSCRIPT)
Cyclohexanone	00108-94-1	ExA[4] PFT[2] U[5] U(EOS)[11] U(ESW)[11]
Demeton (Systox®)	08065-48-3	CH(BL)[10] CH(BL BLP RBC)[16] CH(BLS RBC)[2] CH(RBC)[7,11] U[5]
Diacetone alcohol	00123-42-2	ExA[4] PFT[2]
Diazomethane	00334-88-3	BGa[7] CXr[2,7] Ecg[7] PFT(FVC FEV1)[2,7] SpC[7] WBC[7]
Diborane	19287-45-7	CXr[2,7] PFT(FVC FEV1)[2,7]
Dibromo-3-chloropropane, 1,2-(DBCP)	00096-12-8	BLS(FSH LH TE)[1] GUT(Test Sct)[1] BLS(FSH LH SMA-12 Testo)[7] CBC[7] PFT[2] SCt[7] TFT[7] Ua[7]
Dichlorobenzene, o-	00095-50-1	U[5]
Dichlorobenzene, p-	00106-46-7	ExA[4] LFT[2,7] U[2,3,5,7]
Dichlorobenzidine, 3,3'-	00091-94-1	InR(RIC Ster Preg Cig)[1,7] PFT[2]
Dichlorodiphenyltrichloroethane (DDT)	00050-29-3	AdT[3] BL[5,12] BLS[3,5,9] BT[3] U[3-5,7]
Dichlorophenoxyacetic acid,2,4- (2,4-D)	00094-75-7	BLP[3] U[3] U(24H)[12]
Dichloroethyl ether	00111-44-4	BGa[7] CXr[2,7] Ecg[7] PFT(FVC FEV1)[2,7] SpC[7] WBC[7]
Dichlorvos (DDVP)	00062-73-7	BLS[9] CH (BL)[3,10] CH(BLP RBC)[16] CH(BLS RBC)[2] CH(RBC)[7,11] U[3]
Dieldrin	00060-57-1	BL[3,5,7,15] BLS[9] BT[3] U[5]
Diglycidyl ether (DGE)	02238-07-5	CBC[7] CXr[7] PFT(FVC FEV1)[2,7]
Diisobutyl ketone	00108-83-8	PFT[2]
Dimethyl acetamide	00127-19-5	LFT[2,7] U[5] U(EOS)[18] U(ESW)[11] U(EWW)[18]
Dimethyl sulfate	00077-78-1	BGa[7] BUN[7] CXr[2,7] Ecg[7] LFT[2,7] PFT(FVC FEV1)[2,7] SpC[7] U(Sed)[7] Ua[2,7] WBC[7]
Dimethylaminoazobenzene, 4-	00060-11-7	InR(RIC Ster Preg Cig)[1,7]
Dimethylaniline (N,N-Dimethylaniline)	00121-69-7	BL(MHgb)[2,7,11,16] CBC[2,7]
Dimethyl-1,2-dibromo-2,2-dichloroethyl phosphate (Naled)	00300-76-5	CH(BLP RBC)[16] CH(BLS RBC)[2] CH(RBC)[7,11]
Dimethylformamide (DMF)	00068-12-2	BL[5] ExA[4,5] LFT[2,7] U[5,10,12] U(24H)[3] U(EOS)[11,15,18] U(EWW)[5,18]
Dimethylhydrazine, 1,1-	00057-14-7	BL(BUN Ca CO_2 Glu)[7] CBC[7] LFT[7] PFT[2] U[7]
Dinitrobenzene (all isomers)	Varies	BL(MHgb)[2,7,11,16] BL(MHgb EOS)[15] CBC[2,7] LFT[2,7]
Dinitro-o-cresol	00534-52-1	BLW[3,5]
Dinitrotoluene, 2,4-	00121-14-2	BL(MHgb)[2,7,8,11] CBC[2] LFT[2,7] U[5]
Dioxane (Diethylene dioxide)	00123-91-1	BL(DE)[15] BLP[3] BUN[7] CXr[2] ExA[4] KFT[7] LFT[2,7] PFT[2] U[5] U(Sed)[7] U(EOS)[3,15] Ua[2]
Diphenyl (Biphenyl)	00092-52-4	U[5]
Endrin	00072-20-8	BL[5,7] BL(EOS)[15] BLP[3] BLS[9] U[3,5] U(EOS)[15]
Epichlorohydrin	00106-89-8	CXr[2,7] ExA[4] LFT[2,7] PFT(FVC FEV1)[2,7] Ua[2]
EPN	02104-64-5	CH(BLS RBC)[2] CH(RBC)[7,11]
Ethoxyethanol, 2- (ethyl cellosolve)	00110-80-5	U(ESW)[3]
Ethyl alcohol (Ethanol)	00064-17-5	BL[3,9] ExA[3,4] U(EOS)[3]
Ethyl benzene	00100-41-4	BL(DE)[5] ExA[4,5,11] ExA(EWW)[11] ExA(PNS)[15] U[5] U(EOS)[15,18] U(ESW)[11] U(EWW)[18] U(L2H)[3]
Ethyl ether (Ether)	00060-29-7	BL[3] ExA[3,4] U[3]
Ethyl mercaptan	00075-08-1	BGa[7] Ecg[7] ExA[4] PFT[2] SpC[7] WBC[7]
Ethylene chlorohydrin	00107-07-3	ExA[4] LFT[2,7] Ua[2,7]
Ethylene dibromide	00106-93-4	BUN[7] CXr[2] Ecg40[2] ExA[4] LFT[2,7] PFT[2] U[5] U(Sed)[7]
Ethylene dichloride	00107-06-2	ExA[4] LFT[7] PFT[2]
Ethylene glycol dinitrate	00628-96-6	BL[3,5,12] BL(DE)[5,15] CBC[7] Ecg[7] ExA(DE)[5] U[3,5,12]
Ethylene oxide	00075-21-8	CBC(WBC RBC Hct Hgb)[1] BGa[7] BL[3,5] CBC[7] CXr[7] Ecg[7] ExA[3-5] PFT(FVC FEV1)[2,7] RBC[5] SpC[7] WBC[7]
Ethyleneimine	00151-56-4	InR(RIC Ster Preg Cig)[1] BGa[7] CXr[7] Ecg[7] ExA[4] PFT(FVC FEV1)[2,7] SpC[7] WBC[7]
Ferrovanadium dust	12604-58-9	CXr[2] PFT[2]
Fluoride, inorganic	16984-48-8	BLP[3,12] CXr[2,7] PFT(FVC FEV1)[2,7] PXr[2,7] U[2,4,9,12] U(EOS)[3,11,18] U(EWW)[15,18] U(PNS)[15] U(PPS)[5,7,17] U(PTS)[11] Ua[2]
Fluorine	07782-41-4	BGa[7] CXr[2,7] Ecg[7] PFT(FVC FEV1)[2,7] SpC[7] WBC[7]
Formaldehyde	00050-00-0	PFT(FVC FEV1 FEF)[1,7] BL[3,9] CXr[7] PFT(PPS)[2] PpT[7] U[9] U(PPS)[3]
Furfural	00098-01-1	U[5] U(EOS)[18]
Furfuryl alcohol	00098-00-0	ExA[4] PFT[2] U[5]
Graphite, natural	07782-42-5	CXr[2,7] PFT(FVC FEV1)[2,7]
Hexachloronaphthalene (Halowax 1014)	01335-87-1	LFT[2,7]
Hexane, n-	00110-54-3	BL[3] BL(DE)[5] ExA[3,4] ExA(DE)[5,15] Neur[7] U[3,5,10] U(EOS)[3,15,18] U(ESW)[11] U(EWW)[5,18]
Hydrazine	00302-01-2	CXr[7] PFT(FVC FEV1)[2,7]
Hydrogen chloride (Hydrochloric acid)	07647-01-0	CXr[7] PFT(FVC FEV1)[2,7]
Hydrogen cyanide (Hydrocyanic acid)	00074-90-8	BL(Lac pH)[7] BL(COHb)[4] BLP(Bicarb)[7] Ecg40[2] ExA[4]
Hydrogen fluoride (Hydrofluoric acid)	07664-39-3	BGa[7] CXr[2,7] Ecg[7] PFT(FVC FEV1)[2,7] PXr[2,7] SpC[7] U[2,7] U(PPS)[7] Ua[2] WBC[7]
Hydrogen sulfide	07783-06-4	BL[3] BL(COHb)[4] BLP[3] CXr[2,7] PFT(FVC FEV1)[2,7]
Hydroquinone	00123-31-9	Oph[7] PFT[2]
Iodine	07553-56-2	CXr[2,7] PFT(FVC FEV1)[2,7]
Iron oxide fume	01309-37-1	BL[9,17] BT[9] CXr[2,7] PFT(FVC FEV1)[2,7] U[9,17]
Isophorone	00078-59-1	CXr[2] PFT[2]
Isopropyl alcohol (Isopropanol)	00067-63-0	BL[3] ExA[3-5] U[3,5]
Isopropyl glycidyl ether (IGE)	04016-14-2	PFT[2]
Ketene	00463-51-4	BGa[7] CXr[2,7] Ecg[7] PFT(FVC FEV1)[2,7] SpC[7] WBC[7]
Lead, inorganic	07439-92-1	BL(BLL Hgb Hct ZPP BUN Scr)[1] BLPr[1] Ua(Micr)[1] BL[3-5,7,9,11-13,15-17] BLL[2,7,12,14,16,18] BT[9] CBC[2,7] NCS[7] Neur[7] RBC[5] RBC(ZPP)[2,5,12] U[3-5,9,12-17] U(EOS)[17] Ua(24H)[7] ZPP[2,3,5,9,12,13] ZPP(1M)[2,15,16]
Lindane	00058-89-9	BL[2,3,5,7,12] BLS[15] CBC[2,7] U[4]
Lithium hydride	07580-67-8	BL[9,17] BLS[3] BT[9] U[3]
Magnesium oxide fume	01309-48-4	BL[17]
Malathion	00121-75-5	BLS[9] CH(BL)[3,10] CH(BLS RBC)[2] CH(RBC)[7,11] CXr[2] LFT[2] PFT[2] U[3]
Maleic anhydride	00108-31-6	BGa[7] Ecg[7] SpC[7] U[5] WBC[7]
Manganese and compounds	Varies	BL[4,5,9,13,17] BT[9] CBC[2] CXr[2] PFT[2] U[2,3,5,9,13,17] Ua[2]

TABLE 10-3 (continued)

NAME OF SUBSTANCE	CAS NUMBER	SPECIFIC MEDICAL TESTS OR EXAMINATIONS (REFERENCES IN SUPERSCRIPT)
Mercury, inorganic and compounds	Varies	BL[3-5,12,16] BL(EOS)[18] BL(ESW)[11] BL(EWW)[18] BT[9] BUN[7] NCS[14,16] Neur[7] TFT[7] U[2,3,5,7,9,12,13,16] U(PNS)[15] U(PTS)[11,18] U(Sed)[7] Ua[2,7]
Mercury, organo alkyl compounds	Varies	BH[5] BL[3,5,7,9] Neur[7] U[9] Ua[2,7]
Methoxyethanol, 2- (Methyl cellosolve)	00109-86-4	CBC[2,7] U(EOS)[3]
Methyl alcohol (Methanol)	00067-56-1	BL[5,10,16] BL(pH)[7,16] BL(PPS)[3] BLP(Bicarb)[7,16] ExA[3,4,10] LFT[2] U[3,5,7,10,16] U(EOS)[11,15,18] U(EWW)[15] U(PNS)[15]
Methyl n-amyl ketone	00110-43-0	PFT[2]
Methyl bromide	00074-83-9	BL[5,15,16] BLS[3,5] CXr[2] ExA[4] PFT(FVC FEV1)[2] U[3,5]
Methyl n-butyl ketone (MBK)	00591-78-6	BL[3] ExA[3] Neur[7] PFT[2] U[5] U(EOS)[3] U(ESW)[11] U(EWW)[5]
Methyl chloride	00074-87-3	BL[3] ExA[3,4] U[5]
Methyl chloroform	00071-55-6	BL[5,10,12] BL(DE)[3] BL(EOS)[15,18] BL(ESW)[5,11] BL(EWW)[5,10,15,18] ExA[3,4,10,12] ExA(16H)[5] ExA(EWW)[10,15,18] ExA(PNS)[3,15] ExA(PTS)[18] ExA(PWW)[11] U[5,10,12] U(EOS)[15,18] U(ESW)[5,11] U(EWW)[5,10,11,15,18] U(PNS)[3]
Methyl ethyl ketone (MEK)	00078-93-3	BL[3,5,9,10,17] ExA[3-5,10] PFT[2] U[5] U(EOS)[3,11,15,16,18]
Methyl hydrazine	00060-34-4	BL(MHgb)[16] PFT[2]
Methyl isobutyl ketone (MIBK)	00108-10-1	BL[3] ExA[3,4] PFT[2] U[5] U(EOS)[3,11,18] U(EWW)[18]
Methyl isocyanate	00624-83-9	BGa[7] CXr[2,7] Ecg[7] PFT(FVC FEV1)[2,7] SpC[7] WBC[7]
Methyl mercaptan	00074-93-1	BGa[7] CXr[7] Ecg[7] ExA[4] Neur[7] PFT(FVC FEV1)[2,7] SpC[7] WBC[7]
Methyl propyl ketone	00107-87-9	CXr[2] PFT[2]
Methyl styrene, alpha-	00098-83-9	U[5]
Methylene bisphenyl isocyanate (MDI)	00101-68-8	BGa[7] CXr[2,7] Ecg[7] PFT(FVC FEV1)[2,7] RBC[5] SpC[7] U[5] WBC[7]
Methylene chloride	00075-09-2	Lab(COHb Ecgr Hct LFT Chol)[1] BL[3,5] BL(COHb)[2,4,5,16] BL(COHb EOS)[15] BL(COHb PNS)[3,15] BL(EOS)[15] CBC[2,7] ExA[3-5,16] ExA(EOS)[15] LFT[2] U[4,5,16]
Methylene dianiline, 4,4'-	00101-77-9	Lab(LFT Ua)[1] Skn[1] U[5,16]
Mica	12001-26-2	CXr[2,7] PFT(FVC FEV1)[2,7]
Molybdenum and compounds	Varies	BL[4,9,17] BT[9] U[4,5,9,17]
Monomethyl aniline	00100-61-8	BL(MHgb)[2,7,8,11,16] CBC[2,7] Ua[7]
Naphthalene	00091-20-3	BLP(Hgb)[7] CBC[2,7] LFT[7] RBC(Hem)[2] U[5] U(Hgb)[7] Ua[2,7] WBC[7]
Naphthylamine, alpha-	00134-32-7	InR(RIC Ster Preg Cig)[1,7] Cys[7] Ua[7]
Naphthylamine, beta-	00091-59-8	InR(RIC Ster Preg Cig)[1,7] Cys[7] Ua[7]
Nickel, metal and compounds	Varies	BGa[7] BL[9,17] BLP[5,12,14] BLP(EOS)[3,15] BLS[13,16] BT[9] CXr[2,7] Ecg[7] PFT(PPS)[2,13] SpC[7] U[5,9,12,13,16,17] U(EOS)[3,15] WBC[7]
Nickel carbonyl	13463-39-3	BGa[7] BLP[3] CXr[2,7] Ecg[7] ExA (2H)[3] PFT(FVC FEV1)[2,7] SpC[2,7] U[3,5,7,12,16,17] Ua[2] WBC[7]
Nicotine	00054-11-5	BLP(DE PPS)[3] U[3]
Nitric acid	07697-37-2	CXr[2,7] PFT(FVC FEV1)[2,7]
Nitric oxide	10102-43-9	CXr[2,7] Ecg40[2] ExA[4] PFT(FVC FEV1)[2,7]
Nitroaniline, p-	00100-01-6	BL(MHgb)[2,7,8,11,16] CBC[2,7] LFT[2,7]
Nitrobenzene	00098-95-3	BL[5] BL(COHb)[4,15] BL(MHgb)[2,3,5,7,8,11,16] BL(MHgb EOS)[11] CBC[2,7] U[2,5,7] U(EOS)[15,18] U(ESW)[11] U(EWW)[3,12,18]
Nitrobiphenyl, 4-	00092-93-3	InR(RIC Ster Preg Cig)[1,7] U[2]
Nitrochlorobenzene, p-	00100-00-5	BL(MHgb)[2,7,8,11,16] CBC[2,7]
Nitrogen dioxide	10102-44-0	BGa[7] CXr[2,7] Ecg[7] Ecg40[2] ExA[4] PFT(FVC FEV1)[2,7] SpC[7] WBC[7]
Nitrogen trifluoride	07783-54-2	BL(MHgb)[2,7,8,11] CBC[2,7]
Nitroglycerine	00055-63-0	BL[5] BLP[3] CBC[7] Ecg[7]
Nitropropane, 2-	00079-46-9	PFT[2]
Nitrosodimethylamine, N-	00062-75-9	InR(RIC Ster Preg Cig)[1,7] LFT[7] PFT[2]
Nitrotoluene (all isomers)	Varies	BL(MHgb)[2,7,8,11,16] CBC[2,7] U[7]
Octachloronaphthalene	02234-13-1	LFT[2,7]
Oxalic acid	00144-62-7	U[3]
Oxygen difluoride	07783-41-7	BGa[7] CXr[2,7] Ecg[7] PFT(FVC FEV1)[2,7] SpC[7] WBC[7]
Ozone	10028-15-6	BGa[7] BL(PMN)[16] CXr[2,7] Ecg[7] PFT(FVC FEV1)[2,7] SpC[7] WBC[7]
Paraquat	04685-14-7	CXr[2,7] LFT[2] PFT(FVC FEV1)[2,7] U[3] Ua[2]
Parathion	00056-38-2	BLS[3,9] CH(BL)[3,7,10,15] CH(BLP)[16] CH(BLP RBC)[16] CH(BLS RBC)[2] CH(RBC)[3,7,11] U[3-5,10] U(EOS)[11,15,18] U(EWW)[18]
Pentachloronaphthalene	01321-64-8	LFT[2,7]
Pentachlorophenol	00087-86-5	BL[9,17] BLP[3,5] BLP(EOS)[11,18] BLS[9] U[2,3,5,7,9] U(EWW)[17,18] U(PTS)[18] U(PWW)[11]
Perchloroethylene (Tetrachloroethylene)	00127-18-4	BL[3,12,16] BL(16H)[5] BL(EWW)[15,18] BL(PNS)[15] BL(PTS)[18] BL(PWW)[11] ExA[3,4,10,12,16] ExA(16H)[5] ExA(DE)[5] ExA(EOS)[3,15] ExA(EWW)[15,18] ExA(PNS)[15] ExA(PTS)[18] ExA(PWW)[11] LFT[2,7] U[3,12,16] U(16H)[5,11] U(EOS)[15,18] U(EWW)[5,11,18] Ua[2]

(continues)

TABLE 10-3 (continued)

NAME OF SUBSTANCE	CAS NUMBER	SPECIFIC MEDICAL TESTS OR EXAMINATIONS (REFERENCES IN SUPERSCRIPT)
Perchloromethyl mercaptan	00594-42-3	BGa^7 CXr^{2,7} Ecg^7 LFT^2 PFT(FVC FEV1)^{2,7} SpC^7 Ua^2 WBC^7
Perchloryl fluoride	07616-94-6	BGa^7 BL(MHgb)^{2,7,8,16} CBC^{2,7} CXr^{2,7} Ecg^7 PFT(FVC FEV1)^{2,7} SpC^7 U^{16} WBC^7
Phenol	00108-95-2	LFT^{2,7} U^{2,5,7,9,12} U(L2H)^{3,15} U(EOS)^{11,15,18} U(PPS)^{17} Ua^{2,7}
Phenyl glycidyl ether (PGE)	00122-60-1	PFT^2
Phenylhydrazine	00100-63-0	BLP(Hgb)^7 CBC^{2,7} LFT^7 PFT^2 U(Hgb)^7 Ua^7 WBC^7
Phosdrin (Mevinphos®)	07786-34-7	BLS^9 CH(BLS RBC)^2 CH(RBC)^{7,11}
Phosgene (Carbonyl chloride)	00075-44-5	BGa^{7,16} CXr^{2,7} Ecg^{7,16} ExA^4 PFT(FVC FEV1)^{2,7} SpC^{7,16} WBC^{7,16}
Phosphine	07803-51-2	BGa^7 CXr^7 Ecg^7 PFT(FVC FEV1)^7 SpC^7 WBC^7
Phosphorus (Yellow)	07723-14-0	CBC(Anm)^2 DXr^{2,7} LFT^{2,7}
Phosphorus pentachloride	10026-13-8	CXr^{2,7} PFT(FVC FEV1)^{2,7}
Phosphorus pentasulfide	01314-80-3	CXr^{2,7} PFT(FVC FEV1)^{2,7}
Phosphorus trichloride	07719-12-2	BGa^7 CXr^{2,7} Ecg^7 PFT(FVC FEV1)^{2,7} SpC^7 WBC^7
Phthalic anhydride	00085-44-9	CXr^7 PFT(FVC FEV1)^7 U^5
Picric acid	00088-89-1	LFT^{2,7} Ua^{2,7}
Pindone (2-pivalyl-1,3-indandione)	00083-26-1	BLP(PTT)^{2,7} CBC^7 Ua(RBC)^7
Platinum, metal, and compounds	Varies	BL^{5,9,17} BT^9 U^{9,17}
Polychlorobiphenyl (42%Chlorine)(PCB)	53469-21-9	AdT^{3,5} BL^{3,5,15} BLP^3 BLS^{5,9,17} LFT^{2,7}
Polychlorodiphenyl (54%Chlorine)(PCB)	11097-69-1	AdT^{3,5} BL^{3,5,15} BLP^3 BLS^{5,9,17} LFT^{2,7}
Portland Cement	65997-15-1	CXr^{2,7} PFT(FVC FEV1)^{2,7}
Propiolactone, beta-	00057-57-8	InR(RIC Ster Preg Cig)^{1,7} PFT^2
Propyl nitrate, n-	00627-13-4	BL(MHgb) ^{2,8,11,16} CBC^{2,7} Ua^7
Propyleneimine	00075-55-8	CBC^7 CBC(GL)^2
Pyridine	00110-86-1	ExA^4 LFT^2 Ua^2
Quinone	00106-51-4	Oph^{2,7}
Selenium & compounds, except SeF$_6$	Varies	BH^3 BL^{5,9,13} BLP^{15} BLS^5 LFT^2 U^{3,5,7,9,15} U(24H)^{13} Ua^2 WBC^7
Selenium hexafluoride (SeF$_6$)	07783-79-1	BGa^7 CXr^{2,7} Ecg^7 PFT(FVC FEV1)^{2,7} SpC^7 Ua^2 WBC^7
Silica, amorphous, diatomaceous earth	61790-53-2	CXr^{2,7} PFT(FVC FEV1)^{2,7} Rdg^7 SpC^7
Silica, crystalline cristobalite	14464-46-1	CXr^{2,7} PFT(FVC FEV1)^{2,7} Rdg^7 SpC^7 TBS^7
Silicates — Soapstone	—	CXr^{2,7} PFT(FVC FEV1)^{2,7}
Silicates — Talc (without asbestos)	14807-96-6	CXr^{2,7} PFT(FVC FEV1)^{2,7} Rdg^7 SpC^7 TBS^7
Silver, metal and compounds	Varies	BL^{3,9,17} BLS^5 BT^9 Pgm^2 U^{5,9,17}
Sodium fluoroacetate	00062-74-8	Ecg^{2,7} Ua^{2,7}
Sodium hydroxide	01310-73-2	CXr^2 PFT^2
Stibine	07803-52-3	BLP(Hgb)^7 CBC^{2,7} LFT^{2,7} RBC(Hem)^2 U(Hgb)^7 Ua^{2,7} WBC^7
Stoddard solvent	08052-41-3	CBC^2 ExA^4 LFT^2 Ua^2
Strychnine	00057-24-9	Cnv^{2,16}
Styrene	00100-42-5	BL^5 BL(EOS)^{11,15,18} BL(EWW)^{18} BL(PNS)^{15} BL(PTS)^{18} ExA^{4,5,9,12} ExA(DE)^{15} ExA(PNS)^{15} Neur^7 U^{5,7,9,10,12} U(EOS)^{3,11,15,18} U(EWW)^{18} U(PTS)^{18} U(16H)^5
Sulfur dioxide	07446-09-5	BGa^7 CXr^{2,7} Ecg^7 ExA^4 PFT(FVC FEV1)^{2,7} SpC^7 WBC^7
Sulfur monochloride	10025-67-9	CXr^2 PFT^2
Sulfur pentafluoride	05714-22-7	BGa^7 CXr^{2,7} Ecg^7 PFT(FVC FEV1)^{2,7} SpC^7 WBC^7
Sulfuric acid	07664-93-9	CXr^{2,7} PFT(FVC FEV1)^{2,7}
Sulfotep (TEDP)	03689-24-5	BLS^9 CH(BLS RBC)^2 CH(RBC)^{7,11}
Tellurium hexafluoride	07783-80-4	CXr^7 PFT(FVC FEV1)^7
Tetraethyl pyrophosphate (TEPP)	00107-49-3	BLS^9 CH(BLP RBC)^{16} CH(BLS RBC)^2 CH(RBC)^{7,11}
Tetrachloroethane, 1,1,2,2-	00079-34-5	BUN^7 LFT^{2,7} Ua^{2,7}
Tetrachloronaphthalene	01335-88-2	LFT^{2,7}
Tetraethyl lead	00078-00-2	BL^3 BL(BUN Ca CO$_2$ GLU)^7 BMU^{2,7} U^{3,7,16} U(EOS)^4 Ua^{2,7}
Tetrahydrofuran	00109-99-9	BL^5 Bth^5 U^5 U(EOS)^{11}
Tetramethyl lead	00075-74-1	BMU^{2,7} U^7 U(EOS)^4 Ua^{2,7}
Tetranitromethane	00509-14-8	BL(MHgb)^{2,7,8} CBC^{2,7} CXr^{2,7} PFT(FVC FEV1)^{2,7}
Tetryl	00479-45-8	CBC^7 CBC(Anm)^2 LFT^7
Thallium and compounds	Varies	BL^{3,5,9,17} BT^9 NCS^{15} U^{3,4,5,7,16} Ua^{2,7}
Tin, inorganic compounds (exc oxide)	Varies	BL^{13} CXr^{15} U^{3,9,17}
Tin, organic compounds	Varies	CBC(Hem)^2 Ecg40^{2,15} Glau^2 KFT^{15} LFT^{2,15} U^{3,15} Ua^2
Toluene	00108-88-3	BL^{3,9,10,17} BL(EOS)^{5,15,18} BL(EWW)^{18} BL(PWW)^{11} ExA^{4,5,9,10,11} ExA(DE)^{3,15} U^{2,3,5,7,9,10} U(EOS)^{10,11,15,18} U(L4H)^{15} U(16H)^5 Ua^{2,7}
Toluene, 2,4-diisocyanate (TDI)	00584-84-9	BGa^7 BLP^3 CXr^{2,7} Ecg^7 PFT(FVC FEV1)^7 PFT(PPS)^2 SpC^7 U^{3,5} WBC^7
Toluidine, o-	00095-53-4	BL(MHgb)^{2,5,7,8,11} CBC^2 RBC^5 U^5 U(BL)^7 Ua^2
Trichloroethane, 1,1,2-	00079-00-5	ExA^4
Trichloroethylene	00079-01-6	BL^{3,5,12} BL(EOS)^{15,18} BL(ESW)^{11} BL(EWW)^{5,15,18} ExA^{4,7,11,12} ExA(EWW)^{15} ExA(PNS)^{3,15} ExA(16H)^{15} LFT^2 PFT^2 U^{5,12} U(EOS)^{15,18} U(ESW)^{11} U(EWW)^{11,15,18} U (PNS)^3
Trichlorofluoromethane	00075-69-4	U^5
Trichloronaphthalene	01321-65-9	LFT^{2,7}
Trichlorophenoxyacetic acid (2,4,5-T)	00093-76-5	BLP^3 U^3 U(24H)^{12}
Trichloropropane, 1,2,3-	00096-18-4	ExA^4
Trinitrotoluene, 2,4,6- (TNT)	00118-96-7	CBC^7 CBC(ApA)^2 LFT^{2,7} U^5 Ua^2
Uranium and compounds	07440-61-1	BL^5 CBC^{2,7} CXr^{2,7} U^{2,3,5,7,9} Ua^{2,7}
Vanadium fume (pentoxide)	01314-62-1	BGa^7 BL^{5,17} CXr^{2,7} Ecg^7 PFT(FVC FEV1)^{2,7} SpC^7 U^{3,5,7,13,15} U(EOS)^{18} U(EWW)^{18} WBC^7
Vinyl chloride	00075-01-4	BLS(APh Br GGT SGOT SGPT)^1 APh^9 BLS(LDH)^7 CXr^7 ExA$^{3-5}$ LFT(APh Br GGT SGOT SGPT)^7 LFT^{15} PFT(FVC FEV1)^{2,7} RBC^5 U^{3,5,15} Ua(Alb BL)^7 Usg^9
Warfarin	00081-81-2	BLP^3 BLP(Hyt)^2 BLP(PTT)^{2,7,12} CBC^7 U(BL)^7 Ua(RBC)^{2,7}
Xylene (all isomers)	Varies	BL^{5,12} BL(EOS)^{3,15} CBC^7 CBC(HeD)^2 ExA$^{3-5,10}$ LFT^{2,7} U^{5,9,12} U(EOS)^{3,11,15,18} U(EWW) ^{10,18} U(L4H)^{15} Ua^{2,7}
Xylidine	01300-73-8	BL(MHgb) ^{2,7,8,11} CBC^{2,7}
Zinc chloride fume	07646-85-7	BGa^7 BL^{9,17} BT^9 CXr^{2,7} Ecg^7 PFT(FVC FEV1)^{2,7} SpC^7 U^{2,3,7,9} WBC^7
Zinc oxide fume	01314-13-2	BL^{9,17} BT^9 CXr^2 PFT^2 U^{2,7,9} WBC^7
Zirconium & compounds, except ZrCl4	Varies	BL^{9,17} BT^9 (GraS)^2

TABLE 10-3 (continued)

REFERENCES

1. US DOL OSHA. *Code of Federal Regulations. 29 CFR Part 1910.1000.* Subpart Z. Air Contaminants. US Government Printing Office. July 1, 2004.
2. US DHHS PHS CDC NIOSH and US DOL OSHA. *NIOSH/OSHA Occupational Health Guidelines for Chemical Hazards* DHHS (NIOSH) Pub No. 81 123; 88 118; Suppls. I IV. 1981 1995.
3. Baselt RC. Biological Monitoring Methods for Industrial Chemicals. 3rd Edition. Chemical Toxicology Institute,1997.
4. Linch AL. *Biological Monitoring for Industrial Chemical Exposure Control.* CRC Press. 1974.
5. Lauwerys RR, Hoet P. *Industrial Chemical Exposure. Guidelines for Biological Monitoring.* 3rd Edition. Lewis Publishers. CRC Press, Inc. 2001.
6. Brown KK, Teass AW, Simon S, Ward EM: A Biological Monitoring Method for o Toluidine and Aniline in Urine Using High Performance Liquid Chromatography with Electrochemical Detection. Appl Occup Environ Hyg **10**(6):557 565, 1995.
7. Proctor NH, Hughes JP. *Chemical Hazards of the Workplace.* JB Lippincott Company. 1978.
8. Hathaway GJ, Proctor NH, Hughes JP, Fischman ML. *Chemical Hazards of the Workplace.* 3rd Edition. Van Nostrand Reinhold Publishers. 1991.
9. Kneip TJ, Crable JV. *Methods for Biological Monitoring—A Manual for Assessing Human Exposure to Hazardous Substances.* Am Public Health Association. 1988.
10. Ho MH, Dillon HK. *Biological Monitoring of Exposure to Chemicals. Organic Compounds.* John Wiley & Sons, Inc. 1987.
11. ACGIH. 2004 *Threshold Limit Values for Chemical Substances and Physical Agents and Biological Exposure Indices.* ACGIH Worldwide. 2004.
12. Atio A, Riihimaki V, Vainio H. eds. *Biological Monitoring and Surveillance of Workers Exposed to Chemicals.* Hemisphere Publishing Corporation. 1984.
13. Clarkson TW, Friberg L, Nordberg GF, Sager PR. *Biological Monitoring of Toxic Metals.* Plenum Publishing Corporation. 1988.
14. Dillon HK, Ho MH. *Biological Monitoring of Exposure to Chemicals. Metals.* John Wiley & Sons, Inc. 1991.
15. LaDou J, ed. *Occupational & Environmental Medicine.* 2nd Edition. Appleton and Lange. 1997.
16. Proctor and Hughes' *Chemical Hazards of the Workplace.* 4th Edition. Van Nostrand Reinhold, New York, New York. 1996.
17. US DHHS PHS CDC NIOSH. *NIOSH Manual of Analytical Methods.* 4th Edition. DHHS (NIOSH) Publication No. 94 113.
18. Fiserova Bergerova V, Vlach J: Timing of sample collection for biological monitoring of occupational exposure. Ann Occup Hyg **41**/3:345 353, 1997.

Glossary of Terms Used in Table

Abbrev.	Meaning
1M	= After 1 Month Exposure
2D/2H	= After 2 Hours or 2 Days Exposure
6M	= Every 6 Months
16H	= 16 Hours after End of Exposure
24H	= 24-Hour Collection
AdT	= Adipose Tissue
Alb	= Albumin
Anm	= Anemia
ApA	= Aplastic Anemia
APh	= Alkaline Phosphatase
ß2M	= ß-2 Microglobulin in Urine
BF	= Body Fluid (chemical/metabolite)
BGa	= Blood Gas Analysis
BH/N	= Body Hair/Nail
Bicarb	= Bicarbonate
BL	= Whole Blood (chemical/metabolite)
BLC	= Blood Cyanide
BLL	= Blood Lead Level
BLP	= Blood Plasma
BLPr	= Blood Pressure
BLS	= Blood Serum
BLW	= Blood Analyzed Weekly
BMU	= Biologic Monitoring of Urine every 3 Mos.
Br	= Bilirubin
BT	= Biologic Tissue/Biopsy
Bth	= Breath
BUN	= Blood Urea Nitrogen
Ca	= Calcium
CBC	= Complete Blood Count
CH	= Cholinesterase
Chol	= Cholesterol Level
Cig	= Cigarette smoking
Cnv	= Convulsions
CO_2	= Carbon Dioxide
COHb	= Carboxyhemoglobin
CXr	= Chest X-ray
Cys	= Cystoscopy
Cyt	= Cytology
DE	= During Exposure
DXr	= Dental X-ray/Examination
Ecg	= Electrocardiogram
Ecgr	= Electrocardiogram (resting)
Ecg40	= Electrocardiogram on Workers over 40 yrs
EM	= Every Month
EOS	= End-Of-Shift
Ery	= Erythrocyte Count
ESW	= End of Shift at End of Workweek
EWW	= End of Workweek
ExA	= Expired Air
FEF	= Forced Expiratory Flow Rate
FEV1	= Forced Expiratory Volume (1 sec)
FOB	= Fecal Occult Blood Screening (age>40)
Frg	= Fragility
FSH	= Follicle Stimulating Hormone
FVC	= Forced Vital Capacity

(continues)

TABLE 10-3 (continued)

Glossary of Terms Used in Table (continued)

GGT	= Gamma Glutamyl Transpeptidase		PNS	= Prior to Next Shift
GL	= Granulocytic Leukemia		PPS	= Pre- & Post-Shift
Glau	= Glaucoma		PpT	= Photopatch Testing
Glu	= Sugar/Glucose		Preg	= Pregnancy
GraS	= Granuloma of the Skin		PTS	= Prior to Shift
GUT	= Genito-Urinary Tract		PTT	= Prothrombin Time
Hct	= Hematocrit		PWW	= Prior to Last Shift of Workweek
HeD	= Hematopoietic Depression		PXr	= Pelvic X-ray
Hem	= RBC Hemolysis		RBC	= Red Blood Cells/Count
Hgb	= Hemoglobin		Rdg	= Radiography
Hmt	= Hematuria		ReC	= Reticulocyte Count
Hyt	= Hypoprothrombinemia		RIC	= Reduced Immunologic Competence
InR	= Increased Risk		SCr	= Serum Creatinine
KFT	= Kidney Function Tests		SCt	= Sperm Count
L2H	= Last 2 Hours of 8-Hour Exposure		Sed	= Sediment
L4H	= Last 4 Hours of 8-Hour Exposure		SGOT	= Serum Glutamic Oxalacetic Transaminase
Lab	= Laboratory Surveillance		SGPT	= Serum Glutamic Pyruvic Transaminase
Lac	= Lactic Acid		Skn	= Skin Examination
LDH	= Lactic Dehydrogenase		SMA12	= Serum Multiphasic Analysis
Leuk	= Leukocyte Count		SpC	= Sputum Cytology
LFT	= Liver Function Tests		Ster	= Steroid Treatment
LH	= Luteinizing Hormone		TBS	= Tuberculin Skin Test
LTT	= Lymphocyte Transformation Test		TE	= Total Estrogen (females)
MHgb	= Methemoglobin		Test	= Testicle Size
Micr	= Microscopic Examination		Testo	= Testosterone
Nas	= Nasal Examination		TFT	= Thyroid Function Test/Thyroid Profile
NCS	= Nerve Conduction Studies		Thrm	= Thrombocyte Count
Neur	= Neurologic Examination/Electromyography		U	= Urine (chemical/metabolite)
Oph	= Ophthalmic Examination		Ua	= Urinalysis (routine)
pH	= pH (Hydrogen ion concentration)		Usg	= Ultrasonography of the Liver
PFT	= Pulmonary Function Tests		WBC	= White Blood Cell Count/Differential
Pgm	= Pigmentation Evidence		ZPP	= Zinc Protoporphyrin
PMN	= Polymorphonuclear Leukocytes			

(small hair-like structures in the upper respiratory system designed to catch particulate matter), or by synergistically increasing rates of lung cancers. Individuals who smoke and work with asbestos-containing materials have a risk of lung asbestos-related diseases that is 70 times higher than that of nonsmokers. Consequently, information concerning smoking cessation programs must be provided to employees under certain OSHA standards, such as the Asbestos Standard (**TABLE 10-4**).

In addition, certain occupational groups (such as firefighters) may be required to be "nonsmokers" as a condition of employment due to the connection between toxic materials and lung cancers, or the potential for interference or excessive restrictions with respirator approvals.

Symptomatic Monitoring

In the event an employee is exposed to hazardous environments, when symptoms are present, additional medical monitoring must be performed. In addition to emergency treatment, the LHCP will repeat initial testing for the purpose of determining whether body systems remain functional within recommended limits or whether they have been affected.

A secondary means of triggering symptomatic exposure monitoring may be from toxin characterization (e.g., air monitoring where permissible exposure limits have been exceeded, or where mucous-membrane or open-wound contact with chemicals or bloodborne

TABLE 10-4 OSHA Smoking Cessation Program Information

The following organizations provide smoking cessation information:
1. The National Cancer Institute operates a toll-free Cancer Information Service (CIS) with trained personnel to help you. Call 1-800-4-CANCER to reach the CIS offices serving your area or write: Office of Cancer Communications, National Cancer Institute, National Institutes of Health, Building 31 Room 10A24, Bethesda, Maryland 20892.
2. American Cancer Society, 3340 Peachtree Road, N.E., Atlanta, Georgia 30026, (404) 320-8333:
    The American Cancer Society (ACS) is a voluntary organization composed of 58 divisions and 3,100 local units. Through "The Great American Smokeout" held annually in November, the annual Cancer Crusade in April, and numerous educational materials, ACS helps people learn about the health hazards of smoking and become successful ex-smokers.
3. American Heart Association, 7320 Greenville Avenue, Dallas, Texas 75231, (214) 750-5300:
    The American Heart Association (AHA) is a voluntary organization with 130,000 members (physicians, scientists, and laypersons) in 55 state and regional groups. AHA produces a variety of publications and audiovisual materials about the effects of smoking on the heart. AHA also has developed a guidebook for incorporating a weight-control component into smoking cessation programs.
4. American Lung Association, 1740 Broadway, New York, New York 10019, (212) 245-8000:
    A voluntary organization of 7,500 members (physicians, nurses, and laypersons), the American Lung Association (ALA) conducts numerous public information programs about the health effects of smoking. ALA has 59 state and 85 local units. The organization actively supports legislation and information campaigns for non-smokers' rights and provides help for smokers who want to quit, for example, through "Freedom From Smoking," a self-help smoking cessation program.
5. Office on Smoking and Health, U.S. Department of Health and Human Services, 5600 Fishers Lane, Park Building, Room 110, Rockville, Maryland 20857:
    The Office on Smoking and Health (OSH) is the Department of Health and Human Services' lead agency in smoking control. OSH has sponsored distribution of publications on smoking-related topics, such as free flyers on relapse after initial quitting, helping a friend or family member quit smoking, the health hazards of smoking, and the effects of parental smoking on teenagers.
    In Hawaii, on Oahu call 524-1234 (call collect from neighboring islands).
    Spanish-speaking staff members are available during daytime hours to callers from the following areas: California, Florida, Georgia, Illinois, New Jersey (area code 201), New York, and Texas. Consult your local telephone directory for listings of local chapters.

Source: U.S. Department of Labor. *OSHA Asbestos Standard: 29 CFR 1926.1101.* Appendix J.

---

pathogens has occurred). In these cases, knowledge of an overexposure occurs before the presentation of symptoms.

When initial baseline tests are repeated, the LHCP determines whether an employee remains suitable for work under the original conditions specified. Conversely, regimens of treatment may be provided to return the employee to acceptable work conditions.

In general, prophylactic treatment prior to systemic effects is prohibited (e.g., prophylactic chelation of individuals who will work with lead is prohibited under the OSHA Lead Standard, unless elevated blood lead levels are found).

Routine Monitoring

Routine monitoring is required by many standards and recommended by NIOSH for work in hazardous environments.

At times when routine monitoring is performed, it is important to provide the LHCP with exposure characterization data and employee medical exposure files so that an appropriate test regimen can be developed. Both identities and quantities of hazardous environments need to be provided to the LHCP at these times. Subsequently, the LHCP will determine the employees' continued fitness for work, or restrictions due to exposures.

Close-Out Physicals

In the event an employee will no longer operate within a hazardous environment, close-out medical monitoring is required by many standards and recommended for employer protection. The purpose of the close-out physical examination is to determine employee body functions at the completion of work within hazardous environments. This process protects employers from liability and provides employees with medical procedures at the completion of hazardous work, if required. It is permissible to perform many medical close-out examination protocols within the last several months (e.g., six months) of employment termination. However, examinations subsequent to final exposures are generally recommended.

Chapter Summary

Medical monitoring is essential for individuals who enter hazardous environments. Body systems testing, physical performance testing, or psychological evaluations may be performed. Baselines are determined, and an individual's fitness to enter the hazardous environment is certified. Medical monitoring may subsequently be repeated after exposures (symptomatic or asymptomatic) or on a routine basis, and at the completion of work within the hazardous environment.

Terms

Medical Monitoring: A systematic examination of medical and work histories, generally in conjunction with medical tests and a patient examination provided by a licensed health care provider, designed to identify whether individuals are capable of performing (or continuing) work in hazardous environments without risk to themselves or others in the area. Medical monitoring is performed before initial hazardous work, periodically through the work process, and at the completion of hazardous activities. In addition, it may be performed if symptomatic exposures occur.

References

National Fire Protection Association. Standard on Comprehensive Medical Program for Fire Departments. *NFPA 1582*. Quincy, MA (2013).

———. Standard on Fire Department Occupational Safety and Health Program. NFPA 1500. Quincy, MA (2013).

National Institute for Occupational Safety and Health, Environmental Protection Agency, Occupational Safety and Health Administration, U.S. Coast Guard. *Health and Safety For Hazardous Waste Site Activities, Department of Health and Human Services.* 1989.

U.S. Department of Labor, Occupational Safety and Health Administration. 29 CFR 1910.120/29 CFR 1926.65: Hazardous Waste Operations and Emergency Response. In *General Industry Safety Standards*. Washington, D.C.: Government Printing Office.

———. 29 CFR 1910.134: Respirators. In *General Industry Safety Standards*. Washington, D.C.: Government Printing Office.

———. 29 CFR 1910.1001/29 CFR 1926.1101: Asbestos. In *General Industry Safety Standards and Construction Safety Standards*. Washington, D.C.: Government Printing Office.

———. 29 CFR 1910.1020: Medical Recordkeeping. In *General Industry Safety Standards*. Washington, D.C.: Government Printing Office.

———. 29 CFR 1910.1025/29 CFR 1926.62: Lead. In *General Industry Safety Standards*. Washington, D.C.: Government Printing Office.

U.S. Department of Transportation. *40 CFR, Part 30-40; Federal Motor Carrier Safety Regulations and Medical Standards for Commercial Motor Vehicle Drivers.* 2009.

Exercise 10

Complete the Medical Monitoring Reference Chart, identifying four times when work with hazardous substances may require medical surveillance and a description of why medical surveillance is required. Use another sheet of paper, if necessary.

	Work Timeframe:	Medical Surveillance Rationale:
1.		
2.		
3.		
4.		

11 Personal Safety

Key Concepts

- Confined Spaces
- Permit-Required Confined Spaces
- Excavations and Trenches
- The "811" System
- Work at Heights
- Hot Work
- Powered Industrial Trucks
- Hazardous Energy Sources
- Motor Vehicle Safety

Introduction

Personal safety requirements routinely apply to employees in specialized hazardous environments such as the following:

1. Enclosed or confined spaces
2. Excavations and trenches
3. Work at heights, including work on ladders or scaffolds
4. Operating vehicles and use of **powered industrial trucks**
5. Other generally recognized hazardous environments, such as:
   - Hot work
   - Work with hazardous energy sources

Enclosed or Confined Spaces

Enclosed and confined spaces and **permit-required confined spaces (PRCS)** require special entry procedures due to their potential for hazardous atmospheres and the difficulty of efficiently exiting these spaces. Enclosed spaces, a term used in the shipyard industry, present similar hazards to PRCS.

Low oxygen pressures and high concentrations of flammable vapors or toxic materials may be present in these locations. Typical confined spaces, which may be PRCS and their associated hazards, are pictured in **PHOTOS 11-1** through **11-3**. Employers who have employees who enter confined spaces and PRCS should develop specialized entry protocols by identifying the hazards of each space. Employers must also provide warning and awareness training so that individuals do not enter these identified spaces inappropriately. Finally, if entry is to occur, the employer must provide appropriate testing equipment so that individuals can quantify the hazards within the space and provide appropriate engineering or administrative controls, personal protective equipment (PPE) and emergency rescue, where necessary, for individuals in the spaces. Entry into PRCS is regulated by the U.S. Department of Labor—Occupational Safety and Health Administration (OSHA), as well as many state programs protecting public employees. Some state programs (e.g., the Commonwealth of Virginia: Confined Space Entry at Construction Operations)[1] provide for protection in excess of the OSHA standards. Regulations require that a written safety program be developed

PHOTO 11-1 Tanks and vessels are typically considered permit-required confined spaces.

PHOTO 11-2 Sanitary sewer systems are PRCS and often contain low oxygen levels or flammable atmospheres.

and implemented by the employer before entry into these spaces. Permits, attendants, and on-site rescue personnel are required for entry into PRCS.

TABLE 11-1 identifies components of the OSHA PRCS Standard. All entrants into PRCS are required to wear harnesses and to be attached to lifelines, unless these items are contraindicated. Lifeline contraindications are found in **TABLE 11-2**.

---

[1] Section 16: Virginia Administrative Code (VAC) 16 VAC 25-140-10 et. seq. Confined Spaces Standard for the Construction Industry; AND 16 VAC 25-70-10: Confined Spaces Standard for the Telecommunication Industry.

PHOTO 11-3 Although initial tests of confined spaces and permit-required confined spaces may show "good air," substances brought into or applied in the space may change the air characteristics while it is occupied. Testing requirements are documented on the PRCS permit.

A flowchart identifying confined-space safety protocol and decision making required by the U.S. Department of Labor is found in **TABLE 11-3**. A sample permit, capable of being used for both confined spaces and PRCS, as well as PRCS reclassification, is found at Figure 6-2.

Excavations and Trenching

Excavations are cuts within the surface of the earth. Trenches, a form of excavation, are deep cuts within the surface of the earth. In general, excavations that are 20 feet wide, and wider than they are deep, are seldom considered hazardous. Conversely, trenches (cuts in the ground that are 5 feet deep or more) are considered extremely dangerous and require employee protection in all cases. Employee protection may take the following forms:

- Sloping or benching sides of trenches to reduce collapse potential
- Trench boxes or mechanical protection within trenches (if this operation is chosen, bottoms of protective devices are required to be within 2 feet of the bottom of the trench)
- Soil characterization to ensure appropriate soil stability is in place within trenches, in conjunction with appropriate benching or sloping of sides of trenches

PHOTO 11-4 PRCS entrants must be protected by a rescue team, typically standing by at the scene of the entry. This requirement mimics the requirement of the current OSHA standard for entry with respirators into IDLH environments. Lifeline and retrieval devices, unless contraindicated, are also required.

General Trench Safety Requirements

General trench safety requirements include the following:

- Prenotification of utility locator companies to identify underground services (A one-call system is now in place in all states, whereby underground utilities can be identified if called 24 to 72 hours in advance. **811** is now the designated number to notify utility marking services, and all calls are routed to the local "One-Call" center.[2] **TABLE 11-4** identifies utility marking systems, used universally by utility locator companies)

---

[2] "One-Call" notification standards are codified at 49 CFR 61, and are an extension of the Common Ground Alliance formed in 2000 as an extension of the U.S. DOT Common Ground Study. Along with the 811 number, Common Ground Alliance sponsors the "Call Before You Dig" campaign.

TABLE 11-1 Permit-Required Confined Spaces OSHA Requirements (29 CFR 1910.146)

A. Scope
B. Definitions
 Confined Space:
  - Can be entered, and
  - Limited entry and exit, and
  - Not for continuous occupancy (tanks, vessels, storage bins, hoppers, vaults, pits)
 Engulfment: Surrounding or capturing by aspirable liquid/solid (plugging respiratory system) materials that may strangle, constrict, or crush
 Hazardous Atmosphere: Risk of death or impaired self-rescue due to
  - Flammable vapor >10% LEL
  - Combustible dusts >LFL (less than 5 foot visibility in dusty atmosphere)
  - Oxygen >23.5% or <19.5%
  - >PELs of subpart "g" or "z"
  - Any immediately dangerous to life and health atmosphere (identified from material safety data sheet [MSDS])
 Permit-Required Confined Space: A confined space which:
  - Contains or potentially contains a hazardous atmosphere
  - Contains a material that has the potential for engulfment
  - Has a configuration such that an entrant could be trapped or asphyxiated by inwardly converging walls, or floors that slope downward and taper to smaller cross section
  - Contains any serious safety or health hazard
C. General Requirements
 1. Workplace evaluation required
 2. "Danger" postings of PRCSs required.
 3. "No entry provisions" and exemptions if training is provided
 4. Written program required
 5. a. Alternate procedures only are permissible if
    - Hazardous atmospheres (only) are present, and
    - Sufficient ventilation data is present, and
    - Monitoring and inspection data is present, and
    - Initial entry as a permit space occurs, and
    - Data is available to employees
   b. Alternate procedures
    - Entrance cover restrictions, including cover guards
    - Calibrated testing for
      - Oxygen
      - Combustible vapors
      - Toxins
    - No hazardous atmosphere exists
    - Continuous clean forced air ventilation (directed)
    - Periodic testing
    - If a hazardous atmosphere is detected, the space is evacuated and hazard cause and remedy is determined
    - Written certification is available to employees
    - Reclassify for changes in use or configuration
 6-7. Reclassification of permit to nonpermit space:
    - No potential atmospheric hazards and all hazards eliminated without entry into the space (If entry is necessary to eliminate hazards, this position remains under the PRCS program)
 Note: Forced air ventilation is not hazard elimination, but requires Section 5 (b) documentation.
    - Certification (date, location, signature)
    - Exit and reclassify if hazards are detected
 8. Prime contractor responsibilities:
    - Notification of all permit spaces and standard requirements
    - I.D. hazards in the space(s)
    - Apprise contractor of precautions in or near the space
    - Coordinate multi-employer operations
    - Debriefing
 9. Subcontractor responsibilities:
    - Obtain information
    - Coordinate multi-employer operations
    - Inform prime contractor of program in place and hazards confronted

(continues)

TABLE 11-1 (continued)

D. PRCS Program
   1. Secure spaces against unauthorized entry
   2. Identify and evaluate hazards
   3. Address practices and procedures for safe entry
      - Acceptable conditions
      - Isolation
      - Purging/inerting
      - Providing barriers from external hazards
      - Verifying acceptable conditions
   4. Provide equipment and maintain in calibrated condition
      - Maintain ventilation equipment
      - Maintain communication equipment
      - Maintain PPE
      - Maintain lighting equipment
      - Maintain barriers/ladders
      - Rescue equipment
   5. Evaluate conditions
      - Test, prior to entry
      - Continuously test, if isolation is unfeasible
      - Continually test, as required on the permit
      - Order of testing
         - Oxygen
         - LEL
         - Toxins
   6-7. Provide attendant(s)
      (Note: Single attendant—multiple spaces may be permissible)
   8. Designate individuals trained to fulfill active roles.
   9. Develop alarming/alerting procedures and prevent unauthorized rescue and entry.
   10. Permit system
      - Preparation procedures are written.
      - Issuance of a permit occurs, and the permit remains at the space for employee inspection.
      - Cancellation of the permit occurs at the termination of the activity.
   11. Coordination occurs at multi-employer work sites
   12. Closure procedures are developed.
   13. Permits and programs are reviewed and revised as necessary.
   14. All permits and programs are reviewed annually.

E. Permit System
   1. Prepare and post permit prior to entry
   2. Ensure the permit is signed by the supervisor
   3. Ensure employee positions are identified (entrant, attendant, rescue)
   4. Ensure the permit is issued for one job only
   5. Ensure the permit is cancelled at the completion of work or when non-permitted conditions develop
      Ensure problems with the entry are documented on the permit
   6. Retain the permit for one year and review it at that time

F. Entry Permit—Permit Shall Include:
   - Space
   - Purpose for the entry
   - Date and duration
   - Entrant identification
   - Attendants
   - Supervisor (original current)
   - Hazards
   - Isolation required
   - Acceptable entry conditions
   - Test data (initiated/time/date)
   - Rescue service and means of communication

TABLE 11-1 (continued)

- Communication procedures with entrants
- Equipment requirements
- Other safety considerations
- Other permits in the space

G. Training
  1. Before assignment of duties
  2. Change in assigned duties
  3. Change in operations
  4. Change in permit procedures
     - Proficiencies
     - Training certification
       - Dated, signed by instructor

Entrants must:
- Possess hazards and exposure information
- Understand proper equipment use
- Be in constant communication with attendant
- Alert the attendant when
  - Warning signs or symptoms develop
  - Prohibited conditions develop
- Exit when:
  - Evacuation is ordered
  - Signs of overexposure develop
  - Prohibited conditions develop
  - Evacuation alarm is activated

Attendants must:
- Know the hazards of entry
- Recognize behavioral effects of exposure
- Keep accurate count of individuals in space
- Prohibit entry for unauthorized individuals
- Communicate with entrants
- Monitor areas inside and outside space and evaluates space for
  - Behavior changes
  - Outside endangerment
  - Prohibited conditions
- Summon rescue services, when necessary
- Warn unauthorized persons away from space, out of space, and inform supervisor if they remain in the space
- Perform nonentity rescue
- Enforce noninterference rules of the permit

Supervisors must:
- Know hazards, signs, symptoms of overexposure.
- Verify permit/endorse permit
- Terminate entry as required on the permit or written program
- Verify alarm and rescue services
- Remove unauthorized individuals from PRCS
- Determine responsibilities for transfer of duties

Rescue and Emergency Services:
- PPE and training that are necessary for rescues must be available
- Rescue training is required at least annually, for the type of space being protected
- Entrants training is required for rescue personnel
- First aid/CPR training is required for all team members, and current CPR/first aid certification is required for one person on the rescue team
- Outside rescue services must be:
  - Informed of hazards and facility
  - Provided access to the facility for planning and training

(continues)

TABLE 11-1 (continued)

- Retrieval Equipment Requirements:
  - Full body/chest/back attachment for harnesses used in PRCS entry.
  - Wristlet exemption (wristlet harnesses are permissible for vertical entries)
  - Retrieval lines are required for entrants into PRCS, unless contraindicated
  - A retrieval device must be available for all vertical entries greater than 5 feet
- MSDS requirements (must be on site for substances in the space)
- A written copy of operating and rescue procedures shall be at the work site for the duration of the job

Source: "Permit Required Confined Space." U.S. Department of Labor, Occupational Safety and Health Administrations. General Industry Safety Regulations, 29 CFR 1910.146: Washington, DC. 1997.

TABLE 11-2 Lifeline Contraindications

Lifelines are Contraindicated in a PRCS When:

1. A permit space has obstructions or turns that prevent pulls on the retrieval line from being transmitted to the entrant.
2. An individual being rescued would be injured due to forceful contact with projections in the space.
3. An air line where entanglement with the air line may occur.

Source: "Permit Required Confined Space." U.S. Department of Labor. Occupational Safety and Health Administrations, General Industry Safety Regulations, 29 CFR 1910.146, Washington, DC. 1997."

- Inspection by a competent person prior to initiation of daily activities and subsequent to rain or water infiltration into trenches
- Restriction of entry into trenches by means of fencing or warning devices (two feet back from trench sides) and maintenance of traffic away from trenches
- Air characterization of hazardous atmospheres suspected in trenches
- Placement of ladders at 20-foot intervals for evacuation purposes when trenches deeper than 4 feet are entered; placement of rescue equipment at trench sites where appropriate
- Prohibitions to underpinning

Protection must be provided to employees entering excavations (trenches), unless one of the following apply:

- The cut is in stable rock, or less than 5 feet deep
- A slope or bench ratio of 1.5 horizontal to 1 vertical is maintained
- Soil classifications are performed by a competent person on site, and appropriate sloping installed based on the soil classification (**TABLE 11-5**).
- Additional data, certified by a Registered Professional Engineer, ensures safety of entrants

Soil Classification

Soil classifications are identified in Table 11-5. Tests for soil classifications are found in **TABLE 11-6**. When this option is used, one visual and one manual test must be performed by a competent person for soil classification purposes.

Work at Heights

Work or occupancy at nonprotected heights, generally considered heights in excess of four feet (six feet in construction settings), may produce deadly falls. Consequently, protection is required for this hazardous atmosphere.

Ladders

Ladders may be used for the purpose of access, or for certain types of work, when used in compliance with manufacturer's specifications.

Ladder Rating Systems

OSHA, the American National Standards Institute (ANSI), and the National Fire Protection Association (NFPA) publish ladder-rating systems for portable ladders. The ANSI standard is the most stringent for general ladder use, whereas the NFPA standard (far more stringent) covers the use of ladders in fire-and-rescue situations.

OSHA Ladder Requirements

The general industry OSHA standard requires ladders to support one 200-pound person. The OSHA construction standard requires that ladders sustain three to four times the maximum intended load. Ladders must not be loaded beyond their intended capacity. Practically, this means one person on the ladder at any time. Ladders must be placed at an angle with a 4:1 ratio, and extend three feet beyond the upper landing surface or be

TABLE 11-3 OSHA-PRCS Decision Flow Chart

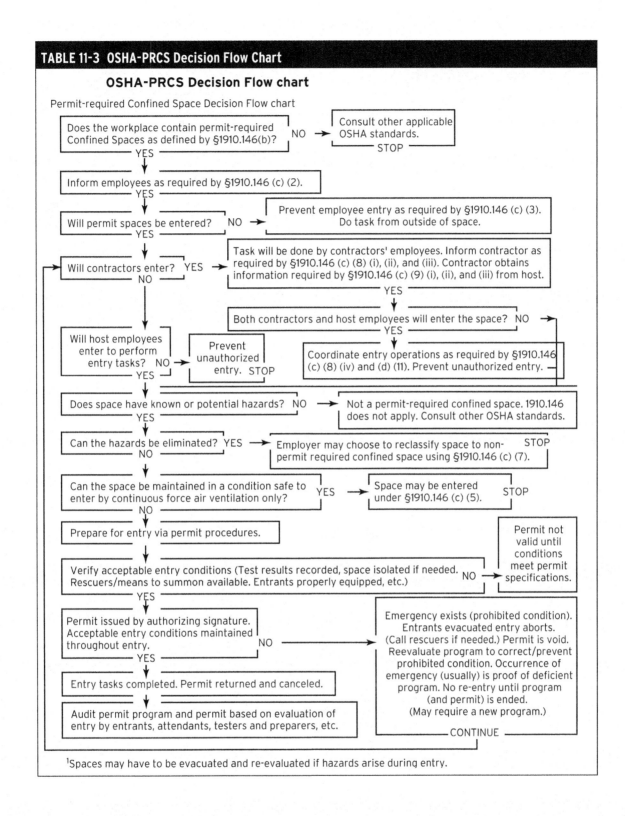

¹Spaces may have to be evacuated and re-evaluated if hazards arise during entry.

secured. Ladders must be secured in any situation where traffic may accidentally displace the ladder. Other OSHA ladder specifications are listed in **TABLE 11-7**.

ANSI Ladder Requirements

A variety of ANSI standards exist, notably Standard A14.1 (2007) for wood ladders, ANSI Standard A14.2 (2007) for portable metal ground ladders, and ANSI Standard 14.5 (2007) for portable reinforced plastic ladders. Ladders meeting the ANSI standard are assumed to meet the OSHA standards.

ANSI ladder types and their duty ratings are found in **TABLE 11-8**. ANSI specifies 5 portable ladder types, with duty ratings between 200 and 375 pounds.

TABLE 11-4 Utility Locator: Uniform Color Code

Color	Description
White	Proposed excavation
Pink	Temporary survey markings
Red	Electric power lines, cables, conduits and lighting cables
Yellow	Gas, oil, steam, petroleum or gaseous materials
Orange	Communication, alarm or signal lines, cables or conduits and traffic loops
Blue	Potable water
Purple	Reclaimed water, irrigation and slurry lines
Green	Sewer and drain lines

Courtesy of American Public Works Association

FIGURE 11-1 Sloping and installation of trench boxes increases occupant safety and is required under the OSHA construction standard. Table 11-5 "OSHA Soil Classification" identifies sloping requirements by soil classification.

NFPA Ladder Requirements

NFPA-compliant ladders must have a minimum duty rating of 750 pounds, have heat sensor labels, and must be able to maintain strength at high temperatures. In addition, they have a minimum width of 16 inches (OSHA/ANSI ladders have a minimum width of 12 inches).

Aerial ladder devices are covered under separate OSHA standards. The NFPA publishes standards for aerial ladders and lifts used in fire and rescue service.

Fall Protection

Work at heights may also be safely accomplished through the provision of fall-protection equipment or practices. Generally required at heights of four feet (six feet in construction industries), multiple options are available to make work at heights safe. Inspection activities,

TABLE 11-5 OSHA Soil Classifications

Stable Rock		90-degree angles permitted in stable rock	90-degree cuts permissible
Type "A" Soil	Cohesive and cemented soils (1.5 ton/ft² compression strength)	Nongranular soils, unless fissured, previously disturbed or subject to traffic and vibrations	3/4 to 1 slope permitted
Type "B" Soil	Cohesive Soil (0.5-1.5 ton/ft² compression strength)	Non-granular soils, which are previously disturbed or fissured or subject to traffic or vibrations, or unstable dry rock	1/1 slope permitted
Type "C" Soil	Granular (<.5 ton/ft² compression strength)	Gravel, sand, wet soil 1 1/2 (v) to 1 (h) slope permitted	1 1/2 (v) to 1 (h) slope permitted

Source: U.S. Department of Labor, Occupational Safety and Health Administrations, Construction Regulations, 19 CFR 1926.

TABLE 11-6 OSHA Soil Classification: Field Testing

1. Visual Test	A.	Visual review and classification as cohesive (fine ground) Type A or B soil
	B.	Visual review and classification as coarse sand/gravel
2. Manual Test	A.	Roll soil to 1/8-inch thickness; if 2 inches can be held, it is cohesive; if it crumbles, it is granular
	B.	Thumb test (ANSI D2488); if soil only indents, it is Type A, if it indents 2 inches or more, it is Type C

Source: U.S. Department of Labor, Occupational Safety and Health Administrations, Construction Regulations, 19 CFR 1926.

PHOTO 11-5 Trench boxes must be within two feet of trench bottoms. A typical trench box is pictured.

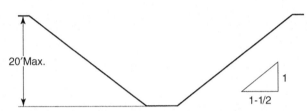

PHOTO 11-6 Type "C" soils (sand or granular soils) have a maximum slope of 1½ (horizontal) to 1 (vertical). This is the default slope when soils are not classified on site by a competent person, or trench boxes are not used.

TABLE 11-7 OSHA Ladder Requirements		
Ladder Type	Maximum Length	Special Requirements
Straight Ladder	30 feet	Width greater than 12 inches
Extension Ladder: Two Section	48 feet	Width greater than 12 inches, overlap stops
Extension Ladder: Three Section	60 feet	Overlap stops
Stepladders	20 feet	Insulating, nonslip pads at bottom of rails and locking device to hold ladder sections

Source: "Ladders." U.S. Department of Labor, Occupational Safety and Health Administrations, Construction Regulations, 19 CFR 1926.

TABLE 11-8 ANSI Ladder Types and Duty Ratings		
Ladder Type	Duty Rating	Description
Type 1AA	375 lbs.	Extra Heavy-duty industrial ladder
Type 1A	300 lbs.	Heavy-duty industrial ladder
Type 1	250 lbs.	Heavy-duty industrial ladder
Type 2	225 lbs.	Medium-duty commercial ladder
Type 3	200 lbs.	Light-duty household ladder

performed before actual work requiring fall protection, is exempt from the duty to provide fall protection. Fifteen construction situations in which fall protection is required have been identified by OSHA, with options for the specific situations listed.

Head protection is required for individuals who work around fall protection and scaffolding areas.

Guardrail Systems and Fall Protection Options

The primary means of protecting individuals at heights is through the use of guardrail systems, which consist of top rails (200-pound strength, 2 inches ± 3 inches in flexibility), a mid-rail system (150-pound strength, crossed members are acceptable), and a kick plate to prevent items such as tools from being kicked over the sides of these areas.

Other options include the installation of a safety net system (which must be placed no more than 30 feet below the occupied area), warning lines flagged and installed six feet from edges, or the use of specialized fall-protection plans. Certain options require the use of more than one system (redundant systems).

Fall protection is now required on low sloped roofs (4/12 angle), in residential construction settings, a recent change in OSHA enforcement policy. Hoist areas and holes are required to be protected, as well as shorter falls (e.g., less than six feet) when individuals are working above dangerous equipment. When guardrail systems are used at openings, a chain to protect the hoist or laddered area is required.

Safety nets, when used, must be drop tested when installed, and openings of more than six inches are prohibited. During certain types of restricted work at heights (leading edge work, roofing, overhand bricklaying), warning line systems, controlled access zones, or safety monitoring systems may be used.

Fall Arrest Systems

Since 1998, body belts have been prohibited by OSHA as personal fall-arrest systems. However, they remain acceptable as positioning devices. Personal fall-arrest systems include sturdy anchor points, full-body harnesses and attachment devices, and any aids that limit falls to six feet. Anchor points, vertical lifelines, and connections to systems must have a minimum tensile breaking strength of 5,000 pounds and be lockable. Self-retracting lifelines (limiting falls to two feet) must have a tensile breaking strength of 3,000 pounds.

Any personal fall-arrest system must limit falls to six feet or prevent contact with lower surfaces. Full-body harnesses with attachment points at the wearer's mid back are required on harnesses. All harnesses are required to be tagged and inspected prior to each use.

Positioning devices (belts), when used, must not allow a fall of greater than two feet, with all strengths rated at 3,000 pounds.

Warning lines, when used, must be flagged at not more than six-foot intervals, and only individuals performing work are allowed in the area between the drop edge and the warning line. Alternatively, control lines with restrictions may be used to restrict access. Safety monitoring, the placement of an individual close enough to communicate with individuals in hazardous areas, is also approved for some applications.

For individuals involved in leading edge work, precast concrete erection, or residential construction who can demonstrate the unfeasibility of other fall protection options, a fall-protection plan authored by a qualified person may be used. Training programs are required for each employee who may be exposed to a fall hazard, which must be verified by a written certification record.

PHOTO 11-8 Ladders are labeled to identify classification and use restrictions.

PHOTO 11-7 This ladder is placed for proper use, with a 4:1 height to base ratio. The worker prepares to climb with a field test to determine whether the ladder angle is appropriate.

PHOTO 11-9 NFPA-compliant ladders are used for lifesaving. These ladders have a heavier duty rating than OSHA-compliant construction ladders.

Scaffolding

Scaffolding, when used, must be designed by a qualified person and be capable of supporting four times the intended load or tipping moment. Scaffold-grade identified materials must be used (**FIGURE 11-2**). Each scaffold or

PHOTO 11-10 Guardrail systems are typically used to protect workers from falls.

PHOTO 11-11 Warning line systems are approved fall-protection systems in certain types of roofing operations.

FIGURE 11-2 Scaffold-grade identified lumber must be used in scaffoldings. Typical scaffolding-grade lumber identifiers are pictured.

walkway must be at least 18 inches wide, and employees must be protected from falls. Front edges of scaffolds may not be more than 14 inches from work faces and scaffolds must be fully planked. Supported scaffolds may not have a height to base ratio of greater than 4:1 or they must be tied to structures at that point.

In general, distances of 10 feet are required between scaffolds and electrical power lines; however, this distance should be increased for lines in excess of 50 kV. Individuals on scaffolds more than 10 feet high must be protected from falls. Multiple requirements exist for specific types of scaffolds, which must be installed under the direction of a competent person.

Training is required for all individuals working on or about scaffolding to identify scaffolding hazards, proper use, and maximum intended loads. Individuals involved with the erection or dismantling of scaffolds must be trained by a competent person in the procedures and design criteria of their work.

Hot Work Procedures

Hot work (burning, cutting, grinding, welding, or other heat or spark producing work) increases fire danger. Consequently, before hot work occurs, the area should be readied through the completion of a "Ready for Hot Work" Permit. A sample permit is found in Figure 6-3 in Chapter 6. Procedures such as those listed below apply to the preparation of areas for hot work.

- Before hot work occurs, the area must be inspected by the fire safety officer for the site. Hot work should not occur in the following environments:
  - In buildings with sprinklers while such protection is impaired
  - In the presence of explosive atmospheres (vapors or dusts)
  - Within 35 feet of combustible materials, or near the storage of large amounts of readily ignitable materials; where these materials may not be removed, shields or wetting materials shall be utilized; openings shall be appropriately protected
- A Ready for Hot Work permit shall be completed for the site. This permit shall specify fire prevention, fire protection, and the watch procedures for the site. It shall be removed 30 minutes after work is closed and the site's final inspection occurs.
  - A fire watch (trained) shall be present, with appropriate extinguishers (minimum of two rated 4A-60B;C extinguishers) while hot

PHOTO 11-12 A well-developed and protected scaffolding system.

PHOTO 11-13 A lockout device with appropriate wording.

work occurs, along with a means of alerting the designated fire response service for the facility. Hose lines (minimum 1 inch, 25 gpm) may be required by the fire safety officer.

- A fire watch shall be maintained for 30 minutes after hot work activities cease.
- Before hot work is initiated, atmospheric tests may be performed to ensure flammable vapors are not present. This test should be performed with a combustible gas indicator, approved by FM/UL for Class 1, Division 1 atmospheres. It must be calibrated prior to use.

Hazardous Energy

Hazardous energy may be transmitted by any electrical, mechanical, hydraulic, pneumatic, chemical, thermal, or other energy source. Consequently, when individuals must work around these energies, protection from routine and unexpected energy start-up must be assured.

An Energy Control Program (written procedures) and associated implementation training must occur for servicing or maintenance where unexpected energy may be released. This written requirement is waived for situations in which a single **lockout device** on a single energy source or machine will completely deenergize or deactivate the machine, and this will place no additional hazards on other employees in the area.

Since 1989, newly installed energy devices have been required to be installed with lockout acceptance capabilities (lockout devices). Alternatively, tagout procedures can be used **(tagout devices)**.

When used, lockout or tagout devices must be individually identifiable and standardized (such as by color) in the facility. They must identify the person applying the device and include a note such as "DO NOT OPERATE."

Training is required for authorized employees who will be the only individuals controlling the hazardous energy.

Prior to starting work in this hazardous area, authorized employees must verify that equipment has been locked out. Lockout devices can be removed after area inspections only by the individuals applying the devices, or under the direction of the employer as written into the Energy Control Program.

Group lockouts must be used for group work, and specific procedures must be developed for shift change times to allow an orderly transfer of lockout devices.

PHOTO 11-14 Typical lockout and tagout equipment. Tagouts remain permissible in older equipment that is incapable of accepting lockout devices.

Motor Vehicle Safety

Motorized vehicles present hazardous atmospheres for individuals and generally are classified as either public roadway or yard vehicles. General motor vehicle safety requirements are found in **TABLE 11-9**. Safety requirements for powered industrial trucks are found in **TABLE 11-10**. Jurisdiction must follow state and federal laws addressing motor vehicles used over the road. However, employers may choose to provide additional training, such as tactical or defensive driving courses. These are essential for individuals who drive for a large portion of work time. Employer powered industrial truck training and certification is required by OSHA, both initially and at three-year intervals.

TABLE 11-9 General Motor Vehicle Safety Requirements

- Licensing as appropriate to jurisdiction
- Commercial licensing (physical [including drug testing] and performance testing) for drivers of Hazmat placarded vehicles, buses, or vehicles in excess of 26,000 pounds GVW (specialized classes)
- Safety restraint policy and procedure in place
- Inspection of vehicle for safety equipment prior to each use

TABLE 11-10 Powered Industrial Truck Safety Requirements

- Training initially on the specific requirements for use of the powered industrial truck
- Retraining of operators in the case of an accident or near miss, observation of unsafe operation, a determination of the need for additional training, changes in the workplaces or changes in types of vehicles.
- Evaluation (testing) of operators and relicensure every three years
- Inspection of vehicle for safety equipment prior to use

Source: OSHA Powered Industrial Truck Standard 29 CFR 1910.178

Chapter Summary

Specialized environments present hazardous atmospheres for employees. These environments include confined spaces, excavations and trenches, work at heights (ladders, scaffolds, and the use of fall protection), heat generating operations, work around hazardous energy sources, and motor vehicle and industrial truck usage. Specialized procedures are required to safeguard each operation, including training in the recognition of hazards and training in the appropriate engineering, administrative, inspectional, and equipment controls that are in place to maintain safety. In some cases, written site permits may be required, as well as licensure or certification.

As with all hazardous environments, the first step in safeguarding individuals in these atmospheres is the recognition by all involved of the potential hazards, and their subsequent ability to implement protective options.

Terms

Hot Work: Work that involves sources of ignition or temperatures sufficient to cause a fire or explosion.

Lockout Device: Advice that utilizes a positive means such as a lock or other key or combination type to hold an energy isolating device in a safe position and prevent the energizing or energy release from a machine or equipment.

Permit Required Confined Space (PRCS): Enclosed and confined spaces that require special entry procedures.

Powered Industrial Truck: Fork trucks, tractors, platform lift trucks, motorized hand trucks, and other specialized industrial trucks powered by electric motors or internal combustion engines. The OSHA Powered Industrial Truck standard does not cover farm vehicles, earth-moving vehicles, or vehicles primarily designed for over-the-road hauling.

Tagout Device: A prominent warning device that can be securely fashioned to an energy-isolating device in accordance with an established procedure, to indicate that the energy-isolating device and the equipment controlled may not be operated until the tagout device is removed.

References

American National Standards Institute. *American National Standard for Powered Industrial Trucks*: ANSI/ITS OF B56.5. New York, NY, 2005.

American National Standards Institute. *Ladders, Wood*. New York, NY: ANSI, 2007.

———. *American National Standard for Ladders, Portable Metal*. New York, NY, 2007.

———. *American National Standard for Portable Re-Enforceable Plastic Ladders*. New York, NY, 2007.

———. *Safety Standard for Low Lift and High Lift Trucks*: ANSI/ITS DF B56.1. New York, NY, 2007.

www.Call811.com: Common Ground Alliance, 2011.

Miller Harness Company. Phone conversation with author, July 2010.

National Fire Protection Association. *Standard for Fire Prevention During Welding, Cutting, and Other Hot Work #51B*. Quincy, MA: NFPA, 2009.

U.S. Department of Health and Human Services, Centers for Disease Control and Prevention, National Institute for Occupational Safety and Health. *Confined Spaces*. Washington, D.C.: Government Printing Office, 1990.

U.S. Department of Labor, Occupational Safety and Health Administration. Construction, Sub Part L: Excavations. In *Construction Industry Safety Regulations*. Washington, D.C.: Government Printing Office, 1999.

———. Fall Protection—Construction Regulations, 29 CFR 1925, Sub Part M, Sections 500-503. In *Construction Industry Safety Regulations*. Washington, D.C.: Government Printing Office, Current.

———. Hazardous Energy Sources—29 CFR 1910.147. In *General Industry Safety Regulations*. Washington, D.C.: Government Printing Office, Current.

———. Ladders—29 CFR 1910. In *General Industry Safety Regulations*. Washington, D.C.: Government Printing Office, 1999.

———. Permit Required Confined Spaces—29 CFR 1910.146. In *General Industry Safety Regulations*. Washington, D.C.: Government Printing Office, Current.

———. Powered Industrial Trucks—29 CFR 1910.178. In *General Industry Safety Regulations*. Washington, D.C.: Government Printing Office, Current.

U.S. Department of Transportation. *One-Call Notification Programs—49 CFR 16*. Washington, DC, February 1, 2010.

VCAS. *Confined Spaces Standard for the Construction Industry*; Commonwealth of Virginia; 16 VAC 25-140-10. 2010.

———. *Confined Spaces Standard for the Telecommunications Industry*; Commonwealth of Virginia; 16 VACS25-70-10. 2010.

———. Welding and Cutting. Fire Prevention; Construction Regulation, Sub Part J. In *Construction Industry Safety Regulations*. Washington, D.C.: Government Printing Office, 1988.

Exercise 11

1–3. Identify the three conditions necessary for the existence of a confined space.

1. _____

2. _____

3. _____

4–7. Identify the four conditions that make a confined space a permit-required confined space.

4. _____

5. _____

6. _____

7. _____

8–10. Identify the three contraindications to life line use in a PRCS.

8. _____

9. _____

10. _____

11. Trench hazards exist and are regulated at depths of _____.

12–13. All trenches must be inspected by a competent person _____ (when), and subsequent to _____ (what events).

14–15. A competent person, as defined by OSHA, is one who exhibits the following two characteristics:

14. _____

15. _____

16. Ladders used for exits must be in place in trenches at distances of no more than _____ feet.

17. A "safe" soil slope (such as for gravel, sand, wet soil) is _____ (vertical) to _____ (horizontal).

18. Fall-protection requirements generally occur at heights of _____ feet.

19. _____ are the most common means of fall protection.

CHAPTER 11: Personal Safety

20. Fall-protection equipment must limit falls to _____ feet and prevent contact with lower surfaces.

21. _____ is a safe distance between scaffolds and electric power lines.

22. A "fire watch" is required during hot work activities and for a minimum of _____ (time) after activities are completed.

23–24. _____ and _____ are two means of controlling hazardous energy sources.

25. Lockout devices can be removed *only* by _____.

26. Powered industrial truck training and certification is required by OSHA, initially and at _____ (when) intervals.

27–31. Identify the type of personal safety hazard and appropriate safety protocol if work will occur in the following areas:

27.

28.

29.

30.

Appendix A
Bloodborne Pathogen Information

Definitions

Bloodborne Pathogen: Microorganisms present in human blood that can cause disease in humans (e.g., HBV/HIV).

Contaminated: Presence or reasonably anticipated presence of blood or other potentially infectious materials on an item or surface.

Exposure Incident: Eye, mouth, other mucous membrane, non-intact skin, or parenteral contact with blood or other potentially infectious materials that results from the performance of an employee's duties.

Universal Precautions: An approach to infection control that assumes that all human blood and certain human body fluids are infectious for HIV, HBV, and other bloodborne pathogens and must be treated accordingly.

Other Potentially Infectious Materials

- Semen
- Vaginal secretions
- Cerebrospinal fluid
- Synovial fluid
- Pleural fluid
- Pericardial fluid
- Peritoneal fluid
- Amniotic fluid
- Saliva in dental procedures
- Any body fluid that is visibly contaminated with blood
- All body fluids in situations in which it is difficult or impossible to differentiate between body fluids
- **Unfixed Tissues (living or dead):** HIV containing organs, cultures or solutions

Exposures and Health Hazards

An exposure is contact with blood or fluids that have the potential to be infectious. Most exposures occur through a needle stick; broken or non-intact skin; or the mucous membranes of the nose, mouth, or eyes.

Workers who are exposed should do the following:
- Wash the affected area.
- Report the incident and be examined by a proper medical authority for assessment and counseling.

Two diseases are prevalent from exposure—Hepatitis and HIV. There are seven known types of Hepatitis. Hepatitis A is passed through contaminated food, water, and feces.

Hepatitis A

Hepatitis A is a contagious liver disease that results from an infection with the Hepatitis A virus. It can range in severity from a mild illness lasting a few weeks to severe illness lasting months. Hepatitis A is usually spread by a person ingesting fecal matter, even in microscopic amounts, from contact with objects contaminated by fecal material of an infected person.

Following are primary risk factors for Hepatitis A:
- Infected person does not properly wash hands after using bathroom and touches objects
- Caregiver does not wash hands properly after changing diaper or cleaning stool of infected person
- Someone engages in certain sexual activities (i.e., oral-anal contact) with infected individual
- Ingesting contaminated food or water

A vaccine for Hepatitis A does exist and is recommended for all children, some international travelers and people with high risk factors or certain medical conditions.

Hepatitis B
Hepatitis B is a viral, incurable liver disease. One-fourth of all infected people develop acute symptoms. Anyone may be a carrier.

Following are the primary risk factors for Hepatitis B:
- Sexual contact with infected persons
- Shared needles and syringes

Because of screening tests, there is little chance, if any, of infection through a blood transfusion.

Occupational risks include the following:
- Accidental needle sticks
- Broken or non-intact skin exposures
- Mucous membrane exposure (eyes, nose, mouth)
- Saliva transmitted by human bite

Hepatitis C
Hepatitis C is a chronic (incurable, but treatable) disease. It is transmitted when blood is exchanged. The latency period is 10 to 30 years, and those who contact the disease may develop cirrhosis or scarring of the liver or liver cancer. People who share needles, razors, and personal hygiene items are at increased risk of Hepatitis C infection.

Acquired Immune Deficiency Syndrome (AIDS)
The cause of AIDS is the HIV virus. It attacks the body's immune system. There is no vaccine and it is incurable. Death is usually a result of other complications from the virus.

HIV transmission usually occurs through sexual contact with infected persons, through sharing needles with an infected person, or from an infected woman to her child at birth.

However, HIV is NOT transmitted through the following means:
- Airborne exposures
- Casual contact
- Nonsexual social situations
- Insects and rodents
- Urine, feces, and nasal secretions
- Sputum, vomitus, saliva (except through a human bite), sweat, and tears

Preventative measures for HIV include the following:
- Do not share needles
- Avoid sexual promiscuity (practice monogamy)
- Avoid sexual relations with individuals known to have multiple sex partners
- Practice safe sex—avoid unprotected sexual contact

How are HBV and HIV similar? Different?

Mode of Transmission	HBV	HIV
Blood	Yes	Yes
Semen	Yes	Yes
Vaginal secretions	Yes	Yes
Saliva (from a human bite)	Yes	Yes
Target in the body	Liver	Immune system
Risk of infection after needle stick from an infected needle	6% to 30%	0.5%
High number of viruses in blood	Yes	No
Vaccine available	Yes	No

The Practice of Universal Precautions

Universal precaution is an attitude that assumes that all blood and body fluids are infected and have the potential to infect. Therefore, practice body substance isolation in all situations. Make use of appropriate personal protective equipment such as gloves, masks, goggles, face shields, and gowns. Properly control and dispose of potentially infectious waste. Hand washing is essential.

Clean-Up/Decontamination
1. Always wear gloves. Wear eye and face protection if there is splash potential. Wear shoe covers as needed.
2. Clean gross materials with disposable towels.
3. Use germicidal or commercial bleach solution (1 part bleach per 9 parts water), and allow it to stand for at least 10 minutes (or 1 part concentrated bleach to 99 parts water).
4. Wipe and rewipe materials to be decontaminated, and air dry.
5. Remove personal protective equipment. Dispose properly of waste (as infectious waste).
6. Wash hands. (Hand washing procedure review: Wash hands with hospital-grade soap, wash all parts of the hands with multiple contact. Ensure a soap/skin residence time of 40 to 60 seconds.) (*Note:* Solution should not be pre-mixed for use.)

Appendix B
Asbestos and the OSHA Asbestos Standards

I. Substance Identification

A. Substance: "Asbestos" is the name of a class of magnesium-silicate minerals that occur in fibrous form. Minerals that are included in this group are chrysotile, crocidolite, amosite, anthophyllite, tremolite, and actinolite.

B. Asbestos is and was used in the manufacture of heat-resistant clothing, automotive brake and clutch linings, and a variety of building materials including floor tiles, roofing felts, ceiling tiles, asbestos-cement pipe and sheet, and fire-resistant drywall. Asbestos is also present in pipe and boiler insulation materials and in sprayed-on materials located on beams, in crawlspaces, and between walls.

C. The potential for an asbestos-containing product to release breathable fibers depends largely on its degree of friability. *Friable* means that the material can be crumbled with hand pressure when dry and is therefore likely to emit fibers. The fibrous fluffy sprayed-on materials used for fireproofing, insulation, or soundproofing are considered to be friable, and they readily release airborne fibers if disturbed. Materials such as vinyl-asbestos floor tile or roofing felt are considered nonfriable if intact and generally do not emit airborne fibers unless subjected to sanding, sawing, and other aggressive operations. Asbestos-cement pipe or sheet can emit airborne fibers if the materials are cut or sawed, or if they are broken.

D. Permissible exposure: Exposure to airborne asbestos fibers may not exceed 0.1 fiber per cubic centimeter of air (0.1 f/cc) averaged over the 8-hour workday, and 1 fiber per cubic centimeter of air (1.0 f/cc) averaged over a 30-minute work period.

II. Diseases Caused by Asbestos

See table on next page.

III. Asbestos Standard Summary

A. Scope

B. OSHA Classes of Asbestos Work: **Class I** asbestos work means activities involving the removal of TSI and surfacing ACM and Pressured Asbestos Containing Materials (PACM).

Class II asbestos work means activities involving the removal of ACM that is not thermal system insulation or surfacing material. This includes, but is not limited to, the removal of asbestos-containing wallboard, floor tile and sheeting, roofing and siding shingles, and construction mastics.

Class III asbestos work means repair and maintenance operations, where "ACM," including TSI and surfacing ACM and PACM, is likely to be disturbed.

Class IV asbestos work means maintenance and custodial activities during which employees contact but do not disturb ACM or PACM and activities to clean up dust, waste, and debris resulting from Class I, II, and III activities.

C. Permissible Exposure Limits:

PEL	0.1 f/cc
Excursion Limit	1.0 f/cc for any 30-minute period

D. Multiemployer worksites: Prime contractor retains responsibility for asbestos identification.

E. Regulated areas: Required for Class I, II, and III work.

Diseases Caused by Asbestos

Disease	Signs and Symptoms	Treatment of Symptoms
Asbestosis [Chronic inflammation and fibrosis of the lung]	Severe shortness of breath (SOB) Dry couth Feeling very tired Clubbed fingers	• Treatment, but no cure. • Stop working with asbestos. • Stop smoking. • Get flu shots. • Treat all chest colds quickly with antibiotics.
Lung Cancer [Uncontrolled cell growth in the tissues of the lung]	Shortness of breath Constant cough Deep chest pain Coughing up blood Weight loss	• Treatments: surgery, radiation and chemotherapy. • 9% to 13% live for 5 years or more. • Poor cure rate. • **Smoking multiplies your chances of getting lung cancer. Stop smoking!**
Mesothelioma [Cancer in the protective lining that covers many of the body's internal organs]	Chest (pleural): lodges in the lining of the chest Shortness of breath Dull chest pain under the ribs Swelling in chest Belly (peritoneal): lodges in the lining of the abdomen. Swollen stomach Belly pain Weight loss	• No treatment, some medical procedures for pain reduction. • Death in 6 months to 2 years after it is discovered. A few people have died 5 years after their mesothelioma was discovered.
Digestive System Cancer	Change in bowel patterns Blood in bowel movement Feeling tired Weight loss	• Treatments: surgery, radiation and chemotherapy. • Chances of living good if colon cancer found early; 80% to 90% live for 5 years or more.

F. Exposure assessment: Required.
G. Methods of compliance:
- HEPA filtration
- Wet methods
- Prompt cleanup
- Reduction to PEL through use of enclosures
- Prohibitions
  o High-speed, abrasive disc saws
  o Compressed air
  o Dry sweeping
  o Employee rotation
- Class I–III specific requirements for roofing

H. Respiratory protection:
- Class I: Required at all times
- Class II–III: Required without negative exposure assessments or wet methods
  o Class II: If not removed intact
  o Class II–III: Where wet methods are not used
  o Class III: Where TSI or surfacing is being disturbed

I. Protective clothing
J. Hygiene facilities and practices:
- Required for Class I activities in excess of 10 ft^2 (0.9 m^2) or 25 linear ft (7.6 m)
- Class I and Class II–III: Equipment room required

K. Communication of hazards:
- Presumption and rebuttal of asbestos
- Signs
- Information and training

L. Housekeeping
M. Medical surveillance: 30 days above the PEL (Note: 1 day equals 1 hour of exposure)
N. Recordkeeping required
O. Competent person requirements: On-site

IV. Where Can Asbestos Be Found?

Asbestos is commonly used as an acoustic insulator, thermal insulation, fireproofing, and in other building materials. Asbestos fibers are incredibly strong and have properties that make them resistant to heat. Many products are in use today that contain asbestos. Most of these are materials used in heat and acoustic insulation, fireproofing, and roofing and flooring. Some of the more common products that may contain asbestos include the following:

Cement Pipes	Laboratory Hoods/Table Tops	Elevator Brake Shoes
Cement Wallboard	Laboratory Gloves	HVAC Duct Insulation
Cement Siding	Fire Blankets	Boiler Insulation
Asphalt Floor Tile	Fire Curtains	Breaching Insulation
Vinyl Floor Tile	Elevator Equipment Panels	Ductwork Flexible Fabric Connections
Vinyl Sheet Flooring	Caulking/Putties	Cooling Towers
Flooring Backing	Adhesives	Pipe Insulation (corrugated air-cell, block, etc.)
Construction Mastics (floor tile, carpet, ceiling tile, etc.)	Wallboard	Heating and Electrical Ducts
Acoustical Plaster	Joint Compounds	Vinyl Wall Coverings
Decorative Plaster	Spackling Compounds	High Temperature Gaskets
Textured Paints/Coatings	Roofing Shingles	Roofing Felt
Ceiling Tiles and Lay-in Panels	Base Flashing	Thermal Paper Products
Spray-Applied Insulation	Fire Doors	Electrical Cloth
Blown-in Insulation	Electrical Panel Partitions	Fireproofing Materials
Taping Compounds (thermal)	Packing Materials (for wall/floor penetrations)	Brake Shoes (Vehicles)
Electrical Wiring Insulation	Chalkboards	Blown In Insulation

V. US EPA Information: National Emissions Standards for Hazardous Air Pollutants (NESHAPs)

A. The Asbestos NESHAP: Definitions:
- Regulated asbestos-containing material (RACM) means (a) Friable asbestos material, (b) Category I nonfriable ACE that has become friable, (c) Category I nonfriable ACM that will be or has been subjected to sanding, grinding, cutting, or abrading, or (d) Category II nonfriable ACM that has a high probability of becoming or has become crumbled, pulverized, or reduced to powder by the forces expected to act on the material in the course of demolition or renovation operations regulated by this subpart. Remove means to take out RACM or facility components that contain or are covered with RACM from any facility.
- Category I nonfriable asbestos-containing material (ACM) means asbestos-containing packings, gaskets, resilient floor covering, and asphalt roofing products containing more than one percent asbestos as determined using the method specified in Appendix A, subpart F, 40 CFR part 763, section 1, Polarized Light Microscopy.
- Category II nonfriable ACM means any material, excluding Category I nonfriable ACM, containing more than one percent asbestos as determined using the methods specified in Appendix A, subpart F, 40 CFR part 763, section1, Polarized Light Microscopy that, when dry, cannot be crumbled, pulverized, or reduced to powder by hand pressure.
- Demolition means the wrecking or taking out of any load-supporting structural member of a facility together with any related handling operations or the intentional burning of any facility.
- Renovation means altering a facility or one or more facility components in any way, including the stripping or removal of RACM from a facility component. Operations in

which load-supporting structural members are wrecked or taken out are demolitions.
- Visible emissions means any emissions, which are visually detectable without the aid of instruments, coming from RACM or asbestos-containing waste material, or from any asbestos milling, manufacturing, or fabricating operation. This does not include condensed, uncombined water vapor.

B. Asbestos NESHAPs Applicability Thesholds

Applicability Thresholds	
Location of Asbestos	Threshold Level of RACM
Pipes	80 linear meters (260 linear feet)
Other facility components	15 square meters (160 square feet)
Asbestos that is already off facility components where the length or area could not be measured previously	1 cubic meter (35 cubic feet)

C. When Must an Asbestos Inspection and Detection Survey Be Completed?
The Asbestos NESHAP requires that a thorough inspection be conducted for all renovations and all demolition. All inspections must be completed before the commencement of a subject renovation and/or demolition activity, and the contractor performing the inspection must be listed on the notification form. Notifications must be postmarked or hand-delivered at least 10 working days prior to renovation or demolition.

D. Is the Activity a Demolition or a Renovation?
A demolition is the wrecking or taking out of any load-supporting structural member of a facility together with any related handling operations or the intentional burning of any facility. A renovation is altering a facility or one or more facility components in any way, including the stripping or removal of RACM from a facility component (excluding operations in which load-supporting structural members are wrecked or taken out).

Appendix C

LEAD

PART 1: Lead and the Lead Standard: A Summary of Occupational Exposure Levels

Source: OSHA 29 CFR 1926.62, Appendices A and B

I. Substance Identification and Health Hazard Data

A. Substance: Pure lead (Pb) is a heavy metal at room temperature and pressure and is a basic chemical element. It can combine with various other substances to form numerous lead compounds.

B. Compounds covered by the standard: The word lead when used in the OSHA standard means elemental lead, all inorganic lead compounds, and a class of organic lead compounds called lead soaps. This standard does not apply to other organic lead compounds.

C. Uses: Exposure to lead occurs in different occupations in the construction industry.

D. Permissible exposure: The permissible exposure limit (PEL) set by the standard is 50 micrograms of lead per cubic meter of air (50 mg/m^3), averaged over an 8-hour workday.

E. Action level: The standard establishes an action level of 30 micrograms of lead per cubic meter of air (30 mg/m^3), averaged over an 8-hour workday. The action level triggers several ancillary provisions of the standard such as exposure monitoring, medical surveillance, and training.

II. Standard Summary 1926.62

Scope and Applicability

The OSHA Lead in Construction Standard applies to construction work where employees may be occupationally exposed to lead. This standard does not apply to those activities that are covered in the OSHA General Industry standard for lead (29 CFR 1910.1025), which provides similar protections. Construction activities covered by this standard include but are not limited to painting, decorating, repair, demolition, salvage, removal or encapsulation of materials containing lead, lead containment and emergency cleanup, maintenance operations, transportation, storage, and disposal of materials containing lead on the site or location at which construction activities were performed. The OSHA lead standard covers:

a. Scope of operations regulated by OSHA
b. Definitions related to lead work
c. Permissible exposure limits:
  - Action level
  - Permissible Exposure Limit
d. Exposure assessments (required for all work generating lead in air):
  - Protection of employees during assessment of exposure: Required task
e. Methods of compliance: Written lead hygiene program
f. Respiratory protection: Required
  - PAPR request option
g. Protective work clothing and equipment: Required
h. Housekeeping
  - HEPA requirements for vacuums
i. Hygiene facilities and practices
  - Changing areas
  - Showers
  - Eating, drinking, smoking prohibitions
  - Hand-washing facilities

j. Medical surveillance
     - Initial monitoring
     - Lead medical monitoring
     - Medical removal protection
  k. Medical removal protection (50 ug/dl)
  l. Employee information and training (annual)
  m. Signs:
     - Warning: Lead Work Area: Poison
     - No Smoking or eating
  n. Recordkeeping
  o. Observation of monitoring by employees

PART 2: Environmental and Public Safety: Lead Based Paint Activities: Prevention of Childhood Lead Poisoning: U.S. EPA: 40 CFR 745, U.S. HUD: 24 CFR 35

I. Historical and Regulatory Perspective

While the phase out of lead as a paint additive was initiated through the Consumer Product Safety Commission in 1977, national strategies to control lead poisoning in children began in 1992, when Congress passed the Residential Lead Based Paint Hazard Reduction Act of 1992 (Title X). Title X's framework allowed for streamlined efforts in eliminating childhood lead poisoning, including protocols for how lead paint hazards are evaluated and controlled. It required real estate transactions that involved targeted structures disclose the presence or potential existence of lead hazards. It required the inclusion of lead paint policies in government-funded properties; and outreach efforts to educate the public on lead paint hazards, as well as steps to minimize exposures. Efforts targeted environments where children were known to spend time, including houses, day cares, and after-school programs.

Primary agencies involved in the efforts were the Environmental Protection Agency, the Department of Housing and Urban Development, the Department of Health and Human Services: Centers for Disease Control and Prevention, and the Department of Labor's Occupational Safety and Health Administration. EPA issued regulations for controlling and evaluating lead hazards and established lead hazard criteria, while HUD developed guidelines for certain activities (e.g., Risk Assessments, Worksite Controls, paint removal via chemical stripping methods, etc.) and regulations for incorporating lead hazard reduction into their housing policies. Additional groups with involvement included state and local labor, environmental and health departments.

Subsequently, in response to studies that indicated potential exposures to lead dust resulting from household renovation activities, the EPA established the **Renovation, Repair and Painting (RRP) Rule (April 2010)**, which required contractors compensated for work in housing or child occupied facilities built prior to 1978 to attend training, obtain licensure, and perform work in accordance with set practices. These practices included controls designed to minimize the spread of dust, use of lead-safe work practices that minimized the amount of dust generation, and specialized cleanup procedures and testing to ensure dust cleanup.

II. Definitions

Target Housing: Housing built prior to 1978 with a bedroom. Target Housing does not include housing built specifically for handicapped or elderly individuals, unless occupied by a child under the age of six.

Child Occupied Facility: A building or portion of a building constructed prior to 1978 that is occupied by a child under 6 for at least 6 hours a week (at least 2 times a week for 3 hours per visit) and 60 hours per year (referred to as the "6-6-60 Rule").

Friction surface: An interior or exterior surface that is subject to abrasion or friction that will speed the break down of coated surfaces.

Impact surface: An interior or exterior surface that is subject to damage by repeated sudden force, such as closing a window or door.

Abatement: Any measure or set of measures specifically performed to eliminate lead-based paint hazards.

The options for abatement include (1) paint removal; (2) component replacement; (3) encapsulation; and (4) enclosure. Abatement does not include renovation activities that are not performed with the intention of addressing lead paint hazards, or interim controls that are temporary in nature.

Elevated blood lead level (EBL): A concentration of lead in whole blood of 20 µg/dl (micrograms of lead per deciliter of whole blood) for a single venous test or of 15–19 µg/dl in two consecutive tests taken 3 to 4 months apart in a child 6 years of age or younger. (*Note: Some health departments utilize a level of 10 µg/dl as an action level.*)

Lead Hazards: A condition that may result in exposure to lead from lead-contaminated dust, lead-contaminated soil, or lead-contaminated paint. Lead hazards are characterized by hazard criteria established by the EPA (*Note: certain state programs may have more stringent criteria*):

Dust: Floors—40 µg/ft^2
Window Sills—250 µg/ft^2
Window Troughs—400 µg/ft^2
Paint: 1.0 mg/cm^2 or 0.5% by weight
Soil: 400 ppm (high contact and play areas)
1200 ppm (areas not frequently contacted)

Paint Condition Classifications: (Damage):
Intact: No damage
Fair: < 2 ft^2 (interior); < 10 ft^2 (exterior); < 10% small component change
Poor: < 2 ft^2 (interior); < 10 ft^2 (exterior); < 10% small component change

III. Lead Disciplines and Certifications

A primary element of the EPA regulations under 40 CFR Part 745 is the requirement for individuals and firms conducting any lead-based paint activity to be properly trained and certified to perform the work. Individuals must attend initial training in EPA curricula provided by an accredited training provider and subsequently apply to the licensing jurisdiction for certification. These individuals must also attend refresher training within set timeframes.

Lead based paint activity disciplines include occupational classifications for evaluating (inspectors, risk assessors) and controlling (worker, supervisor, and project designer) lead hazards.

- **Worker:** Performs lead abatement and hazard controls.
- **Supervisor:** Performs lead abatement and hazard controls, acts as the site competent person, completes reports and Occupant Protection Plans. A licensed supervisor must be onsite for setup and cleaning of the work areas, and must be within 2 hours of the site.
- **Project Designer:** Develops project design and specifications for protecting occupants, completes reports and Occupant Protection Plans.
- **Inspector:** Performs lead inspections and post-abatement clearance assessments.
- **Risk Assessor:** Performs lead inspections, risk assessments, and post-abatement clearance assessments.

For renovation activities under the RRP Rule, each Certified Renovation Firm must have a **Certified Renovator** control each project. Other individuals, identified as **Non-Certified Workers**, on the project must receive training on the site's lead-safe procedures from a Certified Renovator. A Certified Renovator receives 8 hours of training from an EPA authorized training provider or has a lead based paint activity license designated above, with a renovation specific refresher class. This renovator training and licensing is required at 5 year intervals.

IV. Abatement and Control of Lead Hazards

Lead Paint Abatement includes those measures that are performed with the specific intention of eliminating a lead hazard. These activities are to be completed by certified individuals employed by certified companies. These individuals must perform work in accordance with the work practice requirements established by the EPA under 40 CFR 745. Standard practice dictates use of the HUD "Guidelines for the Evaluation and Control of Lead Hazards in Housing." Certain elements of the EPA Regulations and HUD Guidelines are discussed in this section.

1. Specification, Project Design, and Occupant Protection

Many lead abatement projects include a project specification, or project design, that incorporates review of existing lead inspection or risk assessment reports and subsequent development of procedures to address identified hazards. If requested and developed, a written Lead Abatement Project Design must be completed by a certified Project Designer and address all EPA, HUD, and OSHA requirements applicable to the job. This

TABLE C-1 EPA/HUD Lead Abatement Options

Option	Description	Example	Benefits	Disadvantages
Paint removal	Physical or chemical removal of paint from the substrate.	Chemical removal; scraping via heat gun or HEPA shrouded power tool.	Permanent solution for removing a lead paint hazard.	Costly due to extensive preparation, equipment, time, cleaning, etc. Worker exposures depending on method and efficiency.
Component replacement	Replacing leaded building components with new components free of lead based paint.	Door, window, or baseboard replacement	Permanent solution that typically generates less dust than paint removal. May also improve weather resistance, safety and aesthetics.	Cost of removal and new building components, dust must be controlled and cleaned.
Enclosure	Installation of solid, dust-tight materials over leaded components.	Drywall over leaded plaster walls; vinyl siding over leaded wood siding	Minimal dust generated, efficient at prevented an exposure pathway.	Lead dust will still be released after structural damage or renovations.
Encapsulation	Use of a rated encapsulant product to cover a lead paint component. Leaded base coatings do not penetrate the encapsulant layer.	Walls, ceilings, cabinets	Lower cost option, minimal skill and time requirements	Encapsulation is not recommended for doors, windows or other friction/impact surfaces. Lead dust will still be released after structural damage or renovations.

includes development of an Occupant Protection Plan (OPP) within the Project Design. Methods for protection include restriction of occupancy in the building or certain work areas, use of containment barriers, security, and provision of lead safe passageways to bathrooms and exits. Prior to any regulated abatement, the building owners and occupants (or in the case of schools, parents and legal guardians) must be provided with information about the project and the current EPA lead hazard information packet.

2. Agency Notification

Prior to lead abatement, EPA requires notification utilizing the EPA "Notification of Lead Based Paint Abatement Activities" Form (see Figure 6-5) or equivalent state form if applicable. Notification is required a minimum of five business days prior to the anticipated start of the project. The form must specify site information (owner, location, and inspection data), the name/certification information for the abatement company, name and license number of the Certified Supervisor and a description of the project's scope of work, including how the work will be performed.

3. Abatement Options and Work Practices

EPA and HUD recognize four types of abatement options: (1) Paint Removal; (2) Component Replacement; (3) Enclosure; and, (4) Encapsulation. Descriptions, benefits and disadvantages are discussed below.

For lead-contaminated soil hazards, abatement options include removal of soil or installation of a solid barrier, such as pavement or asphalt.

Where abatement options are not selected, or as a temporary measure prior to abatement, Interim Controls may be implemented to minimize lead dust hazards. Examples of Interim Controls include:
- Regular dust cleanup
- Installing doorstops or chair rails at impact points
- Maintenance of slide channels on windows
- Placement of composite or carpeted runners over wood steps
- Mulch, gravel, or fencing that prevents contact with lead-contaminated soil.

4. Worksite Setup

A key step in the abatement project is the development of the work area to minimize the spread of any lead dust that may be generated. It is typically recommended to follow the worksite setup protocols designated by HUD in Chapter 8 of the Guidelines.[1] Worksite setup

---

[1] U.S. Department of Housing and Urban Development. *Guidelines for the Evaluation and Control of Lead Based Paint Hazards in Housing.* June 1995.

procedures are to be listed in the project's Occupant Protection Plan. Elements of an effective worksite safety protocol and occupation protection plan include provisions to keep residents out of the work area (barriers and signs), protect floors, grounds and furniture that remain in the work area, and minimize the spread of dust through critical barriers (doors, windows, ventilation entry points). An effective system will ensure adjacent areas are safe to occupy during work, minimize cleanup outside of the work area, and result in a work area that is most likely to pass clearance testing at the end of the project.

5. Lead Safe Work Practices: Prohibited Activities and Cleanup

Work practices must be selected that minimize the spread of dust outside of the point of generation. These "Lead-Safe Work Practices" range from misting surfaces and scoring paint to use of high-efficiency vacuums, low temperature heat guns and specialized power tools that utilize collection shrouds attached to HEPA vacuums. In certain situations, the use of solvent or caustic-based chemical removal products may be efficient. Due to their propensity to generate high levels of lead dust and fumes, which ultimately settle on surfaces and become difficult to clean, EPA has prohibited certain activities on EPA/HUD regulated projects. These prohibited activities include:

- **Use of open flames or torch cutting**
- **Use of heat guns over 1100°F (593°C)**
- **Extensive dry scraping (over 2 square feet), except near electrical outlets**
- **Use of power tools, grinding, sanding, etc. without dust collection system that include HEPA filtration**
- **Use of compressed air for cleaning purposes (this is an OSHA-prohibited activity).**

Cleaning protocols are to be implemented throughout the project (at a minimum, end of every shift) as well as a final cleaning at the conclusion of all regulated work and before clearance occurs. These protocols should include a cycle of HEPA vacuuming—wet wiping—HEPA vacuuming, and occur in a systemic top-to-bottom, wall-to-door manner. Special attention should be given to areas known to accumulate and trap dust: moldings, registers, upholstery, and gaps in hardwood floors.

Workers are also required to follow standard personal hygiene practices that protect themselves and minimize spreading of dust. These hygiene practices include decontamination, use and changing of protective clothing, and washing hands, face, and hair upon exiting the work area.

6. Clearance of Work

At the conclusion of work, but prior to removal of signs and protective barriers, a clearance assessment is required to ensure the work area has been properly cleaned. The clearance assessment must be completed by a licensed Inspector or Risk Assessor, independent of the abatement contractor. A Clearance Assessment initiates with a visual inspection of the work area with the abatement contractor's supervisor to ensure all work has been completed and that dust, debris, and paint chips are not visible. After the visual inspection is successfully completed, the Inspector/Risk Assessor then collects a series of dust samples from targeted components (floors, sill, troughs) within the contained area, in addition to a sample collected from the floor immediately adjacent to the worker entrance. For projects in which containment barriers were not used, samples may be collected from rooms within the entire unit, regardless if work was performed. The assessment is not considered complete until dust and soil sampling results are found to be below the respective EPA hazard criteria. Only after all results are satisfactory can the warning signs and all protective barriers be removed and occupant entry permitted.

7. Waste Disposal

Unless exempt (e.g., waste generated under RRP activities is exempt, when it is produced from a residential structure, or when less than 220 lbs when the RRP activity effects a child-occupied facility), lead abatement related waste is to be handled as hazardous waste unless testing performed in accordance with EPA's hazardous waste regulations (40 CFR 260-265) proves the waste is not hazardous. This testing protocol—identified as the "Toxicity Characteristic Leaching Procedure" or TCLP—may be performed on various waste streams from the project, such as paint-related debris, containment barriers, or personal protective equipment. If results of the TCLP test are at or above the EPA Toxicity criteria of 5 milligrams per liter (mg/L) of lead, the waste must be handled as hazardous. If results are below this criterion, the waste may be handled as construction debris. Regardless of waste classification, waste should be handled to minimize spread of dust and the threat of lead exposure to the public or occupants. Recommendations include:

- Using containers suitable for the contents, such as double heavy-grade construction bags or fiberboard drums.
- Utilizing sealed containers to prevent accidental release.
- Cleaning waste containers prior to removal from the work area and storage in a secured location.

8. Recordkeeping and Final Reports

After each regulated project, a final report must be developed and signed by a licensed Supervisor or Project Designer. This report must be provided to the building owner and should document all aspects of the project, such as location, scope, setup, methods and clearance results. A copy of the final report and project records are required to be maintained for 3 years by the contractor. Project records include pre-abatement documents (notification, Occupant Protection Plans), abatement activity documents (sign-in sheets, employee and firm licenses) and post abatement documents (waste records, clearance reports and the project final report).

V. Evaluation of Lead Hazards

Lead hazards are assessed in accordance with EPA Regulations and HUD Guidelines by individuals appropriately licensed to perform the evaluation. Options for evaluations include:

Inspection: A Lead Paint Inspection is a surface-by-surface investigation with the objective of identifying

TABLE C-2 Interior Worksite Preparation Levels (Not Including Windows)

Description	Level 1	Level 2	Level 3	Level 4
Typical applications (hazard controls)	Dust removal and any abatement or interim control method disturbing no more than 2 ft^2 (0.2m^2) of painted surface per room.	Any interim control or abatement method disturbing between 2 and 10 ft^2 (0.2 and 0.9 m^2) of painted surface per room.	Same as Level 2	Any interim control or abatement method disturbing more than 10 ft^2 (0.9 m^2) per room.
Time limit per dwelling	One work day	One work day	Five work days	None
Resident location	Inside dwelling, but outside work area. Resident must have lead-safe passage to bathroom, at least one living area, and entry/egress pathways. Alternatively, resident can leave the dwelling during the workday.	Same as Level 1	Outside the dwelling; but can return in evening after day's work and cleanup are completed. Resident must have safe passage to bathroom, at least one living area, and entry/egress pathway upon return. Alternatively, resident can leave until all work is completed.	Outside the dwelling for duration of project, cannot return until clearance has been achieved.
Containment and barrier system	Single layer of plastic sheeting on floor extending 5 feet (1.5 m) beyond the perimeter of the treated area in all directions. No plastic sheeting on doorways is required, but a low physical barrier (furniture, wood planking) to prevent inadvertent access by resident is recommended. Children should not have access to plastic sheeting (suffocation hazard).	Two layers of plastic on entire floor. Plastic sheet with primitive airlock flap on all doorways. Door secured from inside the work area need not be sealed. Children should not have access to plastic sheeting (suffocation hazard).	Two layers of plastic on entire floor. Plastic sheet with primitive airlock flap on all doorways to work areas. Doors secured from inside the work area need not be sealed. Overnight barriers should be locked or firmly secured. Children should not have access to plastic sheeting (suffocation hazard).	Two layers of plastic on entire floor. If entire unit is being treated, cleaned, and cleared, individual room doorways need not be sealed. If only a few rooms are being treated, seal all doorways with primitive airlock flap to avoid cleaning entire dwelling. Doors secured from inside the work area need not be sealed.
Warning signs	Required at entry to room but not on building (unless exterior work is also underway).	Same as Level 1	Posted at main and secondary entryways, because resident will not be present to answer the door.	Posted at building exterior near main and secondary entryways.

TABLE C-3 Exterior Worksite Preparation Levels (Not Including Windows)

Description	Level 1	Level 2	Level 3
Typical applications	Any interim control or abatement method disturbing less than 10 ft² (0.9 m²) of exterior painted surface per dwelling. Also includes soil control work.	Any interim control or abatement method disturbing 10 to 50 ft² (0.9 to 4.6 m²) of exterior painted surface per dwelling. Also includes soil control work.	Any interim control or abatement method disturbing more than 50 ft² (4.6 m²) of exterior painted surface per dwelling. Also includes soil control work.
Time limit per dwelling	One day	None	None
Resident location	Inside dwelling but outside work area for duration of project until cleanup has been completed. Alternatively, resident can leave until all work has been completed. Resident must have lead-safe access to entry/egress pathways.	Relocated from dwelling during workday, but may return after daily cleanup has been completed.	Relocated from dwelling for duration of project until final clearance is achieved.
Containment and barrier system	One layer of plastic on ground extending 10 ft (3 m) beyond the perimeter of working surfaces. Do not anchor ladder feet on top of plastic (puncture the plastic to anchor ladders securely to ground). For all other exterior plastic surfaces, protect plastic with boards to prevent puncture from falling debris, nails, etc., if necessary. Raise edges of plastic to create a basin to prevent contaminated runoff in the event of unexpected precipitation. Secure plastic to side of building with tape or other anchoring system (no gaps between plastic and building). Weight all plastic sheets down with two-by-fours or similar objects. Keep all windows within 20 ft (6 m) of working surfaces closed, including windows of adjacent structures.	Same as Level 1	Same as Level 1
Playground equipment, toys, sandbox	Remove all moveable items to a 20-ft (6-m) distance from working surfaces. Items that cannot be readily moved to a 20-ft (6-m) distance can be sealed with taped plastic sheeting.	Same as Level 1	Same as Level 1
Security	Erect temporary fencing or barrier tape at a 20-ft (6-m) perimeter around working surfaces (or less if distance to next building or sidewalk is less than 20 ft [6 ml]). If an entryway is within 10 ft (3 m) of working surfaces, require use of alternative entryway. If practical, install vertical containment to prevent exposure. Use a locked dumpster, covered truck or locked room to store debris before disposal.	Same as Level 1	Same as Level 1
Signs	Post warning signs on the building and a 20-ft (6-m) perimeter around building or less if distance to next building or sidewalk is less than 20 ft (6 m).	Same as Level 1	Same as Level 1
Weather	Do not conduct work if wind speeds are greater than 20 mph (32 km/h). Work must stop and cleanup must occur before rain begins.	Same as Level 1	Same as Level 1
Cleanup	Do not leave debris or plastic out overnight if work is not completed. Keep all debris in secured area until final disposal.	Same as Level 1	Same as Level 1
Porches	One lead-safe entryway be made available to residents at all times. Do not treat front and rear porches at the same time if there is not a third doorway.	Front and rear process can be treated at the same time, unless unprotected workers must use the entryway.	Same as Level 2

TABLE C-4 Window Treatment or Replacement Worksite Preparation

Appropriate applications	Any window treatment or replacement
Resident location	Remain inside dwelling but outside work area until project has been completed. Alternatively, can leave until all work has been completed. Resident must have access to lead-safe entry/egress pathway.
Time limit per dwelling	None
Containment and barrier system	One layer of plastic sheeting on ground or floor extending 5 ft (1.5 m) beyond perimeter of window being treated/replaced. Two layers of plastic taped to interior wall if working on window from outside; if working from the inside, implement a minimum interior worksite preparation Level 2. Children cannot be present in an interior room where plastic sheeting is located due to suffocation hazard. Do not anchor ladder feet on top of plastic (puncture the plastic to anchor ladders securely to ground). For all other exterior plastic surfaces, protect plastic with boards to prevent puncture from falling debris, nails, etc. (if necessary). Secure plastic to side of building with tape or other anchoring system (no gaps between plastic and building). Weigh all plastic sheets down with two-by-fours or similar objects. All windows in dwelling should be kept closed. All windows in adjacent dwellings that are closer than 20 feet to the work area should be kept closed.
Signs	Post warning signs on the building and at a 20-ft (6-m) perimeter around building (or less if distance to next building or sidewalk is less than 20 ft [6 ml]). If window is to be removed from inside, no exterior sign is necessary.
Security	Erect temporary fencing or barrier tape at a 20-ft (6-m) perimeter around building (or less if distance to next building or sidewalk is less than 20 ft [6 ml]). Use a locked dumpster, covered truck, or locked room to store debris before disposal.
Weather	Do not conduct work if wind speeds are greater than 20 mph (32 km/h). Work must stop and cleanup must occur before rain beings, or work should proceed from the inside only.
Playground equipment, toys, sandbox	Removed items from work area and adjacent areas. Remove all items to a 20-ft (6-m) distance from dwelling. Large, unmovable items can be sealed with taped plastic sheeting.
Cleaning	If working from inside, HEPA vacuum, wet wash, and HEPA vacuum all interior surfaces within 10 ft (3 m) of work area in all directions. If working from the exterior, no cleaning of the interior is needed, unless the containment is breached. Similarly, no cleaning is needed on the exterior if all work is done on the interior and the containment is not breached. If containment is breached, then cleaning on both sides of the window should be performed. No debris or plastic should be left out overnight if work is not completed. All debris must be kept in a secure area until final disposal.

the existence of lead based paint within the structure. The investigation utilizes a process of identifying and sampling/testing combinations based on rooms, components, and substrates with distinct paint histories.

Risk Assessment: A Risk Assessment is an onsite investigation that determines the existence, nature, severity, and locations of lead-based paint hazards. Risk assessments utilize a combination of background information about the property (usually obtained through owner/occupant interviews); visual inspections for damaged paint, dust, and soil hazards, and environmental sampling (typically lead paint, dust wipe, soil sampling). Limited Risk Assessments, referred to as **Hazard Screens,** are typically performed on properties in good condition where a full assessment would be unnecessary.

The Risk Assessor must provide recommendations to the building owner or representative that address hazards identified from the assessment, such as lead dust, soil contamination or lead paint observed in poor or fair condition.

Elevated Blood Lead (EBL) Investigations: A full EBL investigation may be necessary in locations where a child or children (i.e., less than 6 years in age) are identified with blood levels above action or response levels. EBL investigations are typically multi-disciplinary in nature, involving environmental health professionals, health-care professionals, or housing specialists. These investigations may involve a standard risk assessment and investigation into lead exposures outside of the risk assessment's scope (drinking water, toys, utensils, etc.).

VI. Renovation, Repair and Painting Rule

On April 22, 2008, the EPA issued a rule requiring the use of lead-safe practices and other actions aimed at preventing lead poisoning. This rule has been promulgated within 40 CFR 745. Beginning April 22, 2010, any contractor performing Renovation, Repair, or Painting (RRP) activities that disturb lead-based paint in homes and child occupied facilities built prior to January 1, 1978, must be certified and follow specific work practices to prevent lead contamination (some state programs may be more stringent):

RRP REQUIREMENTS:

Training/Certification:
- At least one Certified Renovator for each company must attend an initial course and refresher after five years from accredited training provider.
- The firm must become certified through the EPA and renewed within five years.

Applicability:
- Any contractor, maintenance worker, painter, or other specialty trade disturbing greater than six square feet per room or 20 square feet on the exterior (does not include window replacement, demolition, or prohibited practices) in any 30 day period in any residential structure or child occupied facility constructed prior to January 1, 1978, is required to comply with the RRP rules after April 22, 2010.

Regulatory Summary
- *Pre-Job Requirements*: Prior to the start of any regulated project, the Certified Renovator is required to test the surfaces to be impacted for the presence of lead paint, notify the homeowner/affected tenants, and certify the delivery of the notification. The Certified Renovator is required to test surfaces with an approved chemical spot test, collect paint chips, or hire an EPA-licensed inspector/risk assessor to test by X-ray fluorescence (XRF).
- *Dust Control, Cleaning, and Lead Safe Work Practices:* Specific dust control measures, including plastic barriers, specialized cleaning, and the use of work practices that minimize the generation of dust are required to be used for any regulated job.
- *Project Completion:* At the end of a regulated job, the RRP Rule requires the use of visual inspections and quality assurance measures to ensure the area has been cleaned appropriately.
- *Recordkeeping:* Documentation including training records, cleaning verification/clearance, paint testing, and others are required to be maintained by the firm for a *minimum* of three (3) years.

Additional information may be attained on the web at www.epa.gov/lead or by calling 1-800-424-LEAD (5323).

Appendix D

New York City Department of Health and Mental Hygiene: Fungal Remediation Protocols

Summary of New York City Department of Health Fungal Remediation Protocols					
Remediation Parameter	**Level 1**	**Level 2**	**Level 3**	**Level 4A**	**Level 4B**
Description	Small Isolated Areas (10 ft^2 or less)	Medium-Sized Isolated Area (10–100 ft^2)	Extensive Contamination (Greater than 100 contiguous ft^2)	HVAC Systems (Less than 10 ft^2)	HVAC Systems (Greater than 10 ft^2)
Examples	Ceiling Tiles, Small Areas on Wall	Individual Wallboard Panels	Multiple Wallboard Panels		
Minimum Requirements for Remediation Oversight	Trained Building Staff	Trained Building Staff	Qualified Health and Safety Professional	Trained Building Staff	Qualified Health and Safety Professional
OSHA Regulatory Standards	29 CFR 1910.1200 29 CFR 1910.134	29 CFR 1910.1200 29 CFR 1910.134	29 CFR 1910.1200 29 CFR 1910.134	29 CFR 1910.1200 29 CFR 1910.134	29 CFR 1910.1200 29 CFR 1910.134
Respiratory Protection	N95 Disposable Respirator	N95 Disposable Respirator	Minimum of half-face Respirators with P100 Cartridges	N95 Disposable Respirator	Minimum of half-face Respirators with P100 Cartridges
Gloves	Yes	Yes	Yes	Yes	Yes
Eye Protection	Yes	Yes	Yes	Yes	Yes
Protective Clothing	No	No	Yes	No	Yes
Remediation While Unoccupied	Yes	Yes	Yes	Yes	Yes
Vacation of Adjacent Spaces	Recommended if Occupied by Susceptible Groups*	Recommended if Occupied by Susceptible Groups*	Yes	Recommended if Occupied by Susceptible Groups*	Recommended if Occupied by Susceptible Groups*
Containment Required	No	Critical Barriers	Critical Barriers, Airlocks, Decontamination Room with Critical Barriers	Critical Barriers	Critical Barriers, Airlocks, Decontamination Room Over 30 ft^2 (2.8 m^2)

Remediation Parameter	Level 1	Level 2	Level 3	Level 4A	Level 4B
HEPA Filtered Negative Air	No	No	Yes	No	Yes
Dust Suppression	Misting	Misting	Misting	Misting	Misting
Bag Contamination Materials	Yes	Yes	Yes	Yes	Yes
Post Remediation Cleaning of Work Area and Egress	HEPA Vacuum and Clean with Damp Cloth and/or Mop with Detergent Solution	HEPA Vacuum and Clean with Damp Cloth and/or Mop with Detergent Solution	HEPA Vacuum and Clean with Damp Cloth and/or Mop with Detergent Solution	HEPA Vacuum and Clean with Damp Cloth and/or Mop with Detergent Solution	HEPA Vacuum and Clean with Damp Cloth and/or Mop with Detergent Solution
Clearance Testing	No	No	Yes	No	Yes

*Susceptible groups include infants less than 12 months old, persons recovering from recent surgery, immune-suppressed people or people with chronic inflammatory lung disease (e.g., asthma, hypersensitivity pneumonitis and severe allergies).
Note: Reissued by the New York City Department of Health and Mental Hygiene: 11/2008

Appendix E

U.S. Department of Transportation
Pipeline and Hazardous Materials Safety Administration

DOT CHART 14
Hazardous Materials Markings, Labeling and Placarding Guide

Refer to 49 CFR, Part 172:

Marking - Subpart D

Labeling - Subpart E

Placarding - Subpart F

NOTE: This document is for general guidance only and should not be used to determine compliance with 49 CFR, Parts 100-185.

HAZARDOUS MATERIALS MARKINGS

Package Orientation (Red or Black)

 or

§172.312(a)

Keep Away from Heat

§172.317

§173.25(a)(4)

§172.325

§172.332(a)

Fumigant Marking (Red or Black)

§172.302(g) and §173.9

Biological Substances, Category B

§173.199 (a)(5)

§172.313(a)

§172.316(a)

Excepted Quantity

§173.4a(g)

Marking of IBCs

§178.703(a)(vii)(B)

Marine Pollutant

§172.322

Courtesy of the U.S. Department of Transportation

Hazardous Materials Warning Labels

Actual label size: at least 100 mm (3.9 inches) on all sides

CLASS 1 Explosives: Divisions 1.1, 1.2, 1.3, 1.4, 1.5, 1.6

§172.411

* Include compatibility group letter.
** Include division number and compatibility group letter.

CLASS 2 Gases: Divisions 2.1, 2.2, 2.3

§172.405(b), §172.415, §172.416, §172.417

CLASS 3 Flammable Liquid

§172.419

CLASS 4 Flammable Solid, Spontaneously Combustible, and Dangerous When Wet: Divisions 4.1, 4.2, 4.3

§172.420, §172.422, §172.423

CLASS 5 Oxidizer, Organic Peroxide: Divisions 5.1 and 5.2

Organic Peroxide, Transition-2011

§172.426, §172.427

CLASS 6 Poison (Toxic), Poison Inhalation Hazard, Infectious Substance: Divisions 6.1 and 6.2

§172.323, §172.405(c), §172.429, §172.430, §172.432

For Regulated Medical Waste (RMW), an Infectious Substance label is not required on an outer packaging if the OSHA Biohazard marking is used as prescribed in 29 CFR 1910.1030(g). CDC Etiologic Agent label must be used as prescribed in 42 CFR 72.3 and 72.6. A bulk package of RMW must display a BIOHAZARD marking.

CLASS 7 Radioactive

§172.436, §172.438, §172.440, §172.441

CLASS 8 Corrosive

§172.442

CLASS 9 Miscellaneous Hazardous Material

§172.446

Subsidiary Risk Label

§172.411

Cargo Aircraft Only

NEW
Mandatory January 1, 2013

OLD

§172.448

Empty Label

EMPTY

§172.450

Courtesy of the U.S. Department of Transportation

232   Operating Safely in Hazardous Environments: Second Edition

Hazardous Materials Warning Placards

Actual placard size: at least 273 mm (10.8 inches) on all sides

CLASS 1 Explosives

§172.522
§172.523
§172.524
§172.525

* For Divisions 1.1, 1.2, or 1.3, enter division number and compatibility group letter, when required; placard any quantity. For Divisions 1.4, 1.5, and 1.6, enter compatibility group letter, when required; placard 454 kg (1,001 lbs) or more.

CLASS 2 Gases

§172.528
§172.530
§172.532
§172.540

For NON-FLAMMABLE GAS, OXYGEN (compressed gas or refrigerated liquid), and FLAMMABLE GAS, placard 454 kg (1,001 lbs) or more gross weight. For POISON GAS (Division 2.3), placard any quantity.

CLASS 3 Flammable Liquid and Combustible Liquid

§172.542
§172.544

For FLAMMABLE, placard 454 kg (1,001 lbs) or more. GASOLINE may be used in place of FLAMMABLE placard displayed on a cargo tank or portable tank transporting gasoline by highway. Placard combustible liquid transported in bulk. See §172.504(f)(2) for use of FLAMMABLE placard in place of COMBUSTIBLE. FUEL OIL may be used in place of COMBUSTIBLE on a cargo or portable tank transporting fuel oil not classed as a flammable liquid by highway.

CLASS 4 Flammable Solid, Spontaneously Combustible, and Dangerous When Wet

§172.546, §172.547, §172.548

For FLAMMABLE SOLID and SPONTANEOUSLY COMBUSTIBLE, placard 454 kg (1,001 lbs) or more. For DANGEROUS WHEN WET (Division 4.3), placard any quantity.

CLASS 5 Oxidizer & Organic Peroxide

Organic Peroxide, Transition-2011 (rail, vessel, and aircraft) 2014 (highway)

§172.550, §172.552

For OXIDIZER and ORGANIC PEROXIDE (other than TYPE B, temperature controlled), placard 454 kg (1,001 lbs) or more. For ORGANIC PEROXIDE (Division 5.2), Type B, temperature controlled, placard any quantity.

CLASS 6 Poison (Toxic) and Poison Inhalation Hazard

§172.504(f)(10), §172.554, §172.555

For POISON (PGI) or PGII, other than inhalation hazard) and POISON (PGIII), placard 454 kg (1,001 lbs) or more. For POISON-INHALATION HAZARD (Division 6.1), inhalation hazard only, placard any quantity.

CLASS 7 Radioactive

§172.556

Placard any quantity - packages bearing RADIOACTIVE YELLOW-III labels only. Certain low specific activity radioactive materials in "exclusive use" will not bear the label, but the radioactive placard is required for exclusive use shipments of low specific activity material and surface contaminated objects transported in accordance with §172.504(e) Table 1 and §173.427(o)(6).

CLASS 8 Corrosive

§172.558

For CORROSIVE, placard 454 kg (1,001 lbs) or more.

CLASS 9 Miscellaneous

§172.560

Not required for domestic transportation. A bulk packaging containing a Class 9 material must be marked with the appropriate ID number displayed on a Class 9 placard, an orange panel, or a white square-on-point display.

Dangerous

§172.521

A freight container, unit load device, transport vehicle, or rail car which contains non-bulk packages with two or more categories of hazardous materials that require different placards specified in Table 2 may be placarded with DANGEROUS placards instead of the specific placards required for each of the materials in Table 2. However, when 1,000 kg (2,205 lbs) or more of one category of material is loaded at one loading facility, the placard specified in Table 2 must be applied.

Safety begins with communication!

Courtesy of the U.S. Department of Transportation

General Guidelines on Use of Warning Labels and Placards

LABELS

See 49 CFR, Part 172, Subpart E, for complete labeling regulations.

- The Hazardous Materials Table [§172.101, Col. 6] identifies the proper label(s) for the hazardous material listed.
- Any person who offers a hazardous material for transportation MUST label the package, if required [§172.400(a)].
- Labels may be affixed to packages when not required by regulations, provided each label represents a hazard of the material contained in the package [§172.401].
- For labeling mixed or consolidated packages, see §172.404.
- The appropriate hazard class or division number must be displayed in the lower corner of a primary and subsidiary hazard label [§172.402(b)].
- For classes 1,2,3,4,5,6, and 8, text indicating a hazard (e.g., "CORROSIVE") is NOT required on a primary or subsidiary label. The label must otherwise conform to Subpart E of Part 172 [§172.405].
- Labels must be printed on or affixed to the surface of the package near the proper shipping name marking [§172.406(a)].
- When primary and subsidiary labels are required, they must be displayed next to each other [§172.406(c)].
- For a package containing a Division 6.1, PG III material, the POISON label specified in §172.430 may be modified to display the text PG III instead of POISON or TOXIC. Also see §172.313(d).
- The new ORGANIC PEROXIDE label becomes mandatory on 1 January 2011 and reflects the fact that organic peroxides are highly flammable and eliminates the requirements for a flammable liquid subsidiary label [§172.427]. For information, see §171.14. The color of the border must be black and the color of the flame may be black or white.

PLACARDS

See 49 CFR, Part 172, Subpart F, for complete placarding regulations.

- Each person who offers for transportation or transports any hazardous material subject to the Hazardous Materials Regulations must comply with all applicable requirements of Subpart F [§172.500].
- Placards may be displayed for a hazardous material, even when not required, if the placarding otherwise conforms to the requirements of Subpart F of Part 172 [§172.502(c)].
- For other than Class 7 or the DANGEROUS placard, text indicating a hazard (e.g., "FLAMMABLE") is not required. Text may be omitted from the OXYGEN placard only if the specific ID number is displayed on the placard [§172.519(b)(3)].
- For a placard corresponding to the primary or subsidiary hazard class of a material, the hazard class or division number must be displayed in the lower corner of the placard.
- Any bulk packaging, freight container, unit load device, transport vehicle or rail car containing any quantity of material listed in Table 1 must be placarded [§172.504].
- When the aggregate gross weight of all hazardous materials in non-bulk packages covered in Table 2 is less than 454 kg (1,001 lbs), no placard is required on a transport vehicle or freight container when transported by highway or rail [§172.504(c)].
- Notes: See §172.504(f)(10) for placarding Division 6.1, PG III materials.
- Placarded loads require registration with USDOT. See §107.601 for registration regulations.
- The new ORGANIC PEROXIDE placard becomes mandatory 1 January 2011 for transportation by rail, vessel, or aircraft and 1 January 2014 for transportation by highway. The placard will enable transport workers to readily distinguish peroxides from oxidizers [§172.552]. For information, see §171.14.

PLACARDING TABLES

[§172.504(e)]

TABLE 1

Category of material (Hazard Class or division number and additional description, as appropriate)	Placard name
1.1	EXPLOSIVES 1.1
1.2	EXPLOSIVES 1.2
1.3	EXPLOSIVES 1.3
2.3	POISON GAS
4.3	DANGEROUS WHEN WET
5.2 (Organic peroxide, Type B, liquid or solid, temperature controlled)	ORGANIC PEROXIDE
6.1 (materials poisonous by inhalation (see §171.8 of this subchapter))	POISON INHALATION HAZARD
7 (Radioactive Yellow III label only)	RADIOACTIVE[1]

[1] RADIOACTIVE placard also required for exclusive use shipments of low specific activity material and surface contaminated objects transported in accordance with §173.427(a)(6).

TABLE 2

Category of material (Hazard Class or division number and additional description, as appropriate)	Placard name
1.4	EXPLOSIVES 1.4
1.5	EXPLOSIVES 1.5
1.6	EXPLOSIVES 1.6
2.1	FLAMMABLE GAS
2.2	NON-FLAMMABLE GAS
3	FLAMMABLE
Combustible Liquid	COMBUSTIBLE
4.1	FLAMMABLE SOLID
4.2	SPONTANEOUSLY COMBUSTIBLE
5.1	OXIDIZER
5.2 (Other than organic peroxide, Type B, liquid or solid, temperature controlled)	ORGANIC PEROXIDE
6.1 (Other than materials poisonous by inhalation)	POISON
6.2	(None)
8	CORROSIVE
9	Class 9 (see §172.504(f)(9))
ORM-D	(None)

IDENTIFICATION NUMBER DISPLAYS

 or

§172.332

Appropriate placard must be used with orange panel.

IDENTIFICATION NUMBER MARKINGS ON ORANGE PANELS OR APPROPRIATE PLACARDS MUST BE DISPLAYED ON: (1) Tank Cars, Cargo Tanks, Portable Tanks, and other Bulk Packagings; (2) Transport vehicles or freight containers containing 4,000 kg (8,820 lbs) in non-bulk packages of only a single hazardous material having the same proper shipping name and identification number loaded at one facility and transport vehicle contains no other material, hazardous or otherwise; and (3) transport vehicles or freight containers containing 1,000 kg (2,205 lbs) of non-bulk packages of materials poisonous by inhalation in Hazard Zone A or B. See §§172.301(a)(3), 172.326, 172.328, 172.330, and 172.331.

 Square white background required for placard for highway route controlled quantity radioactive material and for rail shipment of certain explosives and poisons, and for flammable gas in a DOT 113 tank car (§172.507 and **§172.510**).

§172.527

This Chart is available online at the following link:
http://phmsa.dot.gov/hazmat

U.S. Department of Transportation

Pipeline and Hazardous Materials Safety Administration

USDOT/PHMSA/OHMIT/PHH-50
1200 New Jersey Avenue, SE
Washington, D.C. 20590
Phone: (202) 366-4900
Email: training@dot.gov

PHH50-0119-1110

Courtesy of the U.S. Department of Transportation

Index

Page numbers with "t" refer to tables and with "n" refer to footnotes. All government agencies entered under "U.S."

A

accidents
   chemical hazard release incidents, 42, 51–52, 53–54
   hazardous material release incidents, 12, 14, 16–17, 42, 50, 54, 70
   hazardous substance release incidents, 51–52, 53–54, 103, 130, 160, 173, 186
   hazardous waste release incidents, 14, 16, 17, 102
   on-the-job, 161–163
   prevention efforts, 3, 9, 13
ACGHI. *See* American Conference of Governmental Industrial Hygienists (ACGIH)
advanced exterior fire brigades, 167
AEDs (automated external defibrillators), 162–163
Agent Orange, 14, 135n
AIHA. *See* American Industrial Hygiene Association (AIHA)
air purifying respirators (APRs), 135, 141
air quality, 14, 21, 32, 32t, 68–69, 133–134, 136t. *See also* oxygen deficiency hazards
alarming and alerting, emergency actions plans (EAPs), 104, 159
Ambulance and Stretcher Bearer Act of 1864, 4
American Association of Railroads, *Emergency Handling of Hazardous Materials in Surface Transport and Emergency Action Guides,* 56
American College of Occupational and Environmental Medicine, 163
American Conference of Governmental Industrial Hygienists (ACGIH)
   described, 23
   exposure risk levels, 12, 31, 62, 97
   hot environment safety protocols, 31
American fire marks, photo, 3
American Industrial Hygiene Association (AIHA), 62
American National Standards Institute (ANSI), 12, 23, 141
   first aid kit standards, 162, 162t
   ladder standards, 203, 205t
   PPE standards, 126, 128, 129, 130t, 131, 132
   protection factors (PF), respirator, 136
   radiation exposure limits, 31
American Society for Testing and Materials (ASTM)
   described, 141
   PPE standards, 126, 129, 131, 132n
ANSI. *See* American National Standards Institute (ANSI)
APRs (air purifying respirators), 135, 141
asbestos-containing materials
   decontamination area requirements, 148, 150
   historical timeline, 7, 7t
   laboratory analysis procedures, 73t–87t
   legislation, 14
   notification of demolition and renovation form, 118–119
   OSHA standard, 181, 192
   permit for work with, 105
asphyxiant hazards, 31–33, 35. *See also* oxygen deficiency hazards
Association of Iron and Steel Electrical Engineers, 4
ASTM. *See* American Society for Testing and Materials (ASTM)
atmospheres. *See* air quality; hazardous atmospheres; immediately dangerous to life or health (IDLH) atmospheres
ATSDR. *See* U.S. Agency for Toxic Substances and Disease Registry (ATSDR)
automated external defibrillators (AEDs), 162–163

B

backboards, 171
bacteria. *See* etiological hazards
BEEL (biological employee exposure limits), 62
belts, body, 205, 206
Bhophal, India, chemical release (1984), 16, 102
biological employee exposure limits (BEEL), 62
biological hazards
   PPE ensembles, 127, 133, 138
   respiration interfering, 134
bloodborne pathogens, 33, 105, 186t
bomb threats, 161
Boston Fire Department, 3n
Boston Mutual Fire Society, 3
bowline knot, 172
Braidwood, James, 2n
Bubonic Plague, 3
burning (hot work procedures), 207

C

CAER (Chemical Awareness and Emergency Response Program), 102
Cancer Slope Factors (CSFs), 97
carbon dioxide monitoring, 71t. *See also* air quality; oxygen deficiency hazards
carcinogens, 135
cardiopulmonary resuscitation (CPR), 162
carries, 170–171
CBRNE (chemical, biological, radiological, nuclear or explosive) agents, PPE ensembles, 133
CDC. *See* U.S. Centers for Disease Control and Prevention (CDC)
ceiling concentrations, exposure limits, 62, 96
CERCLA. *See* Comprehensive Environmental Response Compensation and Liability Act of 1980 (CERCLA)
CFATS (Chemical Facility Anti-Terrorism Standards), 22–23
Chemical Awareness and Emergency Response Program (CAER), 102

chemical emergency planning zones, 102
chemical energy, tagouts and lockout devices, 208
Chemical Facility Anti-Terrorism Standards (CFATS), 22–23
chemical hazards, 33, 35
   defined, 57
   employee protection, 13–14, 17, 21, 51–52, 105, 150
   labeling, 50
   ozone-depleting, 14
   PPE ensembles, 133, 138, 139t
   release incidents, 42, 51–52, 53–54
   respiration interfering, 134–135
chemical weapons stockpile sites, 103
CHEMTREC/Chemical Manufacturers Association (CMA), 54, 57
Chernobyl Nuclear Power Station accident (1985), 16
Chicago Fire of 1871, 4
Clean Air Act Amendments of 1990, 21, 105
clean media. See laboratory blanks (off-site field sample analysis)
Clean Water Act of 1972 (CWA), 105
clothing, protective. See personal protective equipment (PPE)
clove hitch knot, 172
CMA. See CHEMTREC/Chemical Manufacturers Association (CMA)
Code #704: The Identification of Hazards of Materials for Emergency Response (NFPA), 33, 50, 51t
codes, enforceable, 9. See also Consensus Standards
cold-related disorders, 30, 30t, 31t, 63, 67
combustible gas testing, flash point and explosive ranges, 41–42, 57, 67, 69–70, 70t
Common Ground Alliance, 198n
Communicable Disease Center, 9. See also U.S. Centers for Disease Control and Prevention (CDC)
communities, noise protection, 34. See also Emergency Planning and Community Right-to-Know Act (EPCRA, 1986); local emergency planning commissions (LEPCs)
competent person
   defined, 126n
   PPE selection process, 131
Comprehensive Environmental Response Compensation and Liability Act of 1980 (CERCLA), 14, 16, 103
confined spaces. See also Permit Required Confined Spaces Standard (OSHA)
   rescue, 169, 170
   safety requirements, 197–198, 197n
Consensus Standards, 4, 12. See also American National Standards Institute (ANSI)
   fire suppression and fire response, 168
   medical emergency requirements, 161
   PPE requirements, 126–27
construction industries
   confined space standard, Virginia, 197n
   PPE regulations, 126, 128, 129
contamination. See also decontamination
   cross-contamination, 147, 152
   prevention efforts, 148–149
Control of Hazardous Energy Standard (OSHA), 34
controlled access zones (fall protection), 205
corrosives, 41, 42

CPR (cardiopulmonary resuscitation), 162
cradle-to-grave responsibility, 14
CSFs (Cancer Slope Factors), 97
cutting (hot work procedures), 207
CWA. See Clean Water Act of 1972 (CWA)

D

decibels, 133
decompression sickness (the bends), 134
decontamination, 147–155
   asbestos, 148, 150
   bloodborne pathogens, 105
   chemical, 150
   defined, 147, 152
   emergency, 148, 151, 152
   equipment, 151–152
   exercise, 153–155
   location, 148, 151
   mass (gross), 151, 152
   planning and design procedures, 147, 148–149
   purpose, 147
   terms, 152
   worker protection, 149–151
demolition notification form, 118–119
DHS. See U.S. Department of Homeland Security (DHS)
Disaster Medical Assistance Teams (DMAT), 103–104
disposal, waste. See hazardous wastes, disposal
DMAT (Disaster Medical Assistance Teams), 103–104
DOD. See U.S. Department of Defense (DOD)
"Dover Tower Light" (England), 2
drags, 170–171

E

EAPs. See emergency action plans (EAPs)
ECT (equivalent chill temperatures), 30, 31t. See also wind chill
"811" numbers, 198n
electrical energy, tagouts and lockout devices, 208
emergencies, responding to, 157–179. See also emergency rescue; release and response
   emergency rescue, 136, 168, 169–175
   Emergency Response Planning Guidelines (AIHA), 62
   equipment maintenance, 168–169, 168t
   exercise, 177–179
   fire response and fire suppression, 163–169
   IDLH situations, 169–175
   medical emergencies, 161–163
   National Incident Management System (NIMS), 159–160, 175
   planning, 157–59
   terms, 175
   terrorism, 21–23, 103, 160–161, 175
emergency action plans (EAPs). See also evacuation planning
   alerting and alarming, 104, 159
   defined, 175
   facility emergency coordinator, 159
   hazard substance releases, 173–174, 174t, 175
   ICS, 159, 160, 160t, 167, 174, 175

IDLH situations, 169
requirements, 159t
terrorism, 160–161
warning systems, 105, 159
worksites, hazardous, 105
written, 157
Emergency Broadcasting System, 104
emergency decontamination, 148, 151, 152
Emergency Handling of Hazardous Materials in Surface Transport and Emergency Action Guides (American Association of Railroads), 56
emergency medical services (EMS), 102, 161–162, 169
emergency planning, 157–159. *See also* emergency action plans (EAPs); emergency response plans (ERPs); evacuation planning; first aid services; public protection process
emergency planning zones (EPZs), 102–104, 123, 159
facility emergency coordinator, 159
four-step planning process, 157, 157t, 158t
legislation, 16–17
response options, 157–158
Emergency Planning and Community Right-to-Know Act of 1986 (EPCRA), 17, 51, 102, 157
emergency rescue. *See also* decontamination, emergency; emergencies, responding to; emergency medical services (EMS); first responders; Urban Search and Rescue (USAR) Teams
buddy system, 168, 169, 175
IDLH situations, 136, 168, 169–175
standby teams, 168, 169, 170, 171, 175
tunnel, 105, 169
emergency response coordinator, 173–174
Emergency Response Guidebook (*ERG-2012*, DOT), 12, 54, 151
Emergency Response Planning Guidelines (AIHA), 62
emergency response plans (ERPs), 157, 159
defined, 175
equipment maintenance, 168t
hazardous substances releases, 173–174, 174t
One-Plan format, 174t, 175
worksites, hazardous, 19t–21t, 105
employees. *See also* Hazmat employees, specialists and technicians; health, occupational; personal protective equipment (PPE); safety, occupational; workplaces, hazardous
medical monitoring, 72, 181–195
personal hearing zones, 63
protecting, 21, 102, 105, 106–114, 149–151, 198–213
employers. *See also* workplaces, hazardous
HASPs, 105
HAZCOM compliance, 51–52
OSHA fire response options, 163t
responsibilities for PPE, 126, 127, 133
EMS (emergency medical services), 161–162, 169
enclosed spaces. *See* confined spaces
energy, hazardous
Control of Hazardous Energy Standard (OSHA), 34
Energy Control Program, 208
lockout devices and tagouts, 208–209

engineering and administrative controls, employee protection, 126, 133, 149
ensembles, personal protective equipment. *See* personal protective equipment (PPE), hazardous environment ensembles
environmental hazards, 42, 51. *See also* hazardous environments
environmental protection, 105–122. *See also* hazardous environments; U.S. Environmental Protection Agency (EPA)
asbestos-containing materials, work permit form, 118–119
governmental agencies regulating, 12
hazardous chemical releases, 42, 51–52, 53–54
lead-containing materials, work permit form, 120–122
legislation protecting, 13–14, 16–17
EPCRA. *See* Emergency Planning and Community Right-to-Know Act of 1986 (EPCRA)
epidemiology, history, 3
EPZs (emergency planning zones), 102–104, 123, 159
equipment. *See also* personal protective equipment (PPE)
decontamination, 151–152
emergency response, 168t
explosion-proof, 168–169
guard systems, 34, 126
intrinsically safe, 168–169, 169t
oxygen testing, 68–69
protecting during decontamination, 149
safety, 4, 6t, 9, 12
equivalent chill temperatures (ECT), 30, 31t. *See also* wind chill
ERPs. *See* emergency response plans (ERPs)
etiological hazards, 33, 35
evacuation planning, 54, 102, 104–105, 158, 159, 161. *See also* emergency planning
excavations and trenches
safety requirements, 198, 202
size and danger, 198, 202
sloping and benching trenches, 198
soil classification, 202, 204t
trench boxes, 198
utility locator notification and marking, 198, 198n, 204t
explosive ranges, flammable substances, 41–42, 57, 67, 69–70, 70t
explosives, 67–70
PPE ensembles, 133
terrorist attacks, 161
upper and lower explosive limits (UEL and LEL), 41, 69, 70, 96, 97
exposure(s)
acute, 50, 62
control programs, 105
medical monitoring, 192, 193
noisy environments, 133
occupational exposure limit, 62, 96
PPE requirements, 139
quantifying, 62–63
radiation, 31
recommended levels, 12, 31, 62, 97, 105
toxic materials, 3, 135
eye and face protection, 128–129, 130t

F

face and eye protection, 128–129, 130t
facility emergency coordinator, 159
fall arrest systems, 169, 205
fall protection, 169, 204–205, 207
   controlled access zones (fall protection), 205
   guardrail systems, 205
   safety monitoring systems, 205–206
   safety nets, 205
Federal Insecticide, Fungicide and Rodenticide Act of 1975 (FIFRA), 14, 52–53
field blanks (off-site field sample analysis), 72
field medical services, 4. *See also* first aid services
FIFRA. *See* Federal Insecticide, Fungicide and Rodenticide Act of 1975 (FIFRA)
figure eight knot, 173
fire and pressure releases, 41–42
fire brigades, 167–168
   advanced exterior, 167
   incipient, 167
   London Fire, 2–3
   standard operating procedures, 167
   structural stage, 167–168
fire characteristics and classes, 41–42, 164, 165t
fire extinguishers, 163–166
   inspection, 164, 165t, 166
   outdated and illegal, 166
   ratings, 164
   size and placement, 164t
   use and training, 166
fire protection, historical
   fire service organizations, 3
   Great London Fire of 1666, 2–3
   "Rattle Watch," 3n
   Roman "Vigils" corps, 2
   20th-century developments, 4
fire response and fire suppression, 163–169. *See also* National Fire Protection Association (NFPA)
   alarm systems, 159
   employer options, 163t
   equipment, 168–169, 169t
   fire brigades, 167–168
   fire extinguishers, 163–166, 164t
   PPE requirements, 128n, 132
   standpipe and hose systems, 166–167
   stop, drag, roll, 171
Fire Triangle/Tetrahedron, 41–42, 164, 165t
first aid services. *See also* emergency rescue
   historical practices, 4
   IDLH rescues, 169
   kit types, 162, 162t
first responders. *See also* emergency medical services (EMS); emergency rescue
   hazardous materials releases, initial actions, 54, 57
   medical monitoring, 186
   notifying, 159
   protecting during decontamination activities, 151
fixed facilities, 50–53
   FIFRA labeling requirements, 52–53
   GHS/HAZCOM 2012 requirements, 50, 51–52, 52t, 53, 53t
   hazardous materials information system, 33, 50
   material safety data sheets/safety data sheets (MSDS/SDS), 53, 55t–56t
   NFPA Code #704, 33, 50
   OSHA color coding, 53, 54t
flash points and flammable (explosive) ranges, 41–42, 57, 67, 69–70, 70t
foot protection, 129, 131, 131t
Franklin, Benjamin, 3
fungi, 33, 134. *See also* Federal Insecticide, Fungicide and Rodenticide Act of 1975 (FIFRA)

G

Geiger-Muller sensing tubes, 70, 72
Globally Harmonized System of Classification and Labeling of Chemicals (GHS/HAZCOM 2012), 41, 43, 50, 51–52, 53, 57
   label elements, 52t, 53t
gloves, 132
governmental agencies, 12–13, 25. *See also* specific agencies and departments
Greece
   insurance groups, 2
   physicians, 2
grinding (hot work procedures), 207
guardrail systems, 205
guards, equipment, 34, 126

H

Hamilton, Alexander, 3
hand protection, 132
hard hats, 128
harnesses, 169, 170, 197, 205, 206
HASPs. *See* health and safety plans (HASPs)
Hazard Communications Standard (OSHA). *See* U.S. Occupational Safety and Health Administration (OSHA), Hazard Communications Standard (HAZCOM 2012)
hazardous area emergency planning, 103–104
hazardous atmospheres, 138–141. *See also* hazardous environments; immediately dangerous to life or health (IDLH) atmospheres
   in confined spaces, 197, 202
   mitigating, 2, 9
   PPE ensembles, 42, 138
   quantifying, 62–63, 70
   testing, 68–70
hazardous environments. *See also* environmental hazards; environmental protection; workplaces, hazardous
   defined, 9
   historical activities, 3, 4
   legislation regulating, 18t
   life safety services, 4
   mechanical, 33–35
   medical monitoring for employees, 181–186
   natural, 103

PPE ensembles, 30, 42, 127, 138–141, 139t
remediation efforts, 2, 9, 14, 16, 103, 151
hazardous environments, emergency response, 157–179
  exercise, 177–179
  fire response and fire suppression, 163–169
    equipment maintenance, 168–169
    fire brigades, 167–168
    fire extinguishers, 163, 164–166
    fire triangle/tetrahedron, 164, 165t
    standpipe and hose systems, 166–167
  IDLH situations, 169–175
  medical emergencies, 161–163
  NIMS, 159–160
  planning, 157–59
  terms, 175
  terrorism, 160–161
hazardous environments, quantifying, 62–100. *See also* hazardous atmospheres
  asbestos-containing, 73t–87t
  chain of custody form, 65t
  exercise, 99–100
  IDLH atmospheres, 68–72
  lead-containing materials, 88t–95t
  off-site analysis, 71t, 72
  on-site analysis, 63, 64t, 66t–67t
  physical hazards' testing, 63–67
  process, 62–63
  resampling time frames, 96
  terms, 96–97
hazardous environments, recognizing and identifying, 27–39
  asphyxiant hazards, 31–33, 35
  chemical hazards, 33, 35
  etiological hazards, 33, 35
  exercise, 37–39
  mechanical hazards, 33–35
  radiological hazards, 30–31, 35
  thermal hazards, 27–30, 35
hazardous materials. *See also* exposure(s); transportation of hazardous materials
  defined, 57
  emergencies, 171, 173–175
  PPE ensembles, 139t
  release incidents, 12, 14, 16–17, 42, 50, 54, 70
  reportable quantity (RQ) designation, 43, 52, 159, 175
hazardous materials, recognizing and identifying, 41–60
  chemical releases, initial actions, 42, 51–52, 53–54, 56
  Code #704 (NFPA), 33, 50
  exercise, 58–60
  fixed facilities, 50–53
  GHS/HAZCOM 2012, 41, 43, 50, 51–52, 53, 57
  Hazardous Materials Classifications (DOT), 33, 42–50
  primary dangers, 41–42
  terms, 57
Hazardous Materials Classifications (DOT), 42–50
  examples, 44t–45t, 46t
  hazardous materials information system, 33, 50

Hazardous Materials Shipping Paper (DOT), 45–50, 47t, 49t
  hazardous materials manifest, 43, 45, 47t
  placarding and labeling, 45, 50
  uniform hazardous waste manifest, 45, 48t
hazardous substances
  defined, 42
  PPE guidelines, 139t
  release incidents, 51–52, 53–54, 103, 130, 160, 173, 186
Hazardous Waste Operations and Emergency Response (HAZWOPER) standard (OSHA), 16–17, 19t–21t, 157, 169, 186
hazardous waste sites. *See also* workplaces, hazardous
  defined, 57
  HASPs, 105
  medical monitoring of workers, 70, 186
  PPE ensembles, 127, 138–141
  registration and inspection, 105
  remediation efforts, 151
  uncontrolled, 103
hazardous wastes
  defined, 14, 15t–16t, 57
  disposal, 13–14, 105
  legislation regulating, 14, 16
  release incidents, 14, 16, 17, 102
  uniform hazardous wastes manifest, 45, 48t
HAZCOM 2012. *See* U.S. Occupational Safety and Health Administration (OSHA), Hazard Communications Standard (HAZCOM 2012)
Hazmat employees, specialists and technicians
  communication of materials hazards, 43
  initial actions' guidelines, 53–54, 56
  medical monitoring, 186
  safety regulations, 17
  training requirements, 13t
HAZWOPER. *See* Hazardous Waste Operations and Emergency Response (HAZWOPER) standard (OSHA)
head protection, 128, 129t, 205
health, environmental, liability regulations, 13–14
health, occupational, 2–10, 12–25. *See also* medical monitoring; workplaces, hazardous
  equipment, 4, 6t, 9, 12
  governmental agencies, 12–13, 23
  historical activities, 2–3
  industrialization, 3–4
  legislation, 13–14, 16–17, 21–22
  modern organizations, 4–8
  professional associations, 12, 23
  recent developments, 9
  terms, 9
health, personal, 2–10, 12–25. *See also* medical monitoring; safety, personal
  governmental agencies regulating, 12
  hazard investigations, 12–13
  historical activities, 2–3, 5t–6t
  modern organizations, 4

health, public, 2–10, 12–25. *See also* U.S. Public Health Service
  governmental agencies, 12–13, 23
  historical activities, 2–3, 4, 5t–6t
  industrialization, 3–4, 7
  legislation, 13–14, 16–17, 21–22
  modern organizations, 4–8
  physical performance testing, 186
  professional associations, 12, 23
  recent developments, 9
health and safety plans (HASPs)
  Confined Space Entry Permit, 105, 115–116, 169, 170, 197, 209
    OSHA requirements, 199t–202t, 203t
  decontamination procedures, 147
  defined, 123
  Ready for Hot Work Permit, 105, 117, 207
  site-specific, 102, 105, 106–114
health hazards, 42, 43, 51, 56
hearing, personal. *See also* noisy environments
  monitoring, 63
  protective equipment, 133, 134t, 139, 141
heat indices, 29–30, 29t
heat-related disorders, 27–30, 29t. *See also* hot work environment
heights, 202–207
  controlled access zones, 205
  fall arrest systems, 205–206
  fall protection, 204–205
  guardrail systems, 205
  ladders, 202–204
  safety monitoring systems, 205–206
  safety nets, 205
  scaffolding, 206–207
  warning line systems, 206
helmets, 128
herbicides, 14. *See also* Federal Insecticide, Fungicide and Rodenticide Act of 1975 (FIFRA)
thermal hazards, 27–33
  cold-related disorders, 30, 30t, 31t, 63, 67
  defined, 35
  heat-related disorders, 27–30, 29t
HHS. *See* U.S. Department of Health and Human Services (HHS)
historical activities, 2–10
historical time tables
  asbestos-containing materials, 7t
  health and safety, 5t–6t
  lead-containing materials, 7t–8t
  select safety equipment development, 6t
Homeland Security Act of 2002, 13, 21–23
Homeland Security Appropriations Act of 2007, 22
hot work environment, 27–30, 207–208, 209
  Ready for Hot Work Permit, 105, 114, 207
HUD. *See* U.S. Department of Housing and Urban Development (HUD)
humidity testing, 63, 67, 70
hydraulic energy, tagouts and lockout devices, 208
hydrogen sulfide, 70, 135
hypothermia, 30

I

ICS (incident command system), 159, 160, 160t, 167, 174, 175
IDLH. *See* immediately dangerous to life or health (IDLH) atmospheres
illnesses
  defined, 9
  on-the-job, 161–163
  prevention efforts, 3, 4, 9, 13
  respiration interfering, 134
illumination testing, 63, 67
immediately dangerous to life or health (IDLH) atmospheres, 68–72. *See also* hazardous atmospheres
  combustible gas testing, 68–70
  defined, 68, 96
  emergency rescue situations, 136, 138, 169–175
  fire response and fire suppression, 167–168
  ionizing radiation, 70
  oxygen levels testing, 68–69
  quantifying, 68t
  respiratory protection, 136, 167
  toxic materials testing, 70
incident command system (ICS), 159, 160, 160t, 167, 174, 175
incipient fire brigades, 167
industrial trucks, powered, 209, 209t, 210
industrialization, 3–4, 7, 9, 42. *See also* American Conference of Governmental Industrial Hygienists (ACGIH)
Institute of Electrical and Electronics Engineers, radiation exposure limits, 31
instrument calibration, 63, 63t, 70, 72
insurance groups, historical development, 2, 3
International Labour Organization (ILO). *See* Globally Harmonized System of Classification and Labeling of Chemicals (GHS/HAZCOM 2012)
International Safety Equipment Association (ISEA)
  first aid kit requirements, 162t
  PPE standards, 126, 132
ionizing radiation
  characteristics, 30–31, 30t–31t
  effects, 32t
  protection from during decontamination, 150
  testing, 70, 72
ISEA. *See* International Safety Equipment Association (ISEA)

J

job elimination, employee protection, 126
job matching, psychological testing, 123
job site reviews, 126. *See also* workplaces, hazardous

K

knots, 170
  bowline, 170, 172
  clove hitch, 170, 172
  figure eight, 170, 173
  sheet bend, 170, 171
  square, 170, 171

L

labeling, hazardous materials. *See* placarding and labeling of hazardous materials
laboratories, PPE recommendations, 127, 138
laboratory blanks (off-site field sample analysis), 72
ladders, 202–205, 205*t*
lead-based paint
    decontamination area requirements, 150
    notification of abatement form, 120–122
    permit for work with, 105
    removal, 104, 104*t*
lead-containing materials
    historical timeline, 7*t*–8*t*
    laboratory analysis procedures, 88*t*–95*t*
    legislation, 14
    medical monitoring of employees, 186, 186*t*
    OSHA standard, 96*n*, 186*t*, 192
legislation, 13–14, 16–17, 21–22. *See also* specific acts by title
LEL (lower explosive limit), 69, 70, 96–97
LEPCs (local emergency planning commissions), 17, 102, 175
levels of protection (LEPs), NIOSH guidelines, 139*t*
lifelines, 170, 197, 202*t*, 205–206
lifesaving, historical, 2, 3–4. *See also* decontamination, mass (gross); emergency rescue; first aid services
lifts, 170–171
light levels, testing, 63, 67
light towers and lighthouses, 2
lightning rod, first, 3
LOAEL (Lowest Observed Adverse Effects Level), 62, 97
local emergency planning commissions (LEPCs), 17, 102, 175
lockout devices, 208, 209
London Fire (1666), 2–3
London Fire Brigade, 2–3
Love Canal accident (1978), 14
lower explosive limit (LEL), 69, 70, 96–97
Lowest Observed Adverse Effects Level (LOAEL), 62, 97

M

Manual of Analytical Methods (HHS), 72
maritime lifesaving organizations, history, 2, 3–4
material safety data sheets/safety data sheets (MSDS/SDS), 51, 53, 55*t*–56*t*
Mayan ancestors, 2
mechanical energy, tagouts and lockout devices, 208
mechanical hazards, 33–35. *See also* noisy environments
medical emergencies, 102, 161–163
    automated external defibrillators (AEDs), 162–163
    first aid and CPR, 162
medical monitoring, 181–195. *See also* health, personal
    asbestos workers, 181–185
    close-out physicals, 193
    exercise, 195
    hazardous waste workers, 186
    initial screening, 181
    lead workers, 186, 186*t*, 193
    OSHA Reference Chart, 187*t*–192*t*
    physical performance testing, 186
    preplacement exams, 181
    procedures, 181
    psychological testing, 186
    respiratory protection, 138–141, 168, 181, 182*t*–185*t*
    routine monitoring, 193
    smoking cessation, 186, 193, 193*t*
    symptomatic monitoring, 192
    terms, 194
medical services, historical, 2. *See also* Disaster Medical Assistance Teams (DMAT); field medical services; U.S. Public Health Service
Metropolitan (London) Fire Brigade, 2–3
Middle Ages
    light towers, Italy, 2
    religious activities, 2
Mine Safety and Health Administration (Dept. of Labor), 34, 169
miners, first aid and rescue services, 4, 169
Mission Oriented Protective Posture (MOPP) levels (DOD), 138, 140*t*
molds, 33
MOPP (Mission Oriented Protective Posture levels, DOD), 138, 140*t*
motor vehicle safety, 209, 209*t*
MSDS. *See* material safety data sheets/safety data sheets (MSDS/SDS)
mutagens, 135
mutual aid agreements, 158, 168

N

National Board of Fire Underwriters, 4
National Center for Injury Prevention and Control, 13
National Commission on Terrorist Attacks upon the United States (9/11 Commission), 160
National Council on Radiation Protection and Measurements, radiation exposure limits, 31
National Electrical Code, intrinsically-safe equipment, 169*t*
National Emission Standards for Hazardous Air Pollutants (Clean Air Act), 105
National Fire Protection Association (NFPA), 12, 23
    Code #704: The Identification of Hazards of Materials for Emergency Response, 33, 50, 51*t*
    Consensus Codes, 4
    fire brigade recommendations, 167, 168
    fire extinguisher code, 164
    fire response and fire suppression, 163
    ladder requirements, 204
    lifesaving rope requirement, 169
    PPE requirements, 128*n*, 132
National Incident Management System (NIMS), 159–160, 175
National Institute for Occupational Safety and Health (NIOSH), 9, 12–13
    asbestos exposure guidelines, 181
    fire suppression and fire response recommendations, 168
    hot environment safety protocols, 27
    IDLH atmospheres, 68
    lead analysis procedure, 88*t*–95*t*
    Manual of Analytical Methods, 72
    medical monitoring recommendations, 186, 194–195

National Institute for Occupational Safety and Health (NIOSH) (*Contd.*)
  PBZ sampling, 72
  Pocket Guide to Chemical Hazards, 56
  PPE recommendations, 126, 133, 138, 139t, 140t
  protection factor (PF), respirator, 136
  recommended exposure limit (REL), 62, 97
  respiratory equipment recommendations, 136, 138, 138t
  sound pressure level recommendations, 34
National Oceanic and Atmospheric Administration (NOAA). *See* National Weather Service (NOAA); Weather Radio System (NOAA)
National Pollution Discharge Elimination System (NPDES), 105
National Response Center, 54
National Response Team (NRT), 174–175
National Safety Council, 4, 12, 23
National Weather Service (NOAA)
  heat indices, 29–30, 29t
  wind chill indices, 30, 31t
NFPA. *See* National Fire Protection Association (NFPA)
NIMS (National Incident Management System), 159–160, 175
NIOSH. *See* National Institute for Occupational Safety and Health (NIOSH)
nitrogen narcosis (divers' euphoria), 33, 134
No Observed Adverse Effects Level (NOAEL), 62, 97
NOAA. *See* National Weather Service (NOAA); Weather Radio System (NOAA)
noise reduction ratings (NRR), 133, 141
noisy environments, 34–35. *See also* hearing, personal
  health recommendations, 34t
  hearing protection equipment, 133, 141
  testing, 63
  typical, 34t
nonionizing radiation, 31
NPDES (National Pollution Discharge Elimination System), 105
NRC. *See* U.S. Nuclear Regulatory Commission (NRC)
NRR (noise reduction ratings), 133, 141
NRT (National Response Team), 174–175
nuclear disasters. *See also* U.S. Nuclear Regulatory Commission (NRC)
  early warning process, 105
  power plant accidents, 16, 17, 102
  PPE ensembles, 133, 138
nuclear power plant emergency planning zones, 102

O

Occupational Safety and Health Act of 1970, 12. *See also* health, occupational; safety, occupational; U.S. Occupational Safety and Health Administration (OSHA)
off-site analysis, hazardous environments, 64t, 70–72, 71t
One Call notification standards, 198, 198n
One-Plan format, 174t, 175
on-site analysis, hazardous environments, 63, 64t, 66t–67t
OSHA. *See* U.S. Occupational Safety and Health Administration (OSHA)
oxidation/reduction reaction, 41, 69, 134

oxygen deficiency hazards, 31–33, 133–135. *See also* air quality; respiratory protection
  confined spaces, 197
  physical effects, 32t
  situations producing, 33t, 134–135
oxygen testing, 68–69

P

PASS (personal alert safety system), 168
PBZ (personal breathing zone), 72
PCB (polychlorinated biphenyl), 14, 104
PEL (permissible exposure limit), 62, 97, 181, 186
penetration rates, 132, 152
permeation rates, 132, 152
permissible exposure limit (PEL), 62, 97, 181, 186
Permit Required Confined Spaces Standard (PRCS), 105, 115–116, 169, 170, 197, 209
  OSHA requirements, 199t–202t, 203t
personal alert safety system (PASS), 168
personal breathing zone (PBZ), 72
personal protective equipment (PPE), 126–145
  decontamination workers, 147, 149–151
  effects on decontamination plans, 147
  exercise, 143–145
  firefighting, 128n, 132, 167, 168
  hazardous environment ensembles, 30, 42, 127, 138–141, 139t
  hearing protection, 133, 134t, 139, 141
  penetration and permeation rates, 132
  regulatory requirements, 127t, 137t
  respiratory protection, 133–141
  selection, 126–132
    agencies certifying, 126–127, 140t
    eye and face protections, 128–129, 130t
    foot protection, 129, 131, 131t
    hand protection, 132
    head protection, 128, 129t, 205
    job site review, 126
    MOPP guidelines, 140t
    responsibilities for providing, 126
  standards, 12
  terms, 141
pesticides, 14. *See also* Federal Insecticide, Fungicide and Rodenticide Act of 1975 (FIFRA)
PF. *See* protection factor (PF), respirator
Pharaohs of Alexandria, Egypt, 2
"Philadelphia Contributorship" (Fire Insurance Company), 3
physical hazards
identifying, 41, 51
testing, 63–67
placarding and labeling of hazardous materials, 50–53
  DOT, 45, 50
  FIFRA requirements, 52–53
  GHS label elements, 51–52, 52t, 53t
  NFPA Code 704 marking system, 50, 51t
  OSHA color coding, 53, 54t

planning. *See also* emergency action plans (EAPs); emergency planning; emergency response plans (ERPs); health and safety plans (HASPs); protective planning
   decontamination, 147–149
   evacuation, 54, 102, 104–105, 158, 159, 161
   hazardous area, 103–104
   local emergency planning commissions (LEPCs), 17, 102, 175
   Site Safety Plans, 29
pneumatic energy, tagouts and lockout devices, 208
Pocket Guide to Chemical Hazards (NIOSH), 56
poisons, 33, 42
polychlorinated biphenyl (PCB), 14, 104
posted decontamination plans, 147
post-World War II developments, 9, 42
power plants, nuclear. *See* nuclear disasters; U.S. Nuclear Regulatory Commission (NRC)
powered industrial trucks, 209, 209t, 210
PPE. *See* personal protective equipment (PPE)
PRCS. *See* Permit Required Confined Spaces Standard (PRCS)
preplacement medicals exams, 181
Process Safety Management Standard (OSHA), 21, 22t, 157
professional organizations and associations, 4, 9, 12
property protection, historical development, 2–4. *See also* emergencies, responding to; fire response and fire suppression
protection factor (PF), respirator, 135, 136, 137t, 141
protective equipment, personal. *See* personal protective equipment (PPE)
protective planning, 102–124
   defined, 123
   emergency action and response plans (EAPs), 105
   employee protection, 105
   environmental protection, 105–122
   exercise, 124
   public protection process, 102–105
   terms, 123
psychological testing and job matching, 123
public protection process, 102–105. *See also* emergency planning; health, public; safety, public
   hazardous chemicals releases, 42, 51–52, 53–54, 56
   historical activities, 2–10

Q
quantification processes. *See* hazardous environments, quantifying

R
radioactive materials, 31, 35, 42, 70, 72, 149
radiological hazards
   defined, 35
   ionizing radiation, 30–31, 31t, 32t, 70, 72, 150
   nonionizing radiation, 31
   PPE ensembles, 133, 138
   protection from during decontamination, 150
   respiration interfering, 134
"Rattle Watch," New Amsterdam (New York City), 3n
RBC (Risk Based Concentration), 62, 97

RCRA. *See* Resource Conservation and Recovery Act of 1976 (RCRA)
Ready for Hot Work Permit, 105, 117, 207
recommended exposure limit (REL), 62, 97
Reference Doses (RfDs), 97
REL (recommended exposure limit), 62, 97
release and response. *See also* emergencies, responding to
   chemicals, hazardous, 51–52, 53–54
   materials, hazardous, 12, 70
   substances, hazardous, 103, 130, 160, 173, 186
   wastes, hazardous, 14, 16, 103
renovation notification form, 118–119
reportable quantity (RQ) designation, hazardous materials, 45, 54, 159, 175
resampling time frames, 72, 96
rescue. *See* emergencies, responding to; emergency rescue
Resource Conservation and Recovery Act of 1976 (RCRA), 13–14, 15t–16t, 45, 105
respiration
   interferences, 134–135
   mechanics, 133–134
respiratory protection, 133–141. *See also* oxygen deficiency hazards
   air purifying respirators (APRs), 135
   effects on decontamination plans, 147
   filters, adsorbents, absorbents, 136, 138
   firefighting, 167, 168
   medical approval and monitoring, 138–141, 168, 181, 182t–185t
   OSHA standard, 168, 181, 182t–185t
   PPE ensembles, 138
   protection factors (PF), 136
   self-contained breathing apparatus (SCBA), 135
   supplied air respirators (SARs), 135–136
   written respirator programs, 136
response. *See* emergencies, responding to; release and response
Revenue Cutter Service (U.S. Treasury Department), 3
"reverse 911" systems, 104
RfDs (Reference Doses), 97
Right-to-Know legislation. *See* Emergency Planning and Community Right-to-Know Act (EPCRA, 1986)
Risk Based Concentration (RBC), 62, 97
Risk Management Program (RMP), 21, 21t, 104
risk-based performance standards, hazardous chemical sites, 22–23
Rome
   light towers, 2
   "Vigil"/Vigilante Corps, 2
ropes, lifesaving, 169–170
routine medical monitoring, 193
RQ. *See* reportable quantity (RQ) designation, hazardous materials

S
safety. *See* National Safety Council; Superfund Amendments and Reauthorization Act of 1986 (SARA)

safety, occupational, 2–10, 12–25. *See also* U.S. Occupational Safety and Health Administration (OSHA); workplaces, hazardous
   equipment, 4, 6*t*, 9, 12
   governmental agencies, 12–13, 23
   historical activities, 2–3
   industrialization, 3–4
   information sources, 56
   legislation, 13–14, 16–17, 21–22
   modern organizations, 4–8
   professional associations, 12, 23
   recent developments, 9
   terms, 9
safety, personal, 197–213. *See also* health, personal; personal protective equipment (PPE)
   enclosed or confined spaces, 197–198, 199*t*–202*t*
   excavations and trenching, 198, 202
   exercise, 211–213
   hazardous energy, 208
   historical activities, 2–3, 5*t*–6*t*
   hot work procedures, 207–208
   modern organizations, 4
   motor vehicle, 209
   terms, 209–210
   work at heights, 202–207
safety, public, 2–10, 12–25. *See also* public protection process; U.S. Public Health Service
   equipment, 4, 6*t*, 9, 12
   governmental agencies, 12–13, 23
   historical activities, 2–3, 4, 5*t*–6*t*
   industrialization, 3–4, 7
   legislation, 13–14, 16–17, 21–22
   modern organizations, 4–8
   professional associations, 12, 23
   recent developments, 9
safety data sheets. *See* material safety data sheets/safety data sheets (MSDS/SDS)
safety monitoring systems (fall protection), 205–206
safety nets, 205
sampling and testing, 62–100
   asbestos-containing materials, 73*t*–87*t*
   chain of custody form, 65*t*
   combustible gas, 69–70, 70*t*
   decontaminated materials and equipment, 152
   exercise, 99–100
   hazardous atmospheres, 62–63, 70
   IDLH atmospheres, 68–72
   instrumentation calibration, 63, 63*t*, 70, 72
   ionizing radiation, 70
   lead-containing materials, 88*t*–95*t*
   off-site analysis, 71*t*, 72
   on-site analysis, 63, 64*t*, 66*t*–67*t*
   oxygen in atmosphere, 68–69
   physical hazards, 63–67
   resampling time frames, 96
   sample collection logs, 63, 64*t*

   terms, 96–97
   toxic materials, 62–63, 70
sanitation programs, 33, 105
SARA. *See* Superfund Amendments and Reauthorization Act of 1986 (SARA)
scaffolding, 206–207
SCBA (self-contained breathing apparatus), 135, 167
SDS. *See* material safety data sheets/safety data sheets (MSDS/SDS)
self-contained breathing apparatus (SCBA), 135, 167
sheet bend knot, 171
Shields, Matthew J., 4
shipping hazardous materials. *See* transportation of hazardous materials
shoes. *See* foot protection
short-term exposure limits (STELs), 62, 97
Site Safety Plans, 29. *See also* health and safety plans (HASPs)
sling psychrometer, 70
sloping and benching, trenches, 198
smoking cessation, 186, 193, 193*t*
Snow, John, 3
soil classification, 204*t*, 205
sound pressure levels, 34, 63, 133, 141
spark producing work (hot work procedures), 208
spiked samples (off-site field sample analysis), 72
Spill Prevention Control and Countermeasure (SPCC) sites, 103
square knot, 171
stain detector tubes, 68, 70, 71*t*
Standard for Power Propelled Scaffolds (OSHA), 169
standards. *See also* American National Standards Institute (ANSI); American Society for Testing and Materials (ASTM); Consensus Standards; U.S. Occupational Safety and Health Administration (OSHA)
standpipe and hose systems, fire, 2, 166–167, 175
STELs (short-term exposure limits), 62, 97
Stokes baskets, 171
stop, drag, roll (personal fire suppression), 171
structural stage fire brigades, 167–168
sulfur dioxide monitors, 70
Superfund Amendments and Reauthorization Act of 1986 (SARA), 13, 15, 16–17, 21, 102. *See also* Comprehensive Environmental Response Compensation and Liability Act of 1980 (CERCLA)
symptomatic medical monitoring, 192–193

T

tagout devices, 208, 210
temperatures, extreme, 63, 67, 70, 134. *See also* thermal hazards
teratogens, 135
terrorist attacks, prevention and response, 21–23, 103, 160–161, 175
Texas City, Texas, explosion and fire (1947), 42–43
thermal energy, tagouts and lockout devices, 208
thermal hazards. *See also* temperatures, extreme
Three Mile Island Nuclear Power Generating Station accident (1979), 16, 17, 102
threshold limit value (TLV), 12, 62, 96, 97

timelines
- asbestos-containing materials, 7t
- health and safety, 5t–6t
- lead-contaminated materials, 7t–8t

TLV (threshold limit value), 12, 62, 96, 97

Tox Line/Tox Net, 56

toxic atmospheres. *See* hazardous atmospheres; immediately dangerous to life or health (IDLH) atmospheres

toxic materials. *See also* asbestos-containing materials; exposure(s); lead-based paint; lead-containing materials; U.S. Agency for Toxic Substances and Disease Registry (ATSDR)
- chronological timeline, 7t–8t
- confined spaces, 197
- defined, 57
- employee medical exams, 141
- legislation regulating, 14, 16
- medical evaluation of workers, 139
- remediation efforts, 14, 16
- respiration interference, 135
- safety information, 42t, 56
- stain detector tubes, 70, 71t
- testing, 62–63, 70

Toxic Substances Control Act of 1976 (TSCA), 14

TRACEM mnemonic, 27. *See also* hazardous environments, recognizing and identifying

transportation blanks (off-site field sample analysis), 72

transportation of hazardous materials
- historical overview, 9, 12, 13t
- placarding and labeling, 45, 50
- shipping requirements (DOT), 45–50, 46t, 105
  - hazardous materials manifest, 43, 45
  - Hazardous Materials Shipping Paper, 45–50, 47t, 49t
  - uniform hazardous waste manifest, 45, 48t
  - UN/NA identification number (DOT), 43, 57
  - UN/U.S. DOT hazard classes, 44t–45t

trench boxes, 198

trenches. *See* excavation and trenches

Triangle Shirt Waist Company factory fire, New York City (1911), 4

trucks, powered industrial, 209, 209t

TSCA. *See* Toxic Substances Control Act of 1976 (TSCA)

Tunneling Standard (OSHA), 169

U

UASI (Urban Area Security Initiative), 103

UEL (upper explosive limit), 69, 97

uniform hazardous waste manifest, 45, 48t

Union Volunteer Fire Company, Philadelphia, 3

United Nations Globally Harmonized System of Classification and Labeling of Chemicals. *See* Globally Harmonized System of Classification and Labeling of Chemicals (GHS/HAZCOM 2012)

U.S. Agency for Toxic Substances and Disease Registry (ATSDR), 12, 13, 62, 97

U.S. Army Corps of Engineers, regulated dam sites, 103

U.S. Army Research Institute of Medicine, 30

U.S. Centers for Disease Control and Prevention (CDC), 9, 12, 13
- *Manual of Analytical Methods,* 72
- medical preplacement examination, 186
- PPE guidelines, 138, 140t

U.S. Coast Guard, 3, 4, 138

U.S. Department of Defense (DOD), 16, 31
- MOPP levels, 138, 140t

U.S. Department of Energy, 14

U.S. Department of Health and Human Services (HHS), 9, 12, 13. *See also* U.S. Agency for Toxic Substances and Disease Registry (ATSDR)
- Disaster Medical Assistance Teams (DMAT), 103–104
- *Manual of Analytical Methods,* 72
- PPE guidelines, 140t

U.S. Department of Homeland Security (DHS), 9, 12, 13, 21–23
- emergency planning zones (EPZs), 102
- National Incident Management System (NIMS), 159–160, 175
- Urban Area Security Initiative (UASI) Program, 103

U.S. Department of Housing and Urban Development (HUD), 105

U.S. Department of Labor, 12, 23, 50, 161, 167n1. *See also* U.S. Occupational Safety and Health Administration (OSHA)
- Mine Safety and Health Administration, 34, 169

U.S. Department of Transportation (DOT). *See also* transportation of hazardous materials
- Command Ground Study, 198n
- described, 23
- *Emergency Response Guidebook (ERG-2012),* 12, 54, 151
- Hazardous Materials Classifications, 33, 42–50

U.S. Environmental Protection Agency (EPA), 9, 12, 13, 14, 23
- Acute Exposure Guideline Levels, 62
- hazardous waste site regulations, 103
- hearing protections device testing, 133
- National Emission Standards for Hazardous Air Pollutants, 105
- National Response Team (NRT), 174–175
- PPE ensemble recommendations, 138
- Risk Management Program (RMP), 21, 21t, 104
- Risk-Based Concentration (RBC), 62, 96–97
- Superfund sites, 103
- uniform hazardous waste manifest, 45, 48t

U.S. Federal Communications Commission, radiation exposure limits, 31

U.S. Federal Emergency Management Agency (FEMA), 22, 102, 103, 104

U.S. Lifesaving Service, 4

U.S. Lighthouse Service, 4

U.S. Nuclear Regulatory Commission (NRC), 31, 102

U.S. Occupational Safety and Health Administration (OSHA), 9, 12, 23, 170, 171, 173
- Asbestos Standard, 181, 192
- color coding system, 53, 54t
- Construction Industry regulations, 126
- Control of Hazardous Energy Standard, 34
- EAPs, 105, 157
- enforcement procedures, 96, 205
- fall arrest system requirements, 169, 205
- Fire Brigades Standard, 167, 167n2

U.S. Occupational Safety and Health Administration (OSHA) (*Contd.*)
    fire response and fire suppression, 163*t*, 166
    first aid requirements, 162
    Hazard Communications Standard (HAZCOM, 2012), 41, 50, 51–52, 52*t*, 53, 53*t*, 57
    hazardous materials regulations, 47, 51, 52*t*
    Hazardous Waste Operations and Emergency Response (HAZWOPER) standard, 16–17, 19*t*–21*t*, 157, 169, 186
    Hearing Conservation Program, 133, 134*t*
    hot environment safety protocols, 27–28
    IDLH requirements, 168
    ladder requirements, 202–203, 203*t*, 205*t*
    Lead Standard, 96*n*, 186*t*, 193
    medical monitoring standards, 186, 186*t*, 187*t*–192*t*
    noisy environment protection requirements, 34, 34*t*
    PBZ sampling, 72
    permissible exposure limit (PEL), 62
    Permit Required Confined Spaces Standard, 105, 115–116, 169, 170, 197, 199*t*–202*t*, 203*t*, 209
    PPE requirements, 126, 131, 138, 139, 141
    Process Safety Management Standards, 21, 22*t*, 157
    protection factor (PF), respirator, 136, 137*t*
    radiation exposure protection requirements, 31
    Respiratory Protection Standard, 137*t*, 168, 181, 182*t*–185*t*
    smoking cessation program, 193*t*
    soil classification, 204*t*
    Standard for Power Propelled Scaffolds, 169
U.S. Public Health Service, 3, 12, 33, 186
U.S. Research Innovative Technology (DOT), 9, 12, 43
U.S. Treasury Department, Revenue Cutter Service, 3
UN/NA identification number (DOT), 43, 57
upper explosive limit (UEL), 69, 97
Urban Area Security Initiative (UASI) Program, 103
Urban Search and Rescue (USAR) Teams, 103, 104
utility locator notification and marking, 198, 198*n*, 204*t*

V

vaccinations, 33
vehicle safety, 209, 209*t*
viruses. *See* etiological hazards
vulnerability study, 157

W

warning lines (fall arrest systems), 206
warning systems, emergency action plans (EAPs), 105, 159
wastes. *See* hazardous waste sites; hazardous wastes
Weather Radio System (NOAA), 104
WEEL (workplace employee exposure limit), 62
welding (hot work procedures), 207
Wheatstone bridge, 69, 70
Williams-Steiger Occupational Safety and Health Act of 1970, 9, 12
wind chill, 31*t*, 67. *See also* equivalent chill temperatures (ECT)
workplace employee exposure limit (WEEL), 62
workplaces, hazardous. *See also* hazardous waste sites; safety, occupational
    EAPs, 105
    information sources, 56
    job sites reviews, 126
    medical emergencies, 161–163
    monitoring employees, 62, 72, 141, 181–195
    oxygen deficiency conditions, 33
    protecting employees, 102, 105, 127, 151
    radiation exposure, 31
World War II. *See* post-World War II developments
Wren, Christopher, 3
written respirator programs, 136

Photo Credits

Chapter 1
Opener © Marcin Balcerzak/ShutterStock, Inc.; **1-P1** © Nancy G Fire Photography, Nancy Greifenhagen/Alamy Images.

Chapter 2
Opener © ronfromyork/ShutterStock, Inc.; **2-P2** © Hemera/Thinkstock; **2-P3** Courtesy of Bill Larkin; **2-T3** Source: OSHA General Industry Safety Regulations, 29 CFR 1910.120: Fire Brigades, U.S. Department of Labor, Occupational Safety and Health Administration, 1999.

Chapter 3
Opener © trainman32/ShutterStock, Inc.; **3-F1** Courtesy of OSHA; **3-P1** Courtesy of FEMA; **3-P3** Courtesy of Sandia National Laboratories; **3-P4** © Johnny Habell/ShutterStock, Inc.; **3-P6** © Paul Doyle/Alamy Images; **3-T4** Source: NOAA, National Weather Service.

Chapter 4
Opener © Mark Winfrey/ShutterStock, Inc.; **4-P1** © Lon C. Diehl/PhotoEdit, Inc.; **4-P3** Courtesy of Rob Schnepp; **4-P4** Courtesy of EMD Chemicals, Inc.; **4-P7** © Jonathan Elderfield/Getty Images; **4-P8** Courtesy of National Tank Truck Carriers Association; **4-P9** Courtesy of Sandia National Laboratories; **4-T8** Source: Hazardous Materials Transportation Act, Public Law 93-633, 49 USC 1801 et seq.; **4-UN1 – 4-UN4** Courtesy of the U.S. Department of Transportation; **4-UN5** © Johnny Habell/ShutterStock, Inc.

Chapter 5
Opener © Cheryl Casey/ShutterStock, Inc.; **5-P4** © Drägerwerk AG & Co. KGaA, Lubeck. All rights reserved. No portion hereof may be reproduced, saved or stored in a data processing system, electronically or mechanically copied or otherwise recorded by any other means without our express prior written permission; **5-P5** Courtesy of Berkeley Nucleonics Corp.

Chapter 6
Opener © Colette3/ShutterStock, Inc.; **6-F1** Occupational Safety and Health Guidance Manual for Hazardous Waste Site Activities, U.S. Department of Health and Human Services, Public Health Service, Centers for Disease Control, National Institute for Occupational Safety and Health, DHHS Publication 85-115, U.S. Government Printing Office, October, 1985.; **6-F2** Source: U.S. Department of Labor: Occupational Safety and Health Administration; General Industry Safety Standards; 29CFR1910.146.; **6-F3** NFPA 51B Standard for Fire Prevention During Welding, Cutting, and Other Hot Work, 2009 Edition, Page 51B-11; **6-P1** © Martin Muránsky/ShutterStock, Inc.; **6-P2** © Ulrich Mueller/ShutterStock, Inc.

Chapter 7
Opener © Olivier Le Queinec/ShutterStock, Inc.; **7-T8** U.S. Department of Labor, 29 CFR 1910.134(d)(3)(i)(A).

Chapter 8
Opener © TFoxFoto/ShutterStock, Inc.; **8-UN1** © Bronwyn Photo/ShutterStock, Inc.

Chapter 9
Opener Courtesy of Bill Larkin; **9-P1** © GerryRousseau/Alamy Images; **9-P3** © imageegami/ShutterStock, Inc.; **9-P9** © Cultura RM/Alamy Images.

Chapter 10
Opener Courtesy of Bill Larkin; **10-T3** Data from OSHA.

Chapter 11
Opener © J.D.S./ShutterStock, Inc.; **11-P12** © Tomasz Bidermann/ShutterStock, Inc.; **11-UN1** © Chuck Wagner/ShutterStock, Inc.; **11-UN5** © Igorsky/ShutterStock, Inc.; **11-UN7** © iStockphoto/Thinkstock; **11-UN8** © Keith Muratori/ShutterStock, Inc.; **11-UN9** © PHOTOSTOCK-ISRAEL/Photo Researchers, Inc.

Unless otherwise indicated, all photographs and illustrations are under copyright of Jones & Bartlett Learning or have been provided by the author.

CPSIA information can be obtained
at www.ICGtesting.com
Printed in the USA
BVOW09s0840261016
465759BV00012B/5/P